Application of
Fracture Mechanics
to Design

SAGAMORE ARMY MATERIALS
RESEARCH CONFERENCE PROCEEDINGS

Available from Plenum Press

Forthcoming Volumes

Application of Fracture Mechanics to Design

Edited by

John J. Burke

Army Materials and Mechanics Research Center
Watertown, Massachusetts

and

Volker Weiss

Syracuse University
Syracuse, New York

Springer Science+Business Media, LLC

Library of Congress Cataloging in Publication Data

Sagamore Army Materials Research Conference, 22nd, Raquette Lake, N.Y., 1975.
  Application of fracture mechanics to design.

  (Sagamore Army Materials Research Conference proceedings; 22)
  Includes index.
  1. Fracture mechanics–Congresses. 2. Structural design–Congresses. 3. Rotors–
Design and construction–Congresses. I. Burke, John J. II. Weiss, Volker, 1930-
III. Series: Sagamore Army Materials Research Conference. Proceedings; 22.
UF526.3.S3 no. 22 [TA409] 623'.028s [620.1'126]

ISBN 978-0-306-40040-7   ISBN 978-1-4899-6588-2 (eBook)   78-14819
DOI 10.1007/978-1-4899-6588-2

Proceedings of the Twenty-second Sagamore Army Materials Research Conference
held at Sagamore Conference Center, Raquette Lake, New York, August, 1975

© 1979 Springer Science+Business Media New York
Originally published by Plenum Press, New York in 1979.
Softcover reprint of the hardcover 1st edition 1979

SAGAMORE CONFERENCE COMMITTEE

Chairman
JOHN J. BURKE
Army Materials and Mechanics Research Center

Program Director
VOLKER WEISS
Syracuse University

Secretary
ARAM TARPINIAN
Army Materials and Mechanics Research Center

Conference Coordinator
JOSEPH A. BERNIER
Army Materials and Mechanics Research Center

PROGRAM COMMITTEE

JOHN J. BURKE
Army Materials and Mechanics Research Center

F. R. LARSON
Army Materials and Mechanics Research Center

GEORGE G. MAYER
Army Research Office

R. PELLOUX
Massachusetts Institute of Technology

J. SRAWLEY
NASA Lewis Research Center

C. F. TIFFANY
Air Force Materials Laboratory

VOLKER WEISS
Syracuse University

E. T. WESSEL
Westinghouse Research Laboratories

S. YUKAWA
General Electric Company

Arrangements at Sagamore Conference Center
JAMES REID
Syracuse University

Preface

The Army Materials and Mechanics Research Center has conducted this Twenty-Second Sagamore Army Materials Research Conference in cooperation with the Materials Science Group of the Department of Chemical Engineering and Materials Science of Syracuse University since 1954. The main purpose of these conferences has been to bring together scientists and engineers from academic institutions, industry and government who are uniquely qualified to explore in depth a subject of importance to the Army, the Department of Defense and the scientific community.

This volume, APPLICATION OF FRACTURE MECHANICS TO DESIGN, addresses the areas of Test Methods for Fracture Mechanics Design Data, Pressure Vessels, Structures, Rotating Components, NDE and Failure Analysis and Case Histories of Failures.

We wish to acknowledge our appreciation to Mrs. Helen Brown DeMascio of Syracuse University. The assistance of the Technical Reports Office under the supervision of Mrs. A. V. Gallagher and the Technical Information Office under the supervision of Miss M. M. Murphy of the Army Materials and Mechanics Research Center in reviewing the final manuscript is gratefully acknowledged. Special thanks go to Mr. Harold Laye for final review of the illustrations. The dedicated assistance of Mrs. J. Ayoub of the Army Materials and Mechanics Research Center throughout the stages of the conference planning and finally the publication of this book is deeply appreciated.

Army Materials and Mechanics Research Center   The Editors
Watertown, Massachusetts

Contents

SESSION IV

STRUCTURES
W. E. Anderson, Moderator

SESSION V

ROTATING COMPONENTS
E. T. Wessel, Moderator

SESSION VI

NDE AND FAILURE ANALYSIS
R. Pelloux, Moderator

FRACTURE MECHANICS 1975 - AN OVERVIEW

Volker Weiss

Syracuse University, Syracuse, New York 13210

INTRODUCTION

Catastrophic and unexpected failures, and the need to analyze them in order to prevent future occurrences, were the principal stimuli to the development of Fracture Mechanics. The two original papers [1,2] resulted from strong technological motivation coupled with the desire to apply the latest analytical solutions of the stress distribution near cracks, corners, or slits. The first one is a little known paper by K. Wieghardt [1] who published an extensive paper entitled "On the Clevage and Fracture of Elastic Bodies" in the Zeitschrift für Mathematik und Physik in 1907. He correctly derives the $1/\sqrt{r}$ singularity at the crack tip or corner, Figure 1, but does not attempt to answer the question of the force required to cause fracture. "...the common strength hypotheses would predict fracture at extremely low forces, which is, in fact, not the case". Rather, he treats the problem of the direction of crack propagation. He concludes that experimental results of tests on a specimen designed by Bach to simulate the behavior of "U" shaped supports, Figure 2, and theoretical results agree for a maximum normal stress failure criterion [3].

Griffith [2], undoubtedly stimulated by G. I. Taylor, also cites the inadequacy of the common strength hypotheses to predict the effect of surface scratches or notches on fatigue strength and proceeds to develop the well known energy balance criterion for crack extension, Figure 3.

Concerted efforts in the forties and fifties, primarily in the U.S.A. and in Great Britain, have built the foundation of Fracture

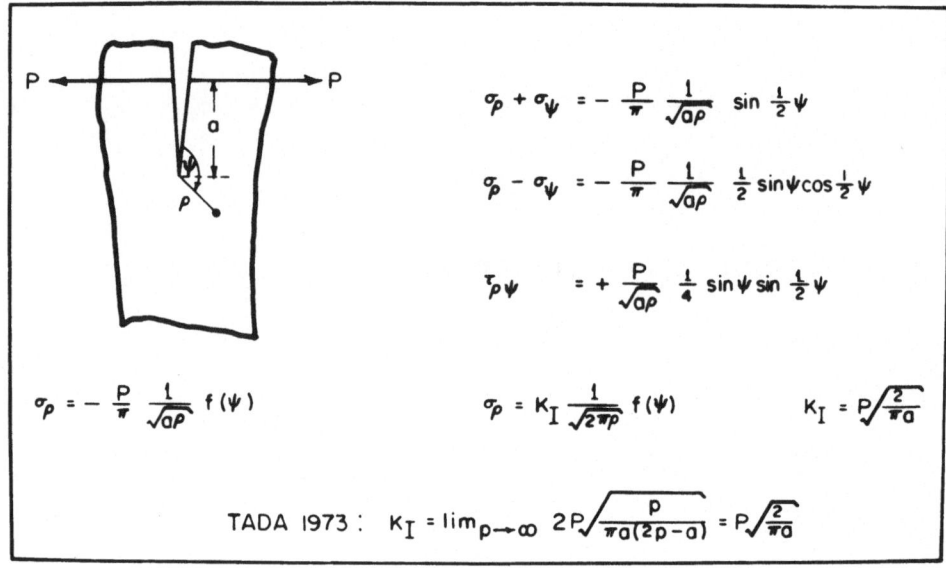

Figure 1.   Wieghardt's stress field solutions.   Published in 1906.

Mechanics, which is now a well recognized sub-discipline of Mechanics
of Solids.   The well known contributions of Orowan, Irwin, Kies,
Sachs, Wells, McClintock, Neuber, Williams, Liu, Paris, Liebowitz,
Soete, Yokobori, Kunio, Myiamoto, Kochendörfer, Kerkhof, Rice, Hahn,
Tetelman, Beeuwkes, Krafft, Bluhm, Srawley, Brown, Low, Beachem,

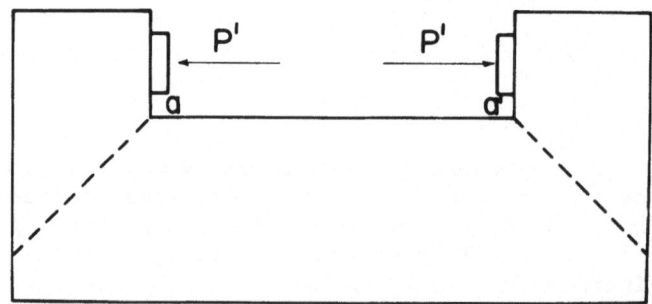

Figure 2.   "Bach" specimen used by Weighardt.

"THE CONDITION THAT THE CRACK MAY EXTEND IS

$$\frac{\partial}{\partial c}\,(W-U) = 0$$

SO THAT THE BREAKING STRESS IS

$$R = \sqrt{\frac{2ET}{\pi c}}\,{}''$$

"-- THE MAXIMUM TENSION --

$$2R\sqrt{\frac{c}{\rho}}$$

--ρ BEING THE RADIUS OF CURVATURE AT THE
CORNERS OF THE ELLIPTIC CRACK."

Figure 3.   Quotes from A. A. Griffith, 1920.

Broek and many others have led to a "Fracture Theory" far beyond
the expectations of the middle fifties and even of those of the
1960 Sagamore Conference [4]. Practical application of this knowledge is certainly warranted in many areas and should be of substantial technological benefit both with respect to new equipment
and devices, improved performance and improved safety. To summarize
the information on how this is and can further be accomplished is
the aim of this conference.

     In this evening session it is my task to give you an overview
of the current status of the experimental and theoretical background
of the sub-discipline we call "Fracture Mechanics". First I shall
review the status of the stress analysis of cracks, then the experimental observations that should lead to a better understanding of
the controlling events in the various fracture processes, and
finally I will attempt to point out some of the questions related
to the practical application of Fracture Mechanics.

                    STRESS DISTRIBUTION NEAR CRACKS

     The two central ingredients of a fracture theory are (1) a
stress analysis for a part containing a crack or crack-like defect

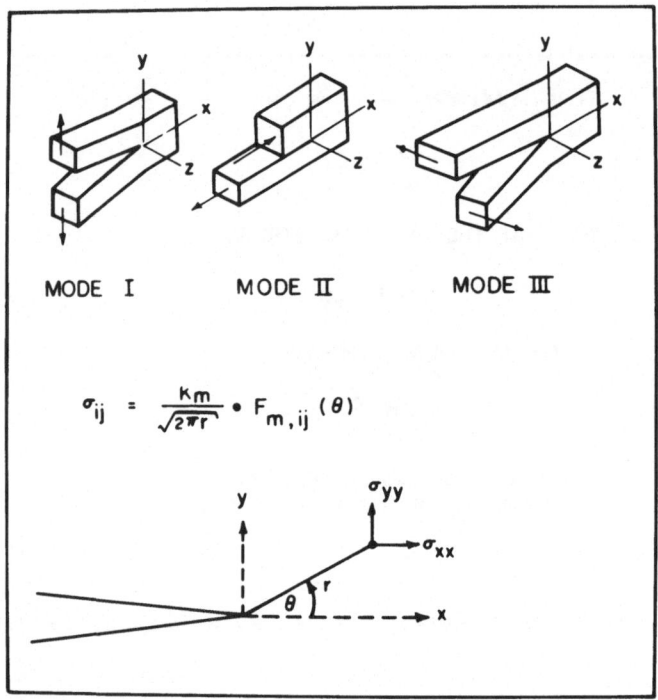

Figure 4.   Crack extension modes (m) and stress field equations.

and (2) a failure hypothesis which somehow defines the events of
critical crack extension. This discussion deals with a review of
the former.

Solutions of the stress and strain distributions at and near
cracks have been obtained for a great variety of loading conditions
and crack geometries, for linear elastic solids as well as for other
stress-strain relationships. Three fundamental modes of crack
extension are defined as illustrated in Figure 4. In the vicinity
of the crack tip, i.e., r<a, the stress distributions for either
mode are uniquely defined through the stress intensity factor K.
This stress intensity factor has the character of a stress or load,
is a linear function of the applied load, and a more complicated
function of the crack geometry and the specimen geometry. It has
the dimension stress times square root of length, Ksi√in or
$MNm^{-3/2}$, which, incidentally, are quite close (within 10%) numerically. The chief problem then is to obtain solutions for K_I, K_{II}

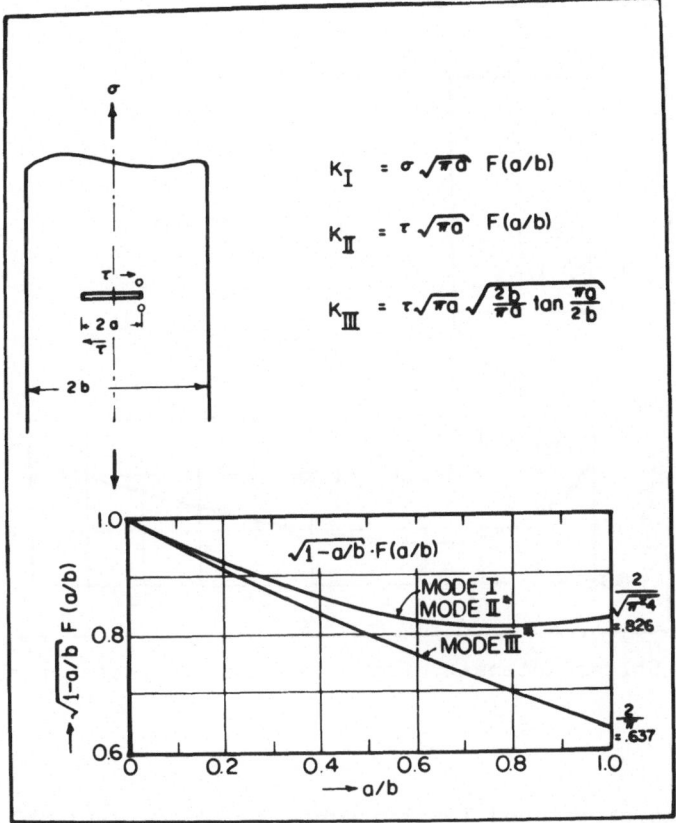

Figure 5.   Stress intensity factor for center cracked test specimen.

and K_{III} as a function of part and crack geometry.

For elastic solids closed form solutions are available for cracks in infinite plates or bodies or for periodic cracks in infinite plates or bodies. A typical example is the Griffith Crack where

$$K_I = \sigma_\infty \sqrt{\pi a}$$

For finite geometries sufficiently accurate solutions can be obtained with boundary collocation techniques, weight function techniques, finite element analysis, compliance calibration, etc. Solutions

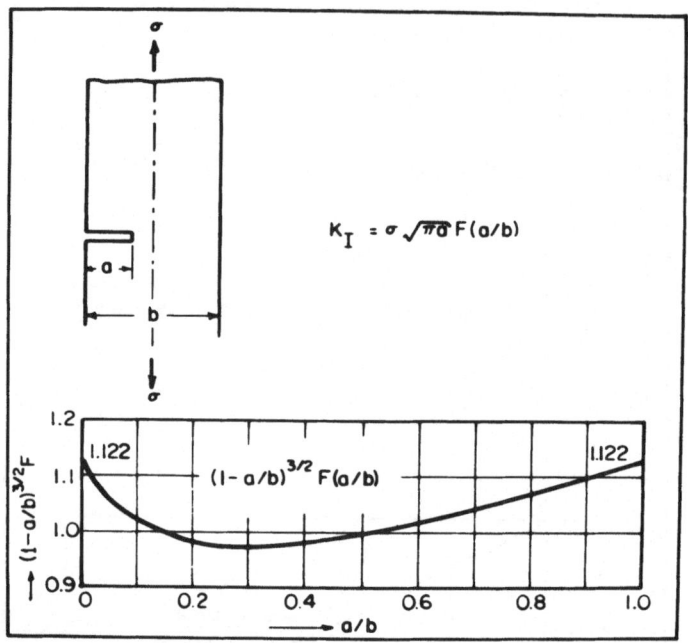

Figure 6.   Stress intensity factor single edge notch test specimen.

for many geometries and loading systems are now available in the literature and even consolidated in some handbooks.   Typical examples of such solutions are shown in Figures 5-7 [3].

In many cases of technical interest the stress state may not be full mode I or full mode II but a mixture of two or all three modes.   For such situations the contribution of each mode to the local stress, $\sigma_{ij}(x,y)$, are additive, as illustrated in Figure 8.

The stress analysis of cracks becomes considerably more intractable for materials that do not follow a linear stress-strain law. If the disturbance due to plasticity is small and limited to a region near the crack tip which itself is small compared to the applicability of the near-tip elastic solutions, plasticity corrections such as that proposed by Irwin at the 1960 Sagamore Conference [4], where the plastic zone size is added to the crack length, usually provide adequate solutions.   The Barenblatt model is another solution.   For the so-called "small scale yielding solutions", it can be argued that specimens or parts having the same K-value must have identical stress and strain distributions inside the plastic zone.

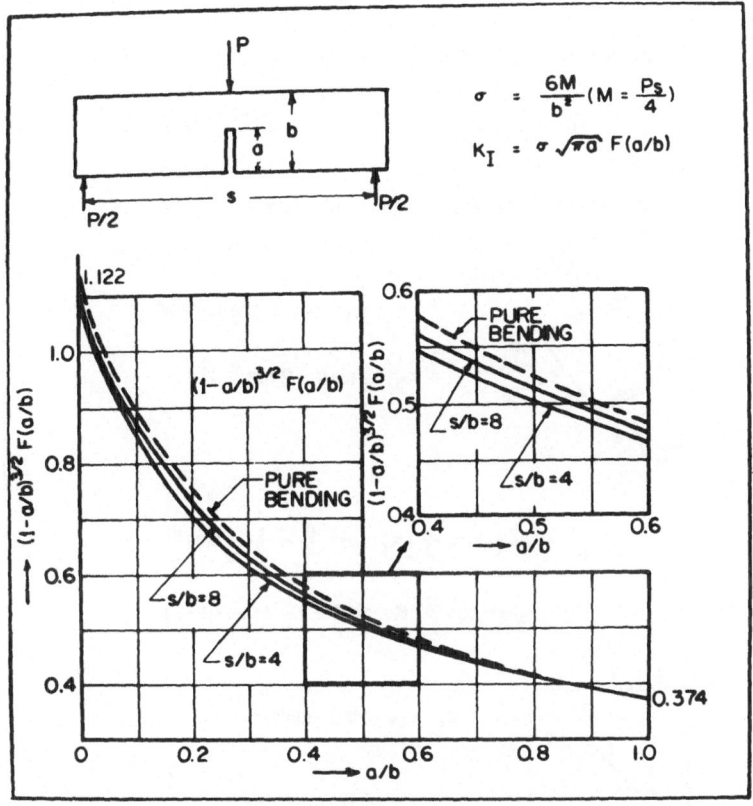

Figure 7.   Stress intensity factor for three point bend test
            specimen.

    The crack opening displacement analysis introduced by Wells and
Burdekin [5] makes use of these concepts. The analysis is roughly
sketched in Figure 9. Evaluated at $r = r_p$, the crack opening
multiplied by the yield strength, the one used to define r_p is
equal to the crack driving force G or K^2/E. Measurement of δ at the
onset of crack extension has therefore been proposed and is used as
an experimental method to characterize fracture toughness.

    Closed form solutions for more extensive plasticity are only
available for mode III. Neuber treated the notch problem for
mode III in 1960 as illustrated in Figure 10. One of the results
of his analysis is the so-called Neuber rule, also illustrated in
Figure 10, which applies to any notch radius for a specific stress-
strain law or to a sharp notch or crack for any stress-strain law.

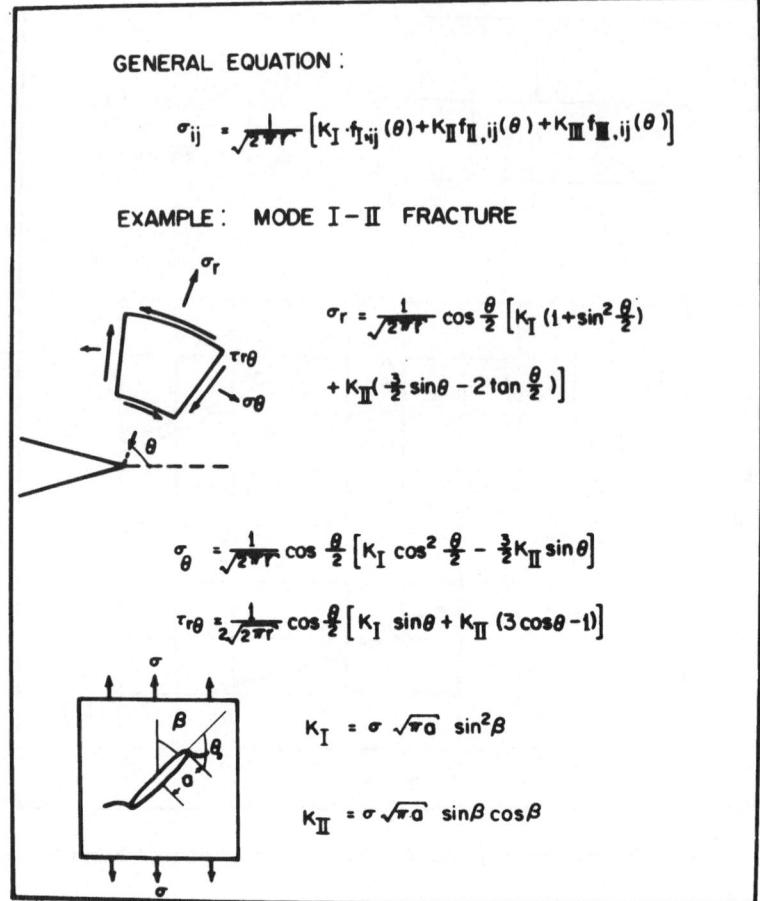

GENERAL EQUATION :

$$\sigma_{ij} = \frac{1}{\sqrt{2\pi r}}\left[K_I \cdot f_{I,ij}(\theta) + K_{II}f_{II,ij}(\theta) + K_{III}f_{III,ij}(\theta)\right]$$

EXAMPLE :   MODE I-II FRACTURE

$$\sigma_r = \frac{1}{\sqrt{2\pi r}}\cos\frac{\theta}{2}\left[K_I\left(1+\sin^2\frac{\theta}{2}\right)\right.$$
$$\left. + K_{II}\left(\frac{3}{2}\sin\theta - 2\tan\frac{\theta}{2}\right)\right]$$

$$\sigma_\theta = \frac{1}{\sqrt{2\pi r}}\cos\frac{\theta}{2}\left[K_I\cos^2\frac{\theta}{2} - \frac{3}{2}K_{II}\sin\theta\right]$$

$$\tau_{r\theta} = \frac{1}{2\sqrt{2\pi r}}\cos\frac{\theta}{2}\left[K_I\sin\theta + K_{II}(3\cos\theta-1)\right]$$

$$K_I = \sigma\sqrt{\pi a}\,\sin^2\beta$$

$$K_{II} = \sigma\sqrt{\pi a}\,\sin\beta\cos\beta$$

Figure 8

For perfect plasticity the strain singularly is characterized by
1/r.

     Hult and McClintock [7] treated a mode III crack for perfect
plasticity and obtained the results sketched in Figure 11.   The
cross section of the plastic zone is a circle and the strain
decreases as 1/r from the crack tip.

     For perfect plasticity following exponential strain hardening
of the type $\overline{\sigma} = k\varepsilon^n$ a number of approximate solutions similar in
character to the "Neuber Rule" are available.   The contributions
of Rice and co-workers deserve special mention here [8].   Some of
the results of these efforts are summarized in Figure 12.   The first

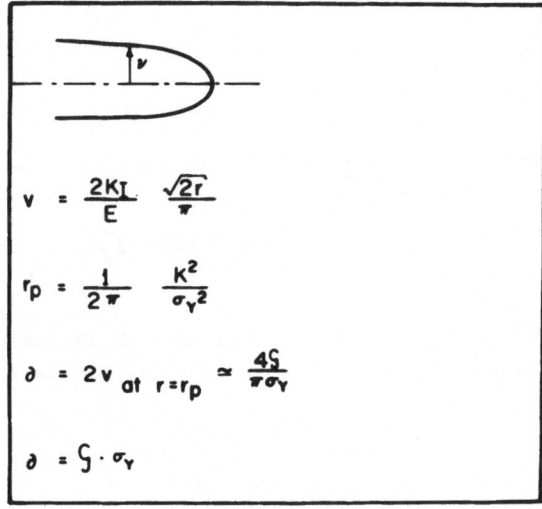

Figure 9.   Crack opening displacement analysis [5].

two equations represent the stress and strain distributions for a
strain hardening material in mode III.   Rice was also able to show
that there exists a path independent integral which assumes the value
of G for the limiting elastic case and is not very sensitive to the
stress-strain law.   Hence it may be useful in characterizing the
stress distribution for non-linear $\sigma - \varepsilon$ behavior.   The J integral
assumes particularly simple forms for the bend bar with a deep crack
and for the center cracked tension bar, which makes these geometries
attractive for J_c and J_{Ic} determinations in tough materials.   The
equations for the limit loads, which one would use for thru-yielding
materials, are also given for comparison in Figure 12.

FRACTURE PROCESSES

As the computer has assumed a key role in the stress analysis
of cracks, so has the scanning electron microscope in the experimental determination of the actual fracturing mechanisms.   From
theoretical considerations and experimental observations there have
evolved a number of fracture criteria (Figure 13).   These criteria
either define or allow us to determine material parameters that
characterize the resistance to crack extension, commonly referred
to as fracture toughness.   The existence of one or several such

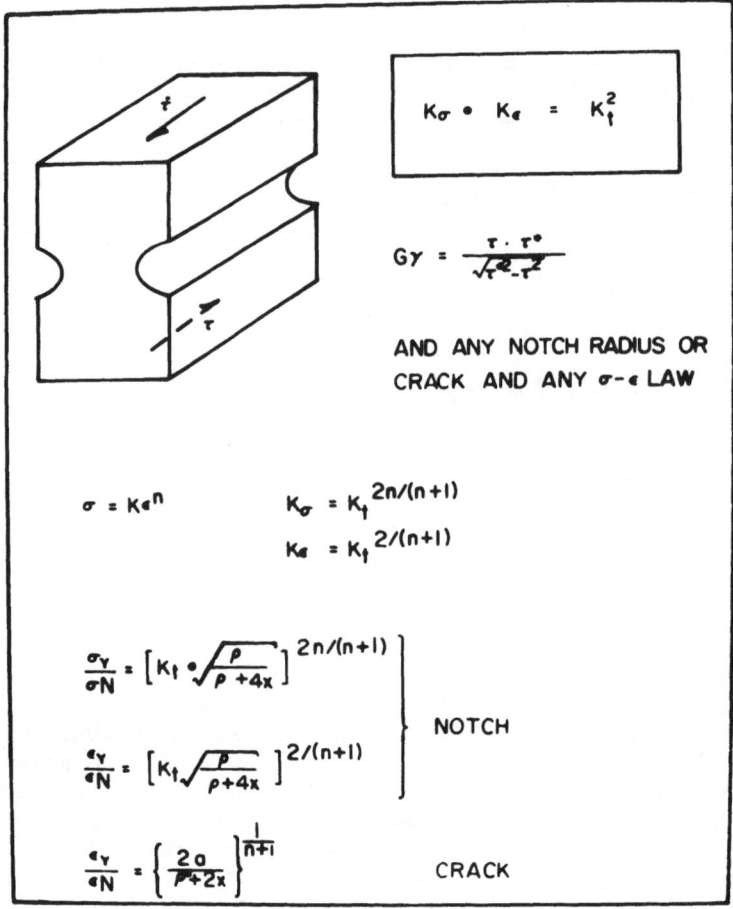

Figure 10.   Notch problem for mode III as treated by Neuber in 1960.

"material constants" and the possibility of their experimental determination is the most essential feature of fracture mechanics. To date we can observe a broad consensus for the existence of K_{Ic}, the plane strain, mode I, fracture toughness, which represents a minimum K_c value for mode I. The majority of cases where fracture mechanics finds practical applications are based on these K_{Ic} numbers.

Let me show some typical examples of fracture surfaces that relate to the actual mechanisms and in turn to the question as to what criteria might be applicable.

Crystallographic clevage, Figure 14, is a case where a critical

surface energy criterion appears most justified. Surface energies
vary with crystallographic direction but range between 1000 and
2000 dyne/cm or 1 to 2 x 10^{-2} in lbs/in^2. G_c would be twice that
value; (K_c between 1 and 2 ksi$\sqrt{in}$ or $MNm^{-3/2}$ for iron). Using a
yield band initiated clevage model Averback obtains G_{Ic} minimum
values of 1 to 3.10^6 dyne cm^{-1}, i.e., about 1000 times the surface
energy values [9]. Nevertheless, G_{Ic} is predicted to be constant
and independent of specimen geometry as long as plane strain is
maintained. Nucleation of the clevage fracture, i.e., by a carbide
particle, must also be considered. Clearly the experimentally
observed toughness is sensitive to the distribution and type of
nucleation sites. A very clear illustration of such a nucleation
event ahead of the crack tip is illustrated in Figure 15, obtained
on Polycarbonate. Control of these microstructural features appears
to be a very promising way towards the practical application of
fracture mechanics to material development.

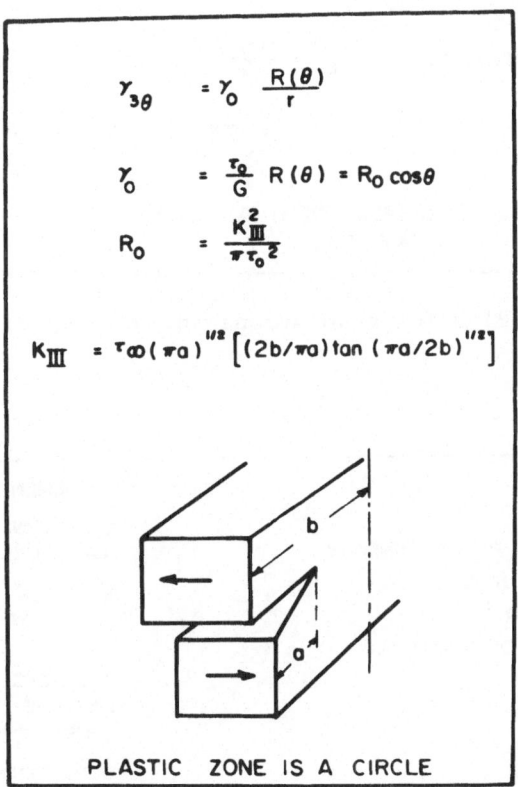

Figure 11. Mode III solution for perfect plasticity by Hult and
            McClintock [7].

RICE $\quad \dfrac{\tau}{\tau_0} = \left[\dfrac{R(\theta)}{r}\right]^{\frac{n}{n+1}}$

$\qquad\qquad \dfrac{\gamma}{\gamma_0} = \left[\dfrac{R(\theta)}{r}\right]^{\frac{1}{n+1}}$

$J = \displaystyle\int_{\Gamma} (Wdy - T \cdot \dfrac{\partial u}{\partial x}\, ds)$

$\quad = \displaystyle\int_0^{\partial} (-\dfrac{\partial P}{\partial a})_{\partial}\, d\partial = \int_0^P (\dfrac{\partial d}{\partial a})_P\, dP$

$J_{ELASTIC} = \mathcal{G} = \dfrac{K^2}{E}$

BEND BAR: $\quad P_L = 1.456\, \sigma_{TEN}\, \dfrac{B}{S}(W-a)^2$

$\qquad\qquad J = 2.912 \cdot \sigma_{TEN}\, \dfrac{\partial}{S}(W-a)$

CENTER NOTCH: $\quad P_L = \dfrac{2\sigma_{TEN}}{\sqrt{3}}(2b-2a)B$

$\qquad\qquad J = \dfrac{2\,\sigma_{TEN}}{\sqrt{3}}\, \partial$

$\partial \cdots$ DISPLACEMENT AT LOAD POINT
$P_L \cdots$ LIMIT LOAD

Figure 12.   Plasticity solutions and J-integral concept.

THEORETICAL:	EXPERIMENTAL:
SURFACE ENERGY	$\sigma\sqrt{a}$ = CONST.
SURFACE ENERGY + PLASTIC ENERGY	σ_{max} = CONST.
MAXIMUM NORMAL STRESS	G_{Ic} OR K_{Ic} = CONST.
CRITICAL TIP RADIUS	COD = CONST.
CRITICAL STRESS DISTRIBUTION	
CRITICAL STRAIN	
CRITICAL COD	TOUGHNESS RELATED TO
J_{Ic}	FRACTURE DUCTILITY
	AS MEASURED UNDER A
VOID COALESCENCE	MULTI-AXIAL STRESS STATE

Figure 13.   Fracture criteria.

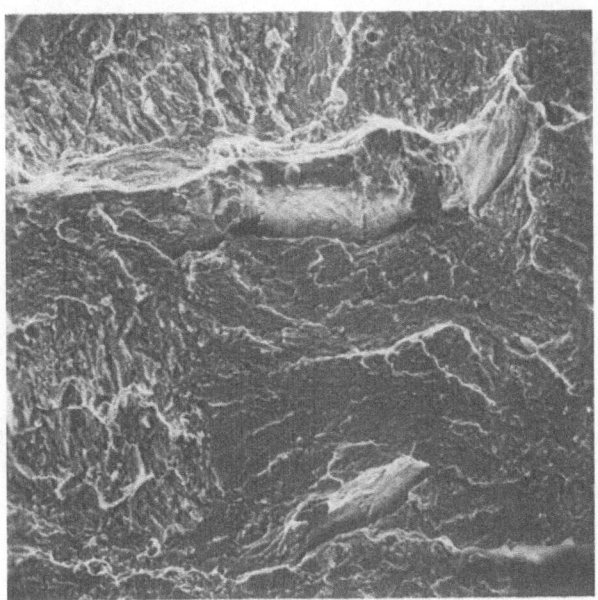

Figure 14.   Scanning electron micrograph of a fracture surface of
             TRIP steel showing cleavage steps.   60,000X

     In materials of higher ductility, crack extension by void
coalescence is a very common fracture mode. This leads to a
fracture surface that contains dimples (Figure 16). Extensive
plastic deformation must be involved in this type of crack extension. Here the surface energy plays a very negligible role. A
constant G_{Ic} value might be expected if the specimen geometry is
such that the conditions for small scale yielding are satisfied.
For more extensive plasticity one could estimate G_c from the work
required to move the plastic zone, as illustrated in Figure 17
[10,11]. According to this very approximate treatment, the
fracture toughness becomes a very strong function of the local
fracture strain at the crack tip. It becomes sensitive to the
stress state, i.e., to the specimen geometry. If the dominant
stress state is preserved during crack advance and if no mode
shift occurs, a constant value of G_{Ic} is expected. Thus a critical
fracture strain criterion and a critical K_c or G_c criterion are
equivalent for this case.

     For many instances of technical interest, fracture occurs
under conditions that are not plane strain, i.e., t, or w, or a
$\leq 2.5 \ (K_{Ic}/\sigma_Y)^2$.

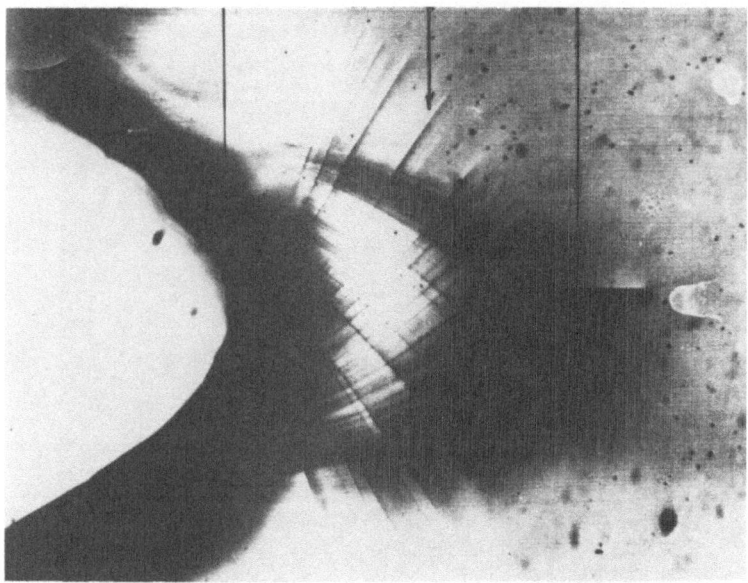

Figure 15.   Nucleation event ahead of crack tip.

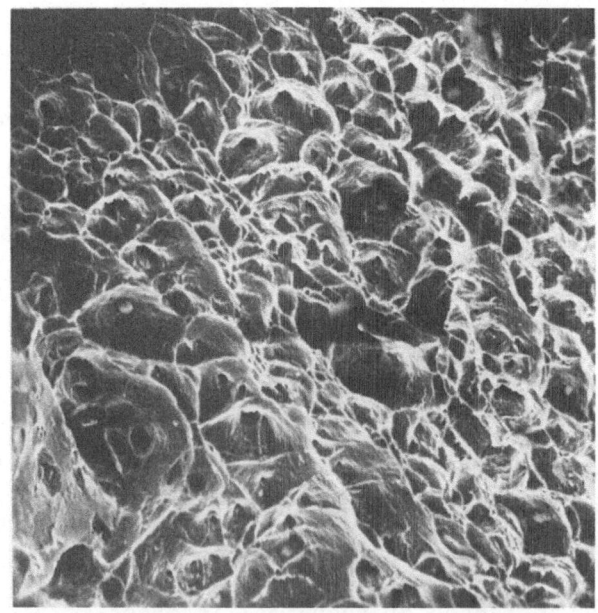

Figure 16.   Scanning electron micrograph of a fracture surface of
             TRIP steel showing dimples.   60,000X

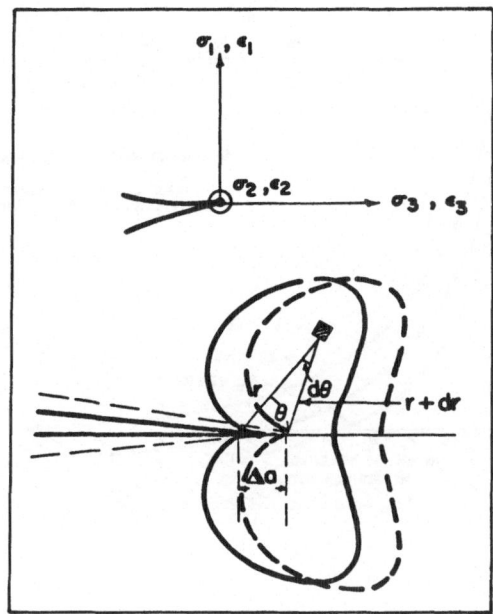

Figure 17. Crack tip coordinate system and schematic for the
            determination of $G_{Ic} = \partial w/\partial A$. This method leads to an
            estimate of G_c in terms of the local multiaxial fracture
            strain at the crack tip corresponding to the stress
            state which exists there, ε_F; $G_{Ic}\alpha\varepsilon_F^2 \cdot f(\rho*,E,n,k,\sigma_Y)$,
            where $\rho*$ is a microstructural size constant and σ_Y is
            the yield strength.

     Experimental data and the preceding analysis show that K_c
becomes a function of these critical dimensions. Fracture toughness
becomes a function of thickness (Figure 18). Crack extension also
occurs under rising load. To characterize a material's fracture
resistance for such cases, the R-curve concept, Figure 19, has been
introduced.

     A number of researchers have studied the interrelationship be-
tween fracture toughness and ductility, as shown in Figure 20. In
recent studies we found a linear relationship between bulge ductility
and plane strain fracture toughness of steels (Figure 21). It is
logical to proceed along these lines to search for the structural
features and mechanisms that control ductility under multi-axial
stress states, particularly void coalesence. Preliminary evidence
indicates that both are strongly stress state dependent [11].
Moreover, the void nucleation strain also depends strongly on the
yield strength and therefore on the test temperature. A schematic
representation of the effect is illustrated in Figure 22. Here the
yield strength dependence of the initiation strain is responsible
for the typical ductile-brittle transition behavior observed in

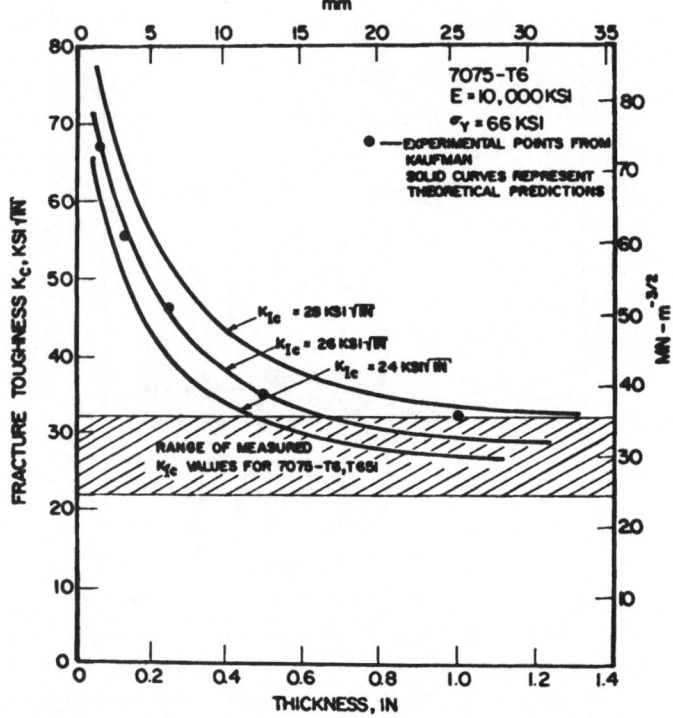

Figure 18.   Experimental data and theoretical curves calculated
according to Ref. [11] for Al 7075-T6 and -T651.

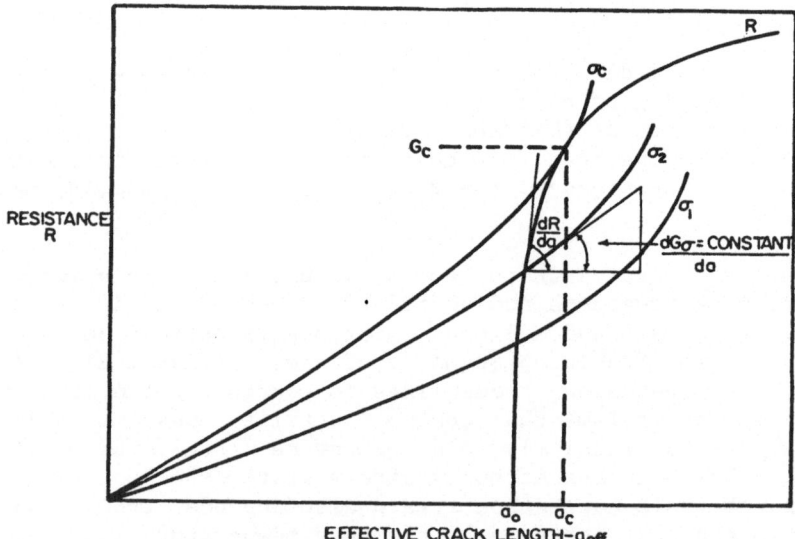

Figure 19.   The R-curve concept.

$$G_c = 2(\gamma_s + \gamma_p)$$

$$2\gamma_p = 2\int_0^\bullet \sigma_{YY}\, dv \qquad \text{COTTRELL}$$

$$G_{Ic} = \partial_{Ic}\, \sigma_Y \qquad \text{WELLS}$$

$$\sigma_{MAX} = \sigma_Y(1+\phi)$$

$$\qquad\qquad\qquad\qquad \text{BEEUWKES}$$

$$K_{Ic} = \sigma_Y\sqrt{\pi s}\cdot\frac{1}{L}$$

$$L = f(\phi) = g(E,\nu,\sigma_Y,\sigma_N,s_0)$$

$$K_{Ic} = En\sqrt{2\pi d_T} \qquad \text{KRAFFT}$$

$$K_{Ic} \approx n\sqrt{\tfrac{2}{3}E\sigma_Y\epsilon_F} \qquad \begin{matrix}\text{HAHN}\\\text{ROSENFIELD}\end{matrix}$$

$$\partial = (\epsilon/\alpha)^{1/m}$$

$$K_{Ic} = A\sqrt{\sigma_Y}\cdot\epsilon_F\cdot ps^{\frac{1}{2m}} \quad \text{BARSOM}$$

$$m \approx 1/4$$

Figure 20. Fracture toughness-ductility relationships proposed.

steels and other bcc materials.

It is well known that certain environments çan cause crack extension under constant loads at K < K_{Ic}. The typical behavior is illustrated in Figure 23. Whether the fracture is transcrystalline or intercrystalline, depends on the material-environment interaction. K_{Iscc} is utilized to characterize the stress condition for the onset of crack extension in a given environment. Final failure occurs when the crack has grown to a size such that K = K_{Ic}.

Of great interest to the practical application of fracture mechanics is our understanding of fatigue crack propagation [12,13]. Most of these are also grounded in experimental findings over larger or smaller ranges of the da/dN vs ΔK curves. The typical curve is shown in Figure 24. The steady state range is bounded by the threshold and the near fracture ranges. The threshold range, if it exists, is very low, approximately 4 - 7 ksi$\sqrt{\text{in}}$ for steels, R = 0. It relates principally to the elastic modulus which has prompted us to propose a model based on the theoretical strength of the material [14]. The threshold level is also affected by

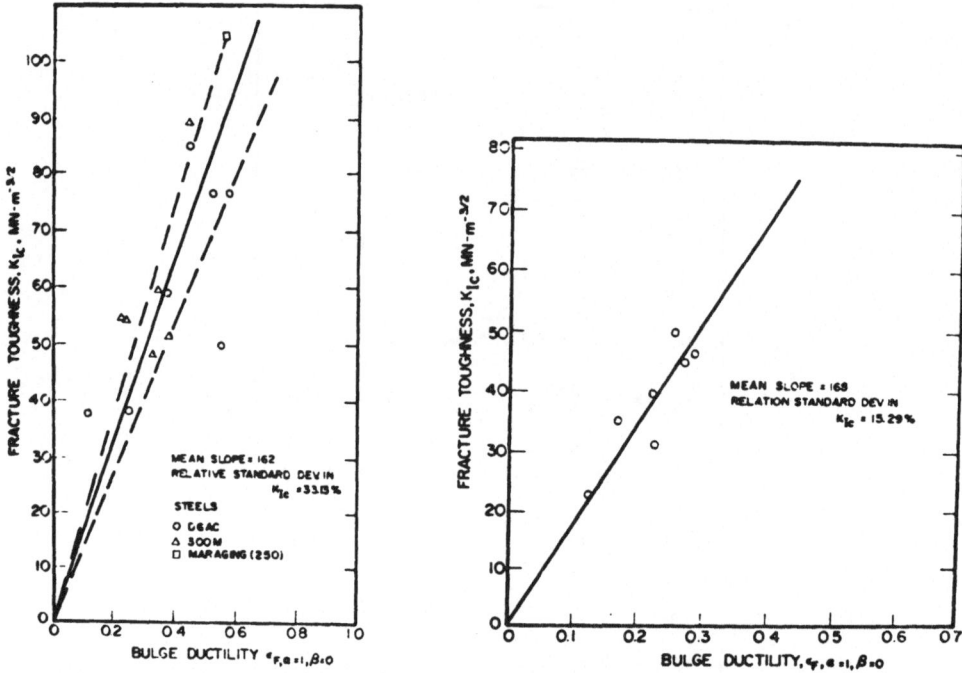

Figure 21.   Correlation between plane strain fracture toughness and
             bulge ductility for high strength steels (a) and
             aluminum alloys (b).   [11]

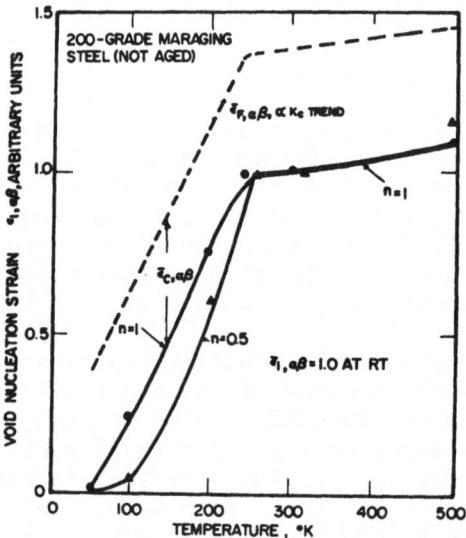

Figure 22.   Schematic of the strain required for void nucleation as
             a function of temperature in unaged 200-grade steel.

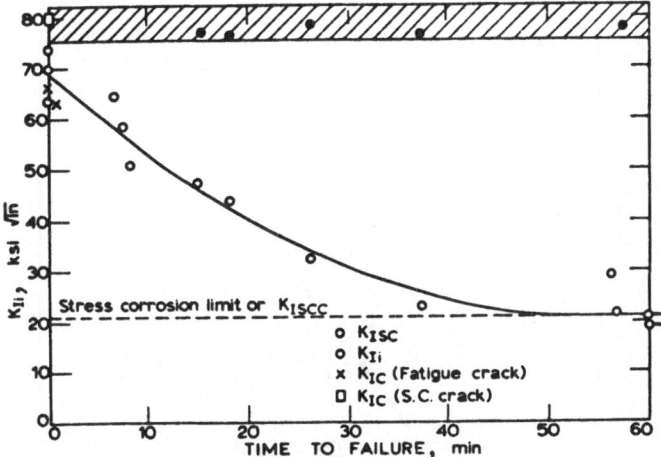

Figure 23.   Effect of initial stress intensity on time to failure
             by stress corrosion for a 4-1/4% Ni-Cr-Mo steel.

environment and by the mean stress.   The environmental effects are
particularly noticeable at the near threshold region.   The mean
stress effects seem to indicate that a $K_{max\ th}$ criterion is operative at low mean stress and a ΔK_{th} criterion at higher R values.
It should perhaps be noted that experimental studies in the threshold range have only recently been initiated.   Nevertheless, it
appears that the results of such studies might well be of great
practical significance, particularly for safe life design situations
[15].

PRACTICAL APPLICATIONS

The greatest success in the practical application of fracture
mechanics can probably be claimed for failure analysis.   These
studies have also done much towards the further development of
fracture mechanics, from the analysis of the Comet failures, turbine rotors and maraging steel rocket motor cases.   Undoubtedly
this trend will continue and more spectacular failures will prompt
more spectacular advances of the theory.

To what extent fracture mechanics has been applied to materials
development is not easily determinable.   One can suspect that notions
of fracture mechanics, particularly notions of micro-structure-
toughness relationships played a role in the development of maraging
steels and TRIP steels, of high strength high toughness aluminum
alloys, and of composites.   We can probably look forward to more
fracture mechanics guided materials design as we learn more about
the micro-mechanisms that control ductility and fracture toughness.

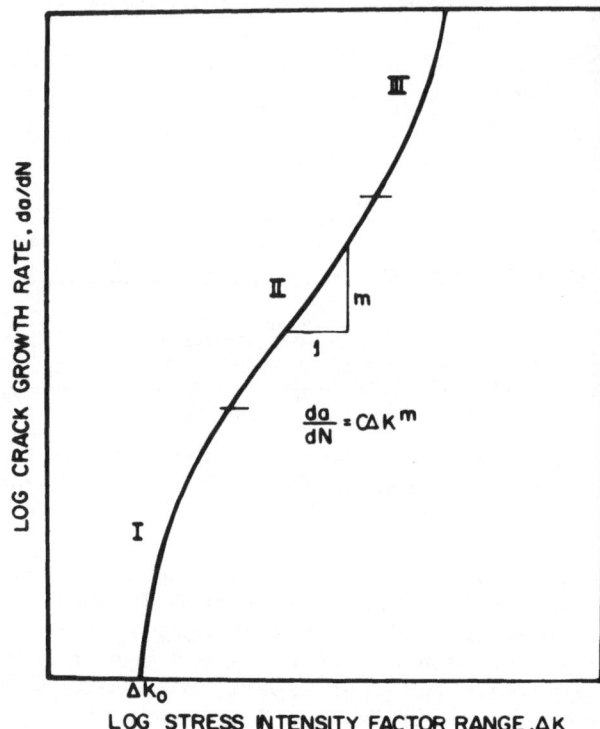

Figure 24. Schematic representation of a typical fatigue crack
            growth curve when the rate da/dN is expressed as a
            function of the stress intensity factor range, ΔK.

        The present conference focuses on applications to design. How
and to what extent this can successfully be accomplished will be
dealt with in detail in the following chapters. Fracture mechanics
tells us that the yield strength and the K_{Ic} value define a critical
flaw size that must be avoided. As flaws of smaller size may grow
in fatigue or under environmental effects to the critical size,
flaws large enough to grow to near critical size during the expected
life span or during service between inspection periods must be
detectable. Thus any practical application must be tied to a flaw
detection capability - by non destructive means or by proof testing.
The safety factor one should apply to this inspection limit will
depend on many parameters, but quite significantly (1) on the
accuracy of the stress analysis of the part of component, (2) on the
accuracy with which the sub-critical crack growth characteristics
are known, and (3) on the accuracy with which the local K_{Ic} value,
which might also be influenced by manufacturing processes, is
known.

REFERENCES

1. Wieghardt, K., Über das Spalten und Zerreissen elastischer Körper. Zeitschr. f. Math. u. Phys., 55 (1907).

2. Griffith, A.A., "The Phenomena of Rupture and Flow in Solids", Phil. Trans. Roy. Soc., A221 (1920).

3. Tada, H., Paris, P. and Irwin, G., "The Stress Analysis of Cracks Handbook", Del Research Corporation, Hellertown, Pa., (1973).

4. Proceedings of the Seventh Sagamore Ordnance Materials Research Conference, Syracuse University (1960).

5. Wells, A.A., "Application of Fracture Mechanics At and Beyond General Yielding", Brit. Weld. J., 10, 563 (1963).

6. Neuber, H., "Theory of Stress Concentration for Shear-Strained Prismatical Bodies with Arbitrary Non-Linear Stress-Strain Law", Trans. Am. Soc. Mech. Eng. J. Appl., 28, 544-50 (1961).

7. Hult, J.A.H. and McClintock, F.A., in "Proceedings of the 9th International Congress on Applied Mechanics", Brussels, Vol. 8, 51 (1956).

8. Begley, J.A. and Landes, J.D., "The J Integral as a Fracture Criterion", ASTM STP 514, 1 (1972).

9. Averbach, B.L., "Some Physical Aspects of Fracture", in Vol. 1, Fracture, ed. Liebowitz, Academic Press, 441 (1968).

10. Weiss, Volker, "Material Ductility and Fracture Toughness of Metals", Mechanical Behavior of Materials, Proceedings of the 1971 International Conference on Mechanical Behavior of Materials, The Society of Materials Science, Kyoto, Japan, 1, 458-74 (1972).

11. Weiss, V., Kasai, Y. and Sieradzki, K., "Microstructural Aspect of Fracture Toughness", Properties Related to Fracture Toughness, ASTM STP 605, 16-33 (1976).

12. Liu, H.W., "An Analysis on Fatigue Crack Propagation", NASA Cr-2032 (1972).

13. Paris, P.C. and Erdogan, F., "A Critical Analysis of Crack Propagation Laws", J. of Basic Engr., Trans. ASME, Series D., 85, 528 (1963).

14. Weiss, V. and Lal, D.N., "A Note on the Threshold Condition for Fatigue Crack Propagation", Met. Trans., 5, 1946 (1974).

15. Mautz, J. and Weiss, V., "Mean Stress and Environmental Effects on Near Threshold Fatigue Crack Growth", Cracks and Fracture, ASTM STP 601, 154-68 (1976).

FRACTURE TOUGHNESS TESTING

J. G. Kaufman

Alcoa Laboratories

ABSTRACT

A tremendous amount of progress has been made in fracture toughness testing over the past 15 years since ASTM Committee E24 was established to focus on the development of fracture test standards [1-6]. From the early leadership of J. R. Low, Jr., and the consistent effort of W. F. Brown, Jr., R. H. Heyer, J. E. Srawley, E. T. Wessel, and many other committee members, several fracture test standards now exist and a number of new methods are being explored for possible standardization. It is the purpose of this chapter to review the available documents, calling attention to those areas of continuing refinement, and to look at new developments likely to lead to new test standards some time in the future.

The variety of methods now available and under consideration are summarized in Figure 1. For purposes of this chapter, the breakdown in Figure 2 will be used, first concentrating on methods for the direct measurement of fracture toughness parameters with potential design implications, and following up with those methods aimed primarily at screening or quality control tests for fracture toughness, which may be most useful because of their relatively lower cost and reasonable correlatability with the more direct fracture toughness parameters.

```
┌─────────────────────────────────────────────────────┐
│                      CURRENT                          │
│   E399  PLANE-STRAIN FRACTURE TOUGHNESS               │
│   E338  SHARP-NOTCH TENSILE TESTING OF SHEET          │
│                                                       │
│                PROPOSED (PUBLISHED)                   │
│   RESISTANCE CURVES                                   │
│   SHARP NOTCH TENSILE TESTING WITH                    │
│   CYLINDRICAL SPECIMENS                               │
│                                                       │
│                  IN DEVELOPMENT                       │
│   SURFACE-CRACK TESTING                               │
│   J INTEGRAL                                          │
│   C-SHAPED K_Ic SPECIMENS                             │
│   PRECRACKED CHARPY SLOW BEND                         │
└─────────────────────────────────────────────────────┘
```

Figure 1.   Fracture Mechanics Test Methods

```
┌─────────────────────────────────────────────────────┐
│   PLANE STRAIN FRACTURE TOUGHNESS TESTING             │
│          REQUIREMENTS (E399)                          │
│          SPECIAL PROBLEMS                             │
│              SIZE REQUIREMENTS                         │
│              FATIGUE CRACKING                          │
│   J INTEGRAL (PLANE STRAIN WITH PLASTICITY)           │
│   THIN SECTION TESTING                                │
│          RESISTANCE CURVES                            │
│          RESIDUAL STRENGTH                            │
│   SURFACE CRACK TESTING                               │
│   SCREENING TESTS                                     │
│          SHEET (E338)                                 │
│          THICK PRODUCTS; CYLINDRICAL SPECIMENS        │
└─────────────────────────────────────────────────────┘
```

Figure 2.   Outline

DIRECT FRACTURE TOUGHNESS MEASUREMENTS

Plane Strain Fracture Toughness Testing

        ASTM Method E399 is the standard method for plane strain
fracture toughness testing to determine K_{Ic}, and has been in the
ASTM Standards sine 1968 [7].   During that period of time it has
undergone a number of modifications to relax some of the more
stringent requirements and make the method more useful. We will
first review the basic steps in making fracture toughness tests,

note some of the areas in which changes have been made, and finally take note of areas of continuing study.

Two types of specimens are presently covered by ASTM Method E399, the bend specimen and the compact specimen (Figure 4 and 5, respectively, of reference [7]). The basic steps in performing fracture toughness tests are illustrated in Figure 3; it is apparent that the test is a relatively complex one and a number of judgments are necessary at several stages along the process, beginning with the selection of the appropriate specimen sizes and ending with judgments of whether or not the data are valid based upon a number of criteria. The major criteria for validity are summarized in Figure 4; they fall into three major categories: size requirements, fatigue cracking requirements, and load displacement curve requirements.

```
SELECT SPECIMEN SIZE
MACHINE SPECIMEN
PRECRACK SPECIMEN BY FATIGUE LOADING
FRACTURE BY MONOTONIC LOADING
      OBTAIN LOAD vs COD CURVE
ANALYZE CURVE; SELECT CRITICAL LOAD
CALCULATE RESULTS
ASCERTAIN VALIDITY OF RESULTS
```

Figure 3. Steps in K_{Ic} Testing

```
A. SPECIMEN SIZE
      1. THICKNESS ⩾ 2.5 [K_Ic/σ_ys]²
      2. CRACK LENGTH ⩾ 2.5 [K_Ic/σ_ys]²

B. FATIGUE CRACKING
      1. K_MAX/K_Ic ⩽ 0.6
      2. CURVATURE ⩽ 5% CRACK LENGTH
      3. TILT ⩽ 10°
      4. LENGTH BETWEEN 0.45 AND 0.55 SPECIMEN WIDTH

C. LOAD-COD CURVE CHARACTERISTICS
      1. P_max/P_Q ⩽ 1.1
```

Figure 4. Major Validity Checks on K_{Ic} Data

The size requirements represent the first major judgment in performing such tests, since the appropriate size of specimens is dependent upon the level of fracture toughness that is to be measured, so that the usefulness of the final result is dependent upon what size of specimen has been used which in turn cannot be determined to be satisfactory until the test result is known.   The present size requirements are those in the initial version of the document, and basically stipulate that the thickness in crack length must be approximately 50 times the radius of the plastic zone as defined by the Irwin [7] approximation.   These size criteria impose rather severe restrictions on the materials for which plane strain fracture toughness can be made, as for relatively tough materials quite large specimens are required.   This has been the subject of a considerable amount of investigation, some seeming to demonstrate that under certain conditions smaller size specimens can be reliably used, but most data, particularly for relatively high toughness materials, suggesting that to meet all requirements for validity, relatively larger sizes are required.

For example, data obtained at Alcoa Laboratories for medium-strength high-toughness alloy 2219-T851 have demonstrated, as

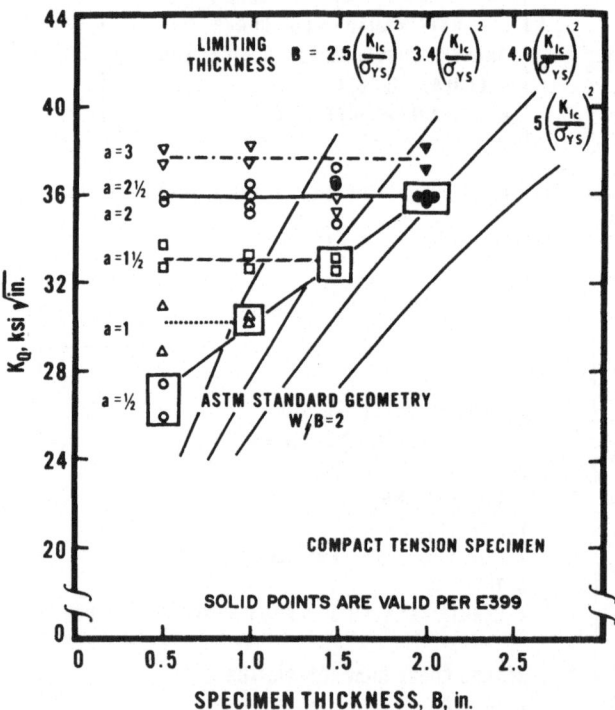

Figure 5.   Influence of specimen thickness in plane-strain fracture toughness tests of 3-in. 2219-T851 plate.

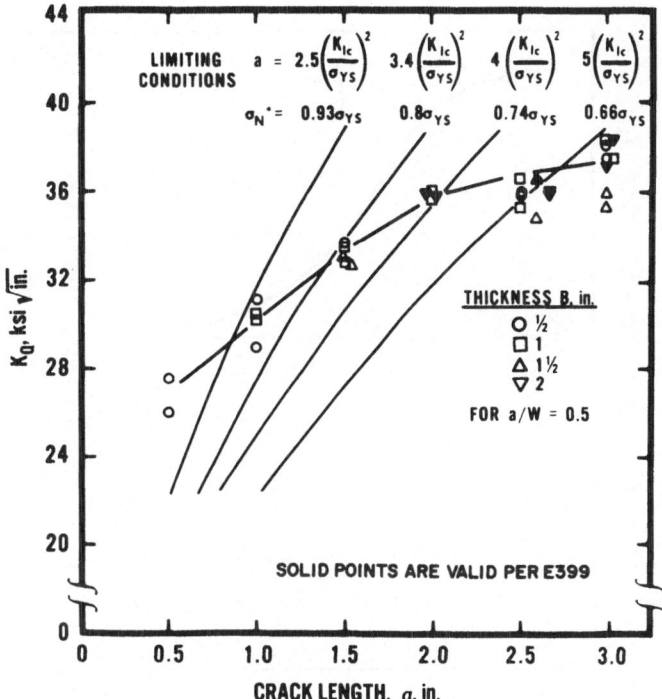

Figure 6.   Influence of crack length independent of thickness in
plane-strain fracture toughness tests of 3-in. 2219-
T851 plate (L-W).

illustrated in Figures 5 and 6, that specimen sizes above the
present limiting criteria may result in load deformation curves
which reflect a relatively rapid rising crack resistance curve and
therefore will not satisfy the criteria on limited plasticity as
expressed by the P_{max}/P_Q ratio to be discussed in more detail later.
For such high toughness materials, these data illustrate that the
selection of specimen sizes where the thickness and crack length
more closely approximate $5(K_{Ic}/\sigma_{YS})^2$ are much more likely to provide
data which satisfy all of the various criteria, and so this guide-
line is used rather regularly in the aluminum industry.   These data
also illustrate that when both thickness and crack length cannot be
made as large as these requirements demand, somewhat less thickness
may be used so long as the plan size is held at that size and the
thickness reduced.   Therefore, when inadequate thickness is available to meet that requirement, the plane size of specimen is selected
to satisfy the criteria with a coefficient of 5, and the maximum
thickness available is utilized.   As the data in Figures 5 and 6
illustrate, the test may not be completely valid, particularly with
regard to the ratio P_{max}/P_Q under these conditions, but values of
K_Q approximately equal to K_{Ic} will be obtained.

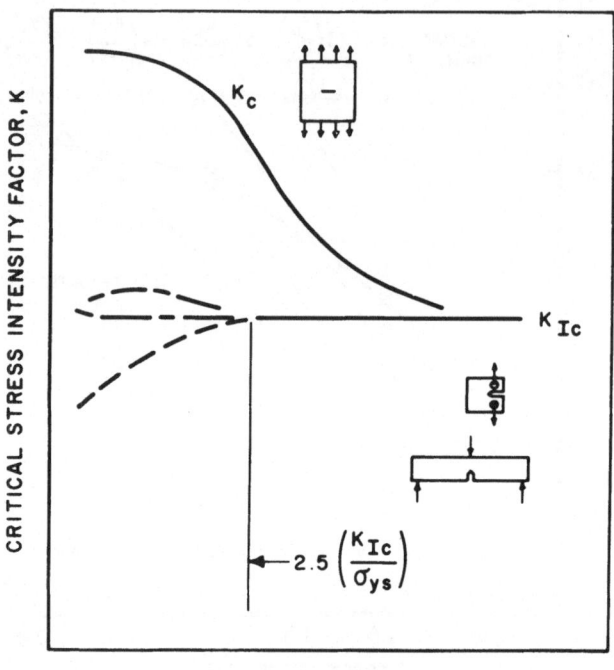

Figure 7. Relationship between K_{Ic} and K_c.

One caution with respect to specimen size requirements: it is often assumed that when undersized specimens are tested and no valid value of K_{Ic} is obtained, the test result may be treated as an estimate of K_c. This is not so as the schematic presentation in Figure 7 illustrates; an invalid number from a hard or compact specimen is likely to far underestimate K_c for any realistic set of conditions.

The seond major set of criteria (Figure 4) involve the fatigue cracking, both the level of stress intensity used to develop the crack, and the characteristics of the fatigue crack which is developed. The limit on stress intensities to be used in the pre-cracking process has been relaxed from 50 percent to 60 percent of K_{Ic} (0.0012 E to 0.002 E), and so this is not a severe requirement, except where relatively low toughness materials are involved. The requirements on fatigue crack front straightness, limiting the center region to variations no more than 5 percent of the average, do cause some problem and study is under way to determine whether or not this criteria can be relaxed. The criteria involving fatigue front flatness and the overall length of the fatigue crack are not usually a problem, though a rather common occurrence is for a crack slightly longer than the 0.55 a/W maximum value to develop. Here again, consideration is being given to relaxation of this requirement, but it seems unlikely for several reasons; first, the stress intensity equation presently in Method E399 is less precise beyond this region; but secondly, even if this were corrected as it will no doubt eventually be, a 5 percent secant offset no longer would provide a relatively precise indication of the load at 2 percent of crack growth (a lesser secant would be required) so that nonconservative (high) values would be achieved unless the secant offset is made variable. This would seen to unnecessarily complicate the method.

The third type of requirement, limiting the P_{max}/P_Q ratio is imposed on the load deformation curve in an attempt to limit the amount of plasticity that takes place during the test and establish whether or not a gradual curvature in the load displacement curve was associated with plastic deformation or to crack growth. The P_{max}/P_Q limit replaces an earlier ineffective requirement limiting the deformation at 80 percent of the 5 percent secant intercept load [8]. The P_{max}/P_Q limiting value has proven quite effective in this case, and while it was established based upon a relatively empirical approach, it does effectively provide assurance that the crack resistance curve is relatively flat for the specimen and, hence, representative of a plane strain test in which plastic deformation is highly constrained (Figure 12).

In summary, ASTM Method E399 is a well established procedure for establishing plane strain fracture toughness, K_{Ic}. Over the seven year life of the method, some improvements have been made, notably in more meaningfull criteria for the judgment of the load displacement curve and relaxation in the fatigue cracking requirements. One added piece of data is also now available in such tests, viz: the specimen strength ratios, designated R_{sb} for the bend specimen and R_{sc} for the compact specimen, and defined in Figure 8. These ratios are essentially ratios of the maximum nominal net-section fracture strength to the tensile yield strength of the material, equivalent to notch-yield ratios, for judgment of the relative toughness of the materials. As a result, in those cases where valid K_{Ic} values cannot be obtained and the significance of of K_Q values is questionable because the criteria for validity are not met, at least relatively useful comparative values of the specimen strength ratios can be obtained. One major caution is noted: specimen strength ratios only suitable for comparison from tests of identically sized specimens both in respect to thickness and plan size.

J Integral

The plane strain fracture toughness test described in Method

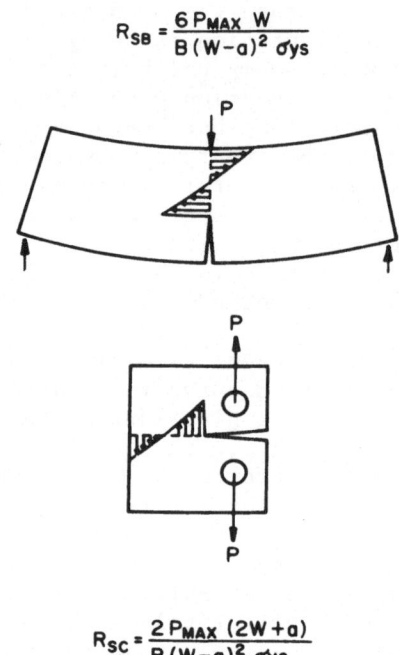

$$R_{SB} = \frac{6\,P_{MAX}\,W}{B\,(W-a)^2\,\sigma_{ys}}$$

$$R_{SC} = \frac{2\,P_{MAX}\,(2W+a)}{B\,(W-a)^2\,\sigma_{ys}}$$

Figure 8.   Specimen Strength Ratios.

E399 assures very limited plasticity by the specimen size requirements.   In testing very high thoughness materials, however, restric-
tively large specimens would be required to measure K_{Ic}.   By
contrast in most real structures of these materials, fracture even
in relatively thick sections is accompanied by relatively large
amounts of plastic deformation.   To handle these cases, an alternate approach to the definition of fracture resistance is needed,
and the J integral approach provides one possible answer.   Though
analytically based upon the path independent J integral of Rice
[9], the J fracture criterion is an energy related criterion [10]
which may be described schematically as in Figure 9, and like G_{Ic}
is based upon the measurement of a critical amount of the energy
associated with initiation of crack growth, but in this case
accompanied by substantial plastic deformation.

     Two approaches to the measurement of this parameter have been
promoted: (1) the multiple specimen approach of Begley and
Landis [11], and (2) the single specimen approximation proposed by
Paris.

     The multiple specimen approach of Begley and Landis is outlined
in Figure 10 and involves monotonic loading of specimens until

PATH INDEPENDENT INTEGRAL OF WORK
TO PRODUCE CRACKING

AS A FRACTURE INDEX

$J_{I_c} = G_{I_c}$ FOR CASE OF BRITTLE FRACTURE OR
SMALL SCALE YIELDING

$J_{I_c} = \dfrac{K_{Ic}^2}{E}$ FOR LARGE SCALE YIELDING

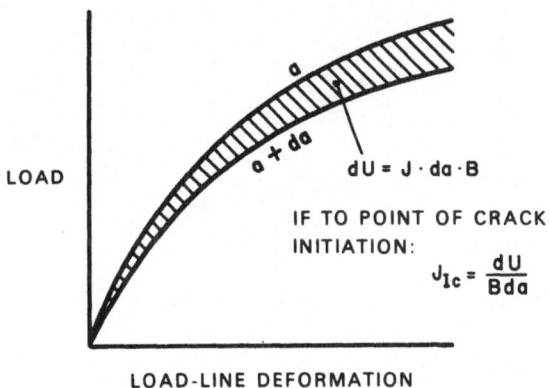

LOAD

$dU = J \cdot da \cdot B$

IF TO POINT OF CRACK
INITIATION:

$J_{Ic} = \dfrac{dU}{Bda}$

LOAD-LINE DEFORMATION

Figure 9.   J Integral

various amounts of crack tip deformation and growth are generated,
measurement of the energy absorbed in developing each of the increments of crack growth, plotting the energy values as a function of
the crack growth and noting the value associated with the initial
substantial crack growth. Begley and Landis noted that such a
plot will frequently have two approximately linear slopes, one
associated with crack blunting prior to real crack growth, and the
second with the crack growth itself. The intersection of these
two lines was designated the J_{Ic} or critical value of the J
integral. From this value of J_{Ic}, values of K_{Ic} are calculated in
much the same way that they are calculated from G_{Ic}. The specimen
size requirements tentatively recommended at this stage for J_{Ic}
testing are:

a(crack length) = B(thickness) = b(remaining ligament or

$$(w - a) = 50 \ (\frac{J_{Ic}}{\sigma_{YS}})$$

The single specimen approximation of J_{Ic} proposed by Paris is

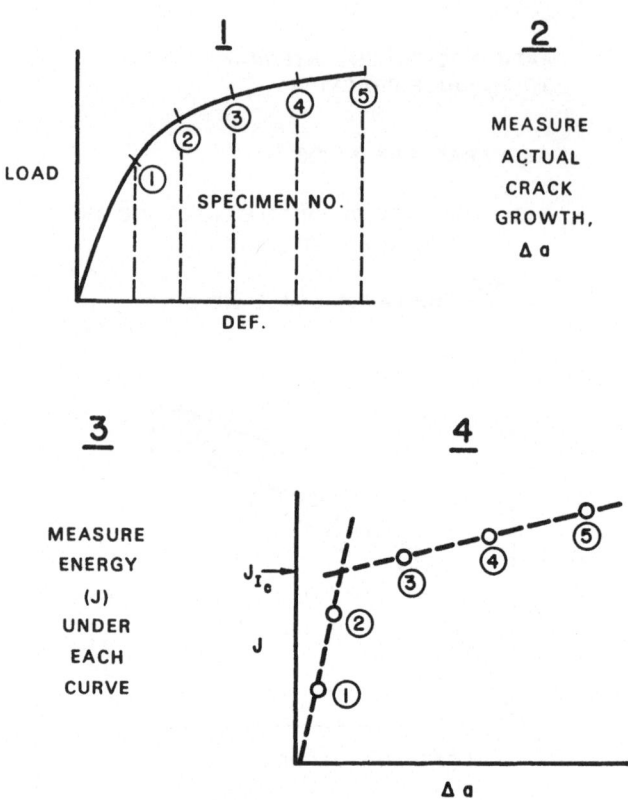

Figure 10. J_{Ic} Testing.

illustrated in Figure 11, and involves the utilization of energy
absorption up the point of crack initiation in a specimen sized to
provide crack initiation at the point of maximum load. All of the
requirements necessary to size the specimen to assure this situation
are not yet completely documented, but additional work may well
illustrate that this procedure is satisfactory.

The current work on the J integral is expected to lead to
some guidelines for a possible recommended test procedure within
the next year. Some discussion still remains about the precision
with which the intersection point is measured and the resultant
relationship between J_{Ic} and G_{Ic} for linear elastic failures.
Since the present guidelines for multiple specimen J_{Ic} testing
involve the selection of a value associated with the initial crack
growth, while G_{Ic} for linear elastic failures is based upon the
stress intensity at an increment of crack growth of 2 percent
(which can be rather substantial for relatively large specimens),
the two parameters may have to be treated separately.

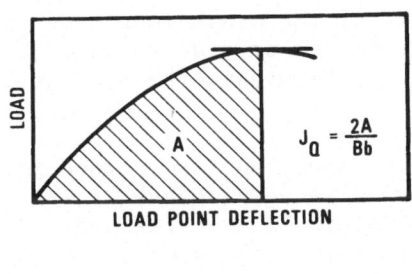

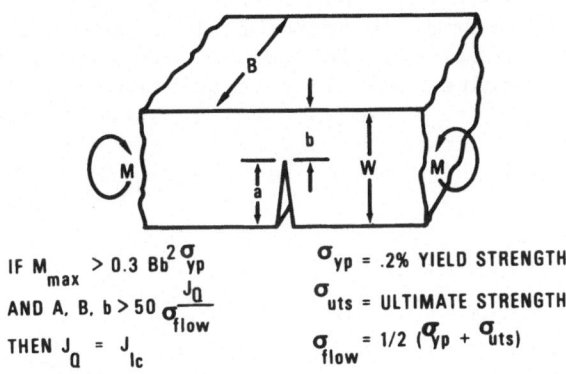

$$\text{IF } M_{max} > 0.3 \ Bb^2 \frac{\sigma_{yp}}{}$$

AND A, B, b $> 50 \frac{J_Q}{\sigma_{flow}}$

THEN $J_Q = J_{Ic}$

σ_{yp} = .2% YIELD STRENGTH

σ_{uts} = ULTIMATE STRENGTH

σ_{flow} = 1/2 $(\sigma_{yp} + \sigma_{uts})$

Figure 11. Maximum load J_{Ic} test method.

Resistance Curves

Moving from relatively thick section fracture problems to those involving a wide range of thicknesses, use of the resistance curve approach originated by Heyer and McCabe [12] is expanding quite rapidly as the method offers promise as the best means of general characterization of fracture behavior. It is applicable to all thicknesses, and is of use to designers as well as materials engineers, as the use of crack resistance curves (K_R vs. Δa) together with driving force curves (K_G vs. Δa) for individual specimens or structures leads to estimates of K_c, the critical stress intensity factor at instability.

Crack resistance curves may be measured from various types of specimens, but the present proposed method [13] involves center-cracked panels and edge-cracked panel of either the wedge load (CLWL) or compact pin load (CS) variety. For the best overall characterization of the crack resistance curve, an edge-cracked specimen offers the advantage in that no instability develops because of the nature of the crack-length dependence of the driving force, whereas with the center-cracked panel and instability does develop and the test is terminated before the complete resistance curve is defined.

In either type of specimen, the stress intensity is measured as a function of the incremental crack growth and curves of the type illustrated in Figure 12 are obtained. Relatively thin or relatively tough materials will be characterized by a gradual curvature of the crack resistance curve and a relatively high plateau value, illustrating that stable crack growth can be tolerated, while relatively thick and/or brittle materials will be characterized by relatively flat curves, suggesting that once any crack growth is obtained indefinite crack growth is likely. Thus a plane strain fracture toughness test is actually the establishment of a point on a crack resistance curve similar to the latter case, and it is for this reason that the P_{max}/P_Q ratio is a useful

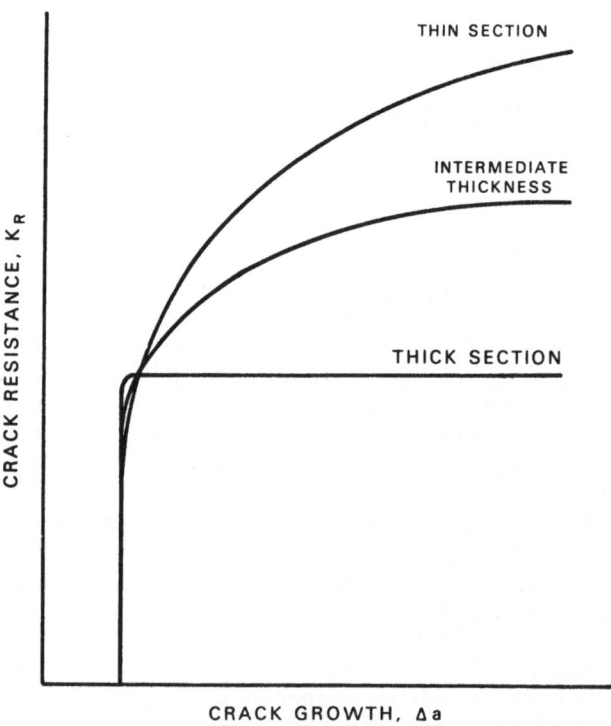

Figure 12.   Crack resistance curves. K_R versus Δa

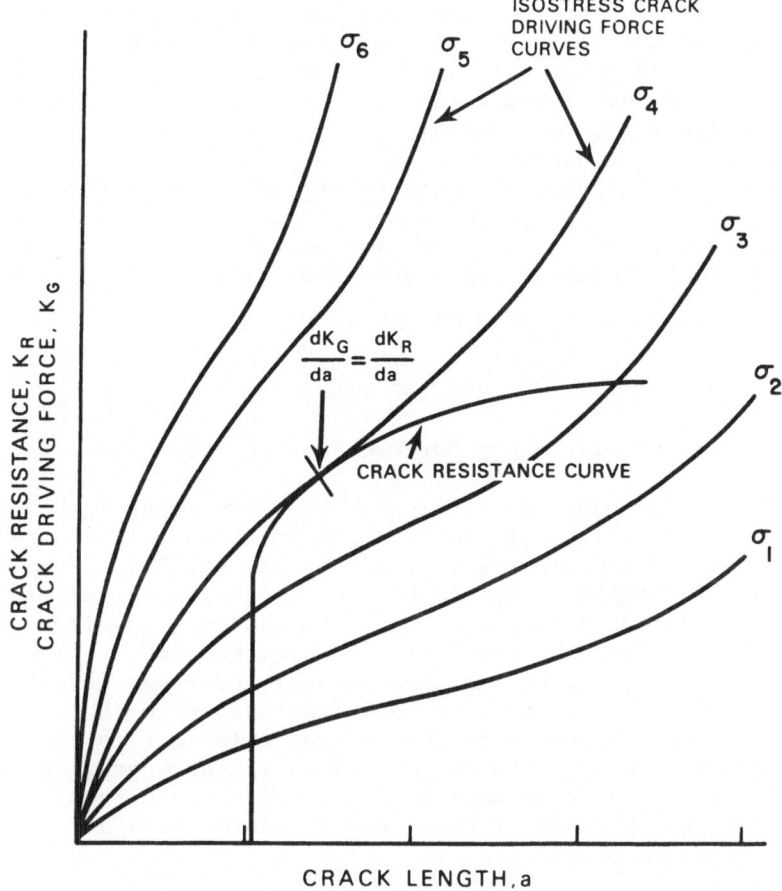

Figure 13. Instability condition defined by tangency of K_R, K_G
versus a curves

parameter in assuring that plane strain conditions have been met
as noted earlier.

The center-cracked panels are normally loaded in tension
testing machines much as the earlier center-cracked panel K_c tests
were conducted. The edge-cracked panels are frequently tested in
horizontal setups of the type designed by Heyer and McCable and a
double compliance method is used to infer both the load and effective
crack length [12].

A substantial round robin test program is presently under way
employing the proposed test method as it appeared in the 1974 ASTM

Standards. It is expected that this method will be advaned to a
tentative standard in the relatively near future and the results of
the current round robin will lead to some refinement and improvement in the method. The usefulness of the data obtained can be
seen from Figure 13. An overlay of the crack resistance curve of
the material at the expected a_0 position, established by non-
destructuve test criteria, on the isostress crack driving force
curves for the structure established the potential conditions for
fracture instability: the stress intensity at which a tengency
(i.e., $dK_G/da = dK_R/da$) is generated between the two types of
curves.

SCREENING TESTS

Rectangular Sharp Notched Tensile Specimens

ASTM Method E338 [14] represents the first test method
available for establishing the relative toughness of materials
based upon notch tensile tests of relatively small and simple
specimens. Two types of specimens are called for, Figure 15 of
reference [14], one a double-edge-machine-notched specimen (EN),
and the other a center-fatigue-cracked specimen (CC). In general,
these specimens are tested in conventional universal tensile
systems and only the maximum nominal fracture strength is measured.
The principal criterion of toughness from this test is the ratio of
the sharp notch strength to the tensile yield strength, as this
parameter has shown to be much more reliable than either the sharp
notch strength by itself or the ratio of sharp notch strength to
ultimate strength [15].

The EN specimen has been rather widely used in the aluminum
industry and provides an economical screening test for toughness
of sheet products. The CC specimen has not been widely used,
probably because of the expense and time required to satisfactorily
fatigue crack such specimens, and the limited data derived from
specimens in which that much is invested. As a result, work is
under way at the present time to come up with a more economically
fatigue cracked specimen, which is fatigue cracked on one side
only. Some type of fatigue cracked specimen is likely to be
necessary for steels and titanium alloys, for which the relatively
sharp notches required for the EN specimen cannot be reproducibly
obtained.

Notched Round Tension Testing

For some time it has been recognized that tensile tests of
sharply notched round specimens provided useful indications of

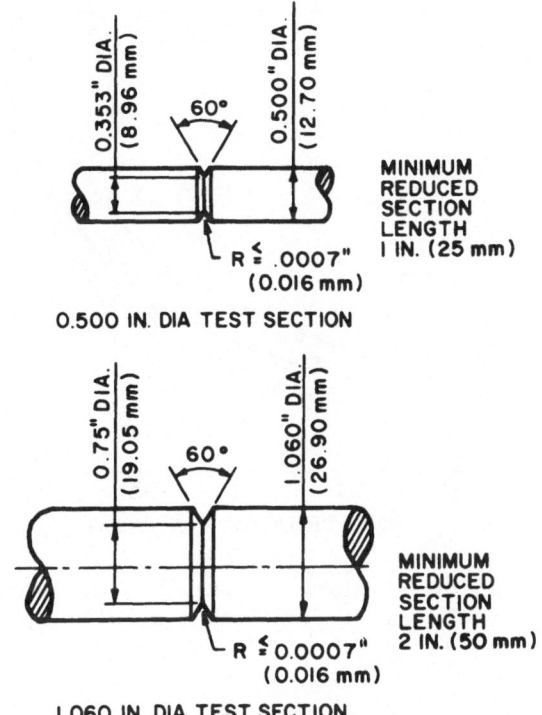

0.500 IN. DIA TEST SECTION

1.060 IN. DIA TEST SECTION

Figure 14.   Standard test sections of notched tensile specimens.

fracture toughness [16], particularly when the net-section fracture
strength was relatively low with respect to the tensile yield
strength.   As a result, this type of specimen has been the basis of
fracture toughness quality assurance systems set up in the aluminum
industry [17].   A proposed method of sharp notched tension testing
has been written and was published in the 1975 ASTM Standards [18];
specimens of the type shown in Figure 14 are included.   The relatively sharp notches required can be quite reproducibly machined
in aluminum and magnesium alloys, and useful correlations of the
sharp notch tensile strength to yield strength ratio (the notch-
yield ratio) to the plane strain fracture toughness K_{Ic} have been
demonstrated, as illustrated in Figure 15.   Such correlations are
in use now by establishing limiting values of the ratio which
provide high assurance that certain minimum values of plane strain
fracture toughness are established.   It is expected that when
larger quantities of data are available, rather stringent statistical approaches to the calculation of the minimum ratios will be
employed.

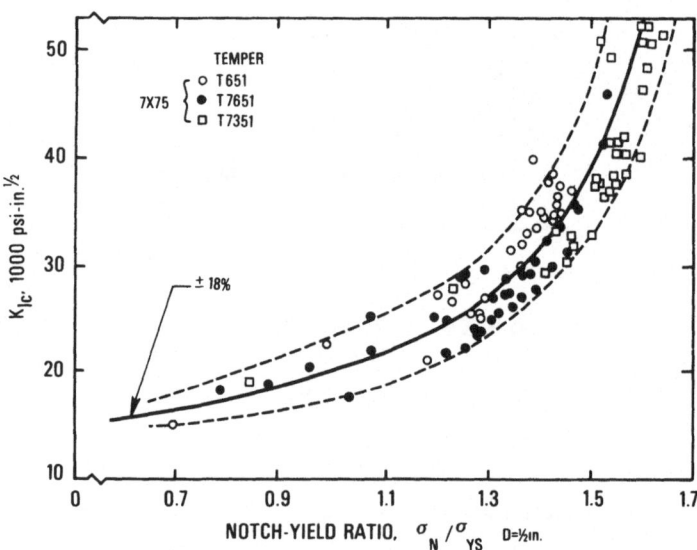

Figure 15.   Correlation of plane-strain fracture toughness with
notch-yield ratio for 7075 and 7475 plate.

Slow–Bend Precracked Charpy Tests

          For a number of years, attempts have been made to relate the
results of various tests of Charpy-size specimens to plane strain
fracture toughness with varying degrees of success [19]. Generally,
impact tests of such specimens either with or without fatigue
cracks have not been found to correlate well with K_{Ic} and it has
been more recently recognized that this is because of the difference between the dynamic and static toughness of many ferritic
materials. However, it also has been fairly well demonstrated
now that if slow bend tests of the Charpy-size specimens are
employed, useful correlations between energy to fracture and K_{Ic}
can be obtained. At this time, work is under way to further
simplify the measurements in instrumented slow-bend precracked
Charpy tests by employing a simpler criteria such as maximum load
for correlation with K_{Ic} and therefore to further reduce the cost.
If this can be accomplished, the relatively small size of specimen
and the relative ease with which it can be fatigue cracked will
unquestionably make it a strong contender for screening and
quality control tests, particularly of steels and titanium alloys
where the sharp edge-notched tensile specimens cannot be readily
machined.

Surface Cracked Specimens

Despite the face that thumbnail, surface or part through cracks are the most common ones obtained in service, relatively little progress has been made in the standardization of test methods for specimens employing this type of crack. This rests primarily upon the problems in accurately characterizing the stress intensity relationships in such specimens. Nevertheless, specimens with surface cracks have been used for a number of years for screening purposes [20] and with various approximate analyses estimating the critical level of stress intensity associated with fracture instability.

As part of an E24 activity, guidelines for surface-cracked specimen testing have been summarized [21]. Data available to date indicate that the guidelines from that study, shown in Figure 16, are likely to be incorporated into a future recommended practice or test method.

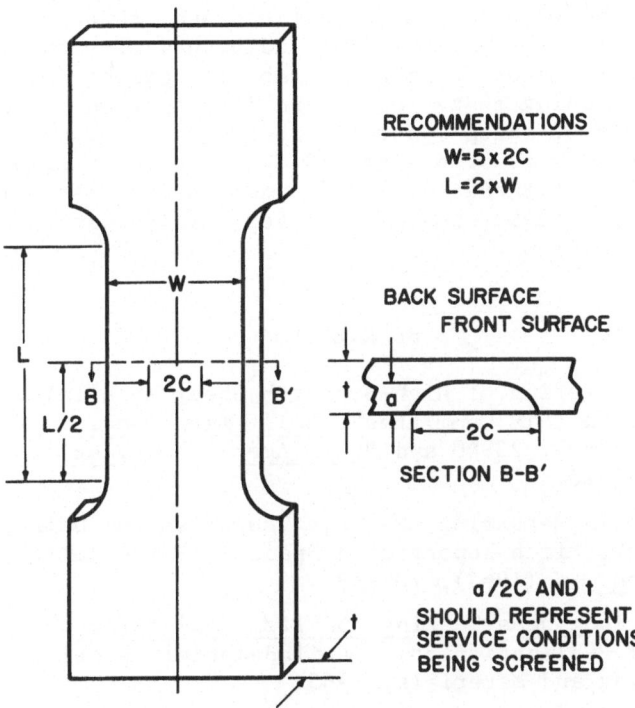

Figure 16.   Typical surface-crack specimen, nomenclature, and conventions (grip details omitted).

For specimens in which the surface cracks are relatively shallow (less than half the thickness of the material) the stress intensity relationships developed early by Irwin [20] seem to provide quite useful estimates of the fracture instability stress intensity factors. For relatively tough materials and deeper cracks, the analysis of Collipriest [22] and Kobayashi [23] should be considered.

SUMMARY

The status of work in the establishment of fracture toughness test methods both of the direct measurement and the screening variety have been summarized. ASTM Method E399 is well established for the determination of plane strain fracture toughness, K_{Ic}, for relatively thick sections. Work is under way to broaden the applicability of such testing to larger amounts of plasticity through the J integral approach. For thin sections and those of intermediate thicknesses with mixed mode fracture, the original K_c approach has been shown to be limited and the crack resistance curve has been recognized as the most versatile and meaningful toughness index. This has the advantage that the complete crack resistance of the material can be defined and, through comparison with crack driving force curves for individual structures (or specimens), a broad range of fracture instability can be described. In the area of screening tests, considerable attention is being given to tension tests of sharply notched rectangular and round specimens, and to slow-bend tests of precracked Charpy-size specimens; correlations of results of these relatively inexpensive tests with K_{Ic} measurements, per E399, show promise for quality assurance and screening applications.

REFERENCES

1. "Fracture Testing of High-Strength Sheet Materials: A Report of a Special ASTM Committee", Bull. Amer. Soc. Test. Mater., No. 243 (1960), 23-40 and Bull. Amer. Soc. Test Mater., No. 244 (1960), 18-20.

2. "Progress in Measuring Fracture Toughness and Using Fracture Mechanics", Fifth Report of a Special ASTM Committee, Mater. Res. Stand., 4 (1964), 107-19.

3. Fracture Toughness Testing and Its Applications. Special Technical Publication 381. Philadelphia: American Society for Testing and Materials, 1965.

4. Brown, W.F., Jr., and Srawley, J.E., Plane Strain Crack Toughness Testing of High Strength Metallic Materials. Special Technical Publication 410. Philadelphia: American Society for Testing and Materials, 1967.

5. Kaufman, J.G., "Progress in Fracture Testing of Metallic
    Materials", in Review of Developments in Plane Strain Fracture
    Toughness Testing, ed. by W.F. Brown, Jr. Special Technical
    Publication 463. Philadelphia: American Society for Testing
    and Materials (1970), 3-21.

6. Fracture Toughness. Special Technical Publication 514.
    Philadelphia: American Society for Testing and Materials, 1972.

7. "Standard Method of Test for Plane Strain Fracture Toughness of
    Metallic Materials", Designation: E399-74, in 1975 Annual
    Book of ASTM Standards, Pt. 10. Philadelphia: American Society
    for Testing and Materials (1975), 561-80.

8. Srawley, J.E., Jones, M.H. and Brown, W.F., Jr., "Determination
    of Plane Strain Fracture Toughness", Mater. Res. Stand., 7
    (1967), 262-66.

9. Rice, J.R., "A Path Independent Integral and the Approximate
    Analysis of Strain Concentration by Notches and Cracks",
    Trans. ASME, Ser. E, J. Appl. Mech., 35 (1968), 379-86.

10. Begley, J.A. and Landes, J.D., "The J Integral as a Fracture
    Criterion", in Fracture Toughness. Special Technical Publica-
    tion 514. Philadelphia: American Society for Testing and
    Materials (1972), 1-23.

11. Landes, J.D. and Begley, J.A., "The Effect of Specimen Geometry
    on J_{Ic}", in Fracture Toughness. Special Technical Publication
    514. Philadelphia: American Society for Testing and Materials
    (1972), 24-39.

12. Fracture Toughness Evaluation by R-Curve Methods. Special
    Technical Publication 527. Philadelphia, Pa.: American
    Society for Testing and Materials, 1973; in particular,
    McCabe, D.E. and Heyer, R.H., "R-Curve Determination Using a
    Crack-Line-Wedge-Loaded (CLWL) Specimen", 17-35.

13. "Proposed Recommended Practice for R-Curve Determination", in
    1975 Annual Book of ASTM Standards, Part 10. Philadelphia:
    American Society for Testing and Materials (1975), 811-25.

14. "Standard Method of Sharp Notch Tension Testing of High
    Strength Sheet Materials", Designation: E338-68, in 1975
    Annual Book of ASTM Standards, Part 10. Philadelphia: Ameri-
    can Society for Testing and Materials (1975), 550-57.

15. Kaufman, J.G. and Johnson, E.W., "The Use of Notch-Yield Ratio
    to Evaluate Notch Sensitivity of Aluminum Alloy Sheet", Proc.
    Amer. Soc. Test. Mater., 62 (1962), 778-91.

16. Kaufman, J.G., "Sharp Notch Tension Testing of Thick Aluminum
    Alloy Plate with Cylindrical Specimens", in Fracture Toughness.
    Special Tehcnical Publication 514. Philadelphia: American
    Society for Testing and Materials (1972), 82-97.

17. "The Aluminum Association Postion on Fracture Toughness
     Requirements and Quality Control Testing", Interim Report
     T-5, September 1974.

18. "Proposed Method for Sharp Notch Tension Testing of Thick
     High-Strength Aluminum Magnesium Alloy Products with Cylindri-
     cal Specimens", in 1975 Annual Book of ASTM Standards, Part 10.
     Philadelphia: American Society for Testing and Materials
     (1975), 799-810.

19. Sailors, R.H. and Corten, H.T., "Relationship Between Material
     Fracture Toughness Using Fracture Mechanics and Transition
     Temperature Tests", in Fracture Toughness. Special Technical
     Publication 514. Philadelphia: American Society for Testing
     and Materials (1972), 164-91.

20. Orange, T.W., "Fracture Testing with Surface Crack Specimens",
     J. Test. Eval., 3 (1975), 335-42.

21. Collipriest, J.E., Jr., "An Experimentalist's View of the
     Surface Flaw Problem", in The Surface Crack: Physical Problems
     and Computational Solutions. New York: American Society of
     Mechanical Engineers (1972), 43-61.

22. Kobayashi, A.S. and Moss, W.L., "Stress Intensity Magnification
     Factors for Surface Flawed Tension Plate and Notched Round
     Tension Bar", in Fracture 1969, Proceedings of the Second
     International Conference on Fracture, Brighton, April 1969.
     London: Chapman and Hall Ltd. (1969), 31-45.

METHODS FOR DYNAMIC FRACTURE RESISTANCE TESTING

E. A. Lange

Naval Research Laboratory, Washington, D. C.

ABSTRACT

The principal need for a dynamic fracture tests is to define
the ductile-brittle transition in the fracture resistance of conventional ferritic steels and irons as a function of temperature
and size. Dynamic criteria of fracture resistance have also been
used for structural metals with a limited sensitivity to strain
rate due to the economy of conducting an impact test. Because of
the general availability of impact machines for conducting a
Charpy test, much research effort has been expended on attempts to
make that test a more quantitative tool. For some materials, Charpy
test results are difficult to interpret, and new tests have been
developed and standardized that more readily provide information
related to structural performance. Modified Charpy tests, the Drop-
Weight NDT test, the Drop-Weight Tear test, and the Dynamic Tear
test are discussed with respect to their attributes and their
limitations.

INTRODUCTION

The recent maturation of linear elastic fracture mechanics
(LEFM) technology was timely because of the growing need to understand and improve the empirical tests that have been in general use
for the routime measurement of fracture resistance. An understanding of the basic limitations of the two most widely used conventional
impact tests, the Charpy V-notch (C_V) test, and the Drop-Weight Nil-
Ductility-Transition (DWT-NDT) test, prompted the development of a
new test with broad range measuring capability [1,2,3]. In the new

test method, which is called the Dynamic Tear (DT) test, the
simplicity of a three-point bend specimen and an energy criterion
is maintained. Although empirical correlations have been made
between DT energy (DTE) and K_{Ic} values for various high-strength
alloy systems, the primary purpose of the DT test is to provide
information on materials for structures where some ductility is
required for reliability even though flaws may be present [4].

Effects of Strain Rate

The question of applicability of fracture resistance data
generated by the high strain rates associated with impact tests has
been raised for many structural applications. Stress-intensity
rates from 3×10^4 to 1×10^6 ksi$\sqrt{in.}$/s (3×10^4 to 1×10^6 MN·m$^{3/2}$/s)
causing fracture to occur in 1 to 0.1 ms are common for conventional
impact loading rates. This relatively high strain rate may be
several orders of magnitude higher than that encountered in
elements of bridges, ships, or pressure vessels. Although the
loading rates on a structure may be intermediate, service experience has shown that the catastrophic fractures of large welded
structures are related to the dynamic, high-strain rate properties
of the materials. A flaw or crack may be dormant for years with
unstable crack extension controlled by the static fracture resistance properties, but when the conditions at the crack tip cause a
few grains to fracture by cleavage, the local mechanics change
drastically. The local strain rate associated with the "pop-in"
is of the order of that in an impact test and continued extension
of the crack is then dependent upon the dynamic properties of the
material even through the structure is under a static load. Therefore, dynamic tests are not only conducted for economic reasons,
but they are needed to provide information related to potential
structural performance under emergency conditions which may be a
small pop-in crack or a plastic overload.

The high loading rates associated with impact tests impose a
complex mechanical condition in the specimen. Recent analyses of
the mechanical conditions in impact tests have shown that for
loading rates causing fractures to occur in less than 1 ms, the
stricker force may not be in phase with the bending moment. This
complicates the generation of dynamic K_{Ic} or J_{Ic} values, frequently
referred to as K_{Id} or J_{Id} values [5,6]. To obtain K_{Id} values under
impact conditions, the impact velocity is frequently decreased to
the point where a static analysis can be employed or the specimen
is instrumented to obtain an accurate measure of the load or strain
level associated with fracture [6,7].

Caution must be exercised when any generalities are proposed
for the effects of intermediate loading rates that cause fracture

to occur in less than 1 ms. A recent study involving a variety of bridge steels showed that the shift in the temperature of the transition region due to strain rate is complex and not a predictable characteristic for a broad range of steels [8].

Effects of Constraint

Constraint has a very significant effect on the initiation of cleavage fracture in ferritic steels. Therefore, constrain has an important influence on the temperature at which the ductile to brittle transition in fracture resistance occurs. Constraint is related to any dimensional parameter that tends to restrict plastic flow at the notch tip. This includes the depth of the notch, the sharpness of the tip of the notch, and the thickness of the section. Because of the unpredictable effect of side grooves on fracture resistance in the elastic-plastic regime, side grooves are not generally used on impact specimens.

Constraint is developed by using a notch with a sharp tip and a notch depth sufficient to prevent the plastic zone from extending to the top surface of the specimen. The level of constraint developed by a specimen can be defined as the capability of the specimen to cause fracture to occur under the conditions defined as "plane-strain" in ASTM E399. For "plane" strain fracture to occur, the thickness of the specimen must comply with the following conditions:

$$B \geq 2.5 \ (K_{Ic}/\sigma_{ys})^2 \tag{1}$$

where:

B = thickness, in.

K_{Ic} = critical stress-intensity factor ksi$\sqrt{\text{in}}$.

σ_{ys} = yield strength, ksi

The relationship in Equation (1) is useful for evaluating the effects of section size on the transitions in fracture resistance that can occur with relatively small changes in yield strength or temperature. The "strength transition" is an important characteristic in the fracture properties of high-strength metals, and the "temperature transition" is an important characteristic in the fracture properties of the ferritic irons and steels [9].

The relationship between specimen size and DT energy for fully plastic fracture has been investigated, and this subject is discussed in a later section concerning the DT test. The point here is that the section size effect can be complicated when the transition features in the fracture resistance of a material are

being determined with a subthickness specimen. Linear elastic
fracture mechanics is most useful in defining fracture resistance
in the plane-strain regime and into the elastic-plastic regime.
However, when extensive through-thickness yielding occurs, the
size of the plastic zone is limited by the width of the specimen
in addition to its thickness and the criterion used as a measure
of fracture resistance becomes a three-dimensional parameter. In
this case, the fracture resistance of the material is a function
of the geometry of the test piece.

Effects of Stable Crack Extension

The translation of empirical measures of fracture resistance
to structural performance involves more than simple relationships
between stress and flaw size when extensive through-thickness
yielding precedes fracture. At such high levels of fracture
resistance, a small crack front extension does not lead to an
unstable condition, and continued extension of the crack is dependent upon continued input of plastic overload energy. Therefore,
the fracture mode and propagation rate becomes dependent upon the
design of the structure and the nature of the overload being
experienced.

In the plastic regime of high fracture resistance, pop-in flaws
can be arrested and any flaws can grow until the stress in net
section of the load bearing member exceeds the yield strength or the
limit load condition. A guarantee of limit load performance in the
presence of a through-thickness flaw would be sufficient evidence
to certify a high level of integrity for most structures. Although
it is of primary importance, material selection is only one aspect
of a structural integrity analysis which also includes trade-offs
in design refinement, fabrication quality, and inspectability.

If a structure, such as a gas line pipe, requires the material
to withstand plastic overload in order to attain an acceptable level
of structural integrity, the correlation between a dynamic fracture
test result and the performance of the material in the structure
can only be reliably developed with full-scale tests. Depending
upon the consequences of limited, stable crack extension, fracture
in a critical region may be load, strain, or energy dependent, and
the level of fracture resistance of a material that is needed for
acceptable reliability can be determined most economically by
empirical correlation. For some cases, finite element analysis can
provide useful and economical guidance especially where the design
of a detail being considered is repetitive. Although the available
analytical tools are useful, translation of fracture resistance
criteria to structural performance outside of the plane-strain
regime remains an empirical exercise.

CHARPY V-NOTCH TEST

Two of the most overworked words that make the semantics of impact testing confusing to many people are "Charpy" and "Drop-Weight". Because of the proliferation of specimens and criteria for fracture resistance using these generic terms, some clarification is in order. About the only thing that the impact tests with these prefixes have in common is a three-point bending load. The Drop-Weight term indicates, however, that testing is by impact, and the parameter used for fracture resistance is the appearance of the fracture, ductile or brittle.

The Charpy test is the oldest standardized impact test in general use, E23-33T. Although there are several specimen designs, the most commonly used specimen today is the 10 mm square specimen with a 2 mm deep notch having a 0.25 mm root radius. The C_v test has served well for comparing certain materials, but translation of C_v energy to structural parameters is on a case-by-case basis.

The C_v energy criterion provided the first correlation between the results of a laboratory fracture test and the service performance of the material in a structure. The correlation of C_v energy and the initiation, propagation, and arrest of fractures in the WW II Liberty ships and T-2 tankers is shown in Figure 1. Fractures initiated from small flaws when the C_v energy was less than 10 ft-lb (14 J) and fractures arrested when the C_v energy was more than 20 ft-lb (27 J). These correlations have been used to justify the use of a 15 ft-lb (20 J) criterion to preclude brittle fracture in all types of welded steel structures. Unfortunately, this 15 ft-lb (20 J) criterion was found to be a unique criterion that could predict a certain level of structural performance in only a few steels.

All steels have a very low level of fracture resistance when their C_v energy value is less than 15 ft-lb (20 J). However, the temperature associated with this criterion is not necessarily the start of the transition region to higher levels of fracture resistance which was the case for the ship steels in Figure 1. For this reason, other criteria of fracture resistance, such as percent shear and lateral expansion have been promoted as parameters that are more generally indicative of structural performance [3,10].

Numerous papers have been written on the problems that arise when attempts are made to develop a correlation between structural performance and any criterion of fracture resistance generated with the Charpy test [3,4,8,11,12,13]. The difficulty does not stem from the criterion used for fracture resistance, because the amount of ductility that occurs prior to fracture is related to the geometry of the test piece. Therefore, the various criteria that

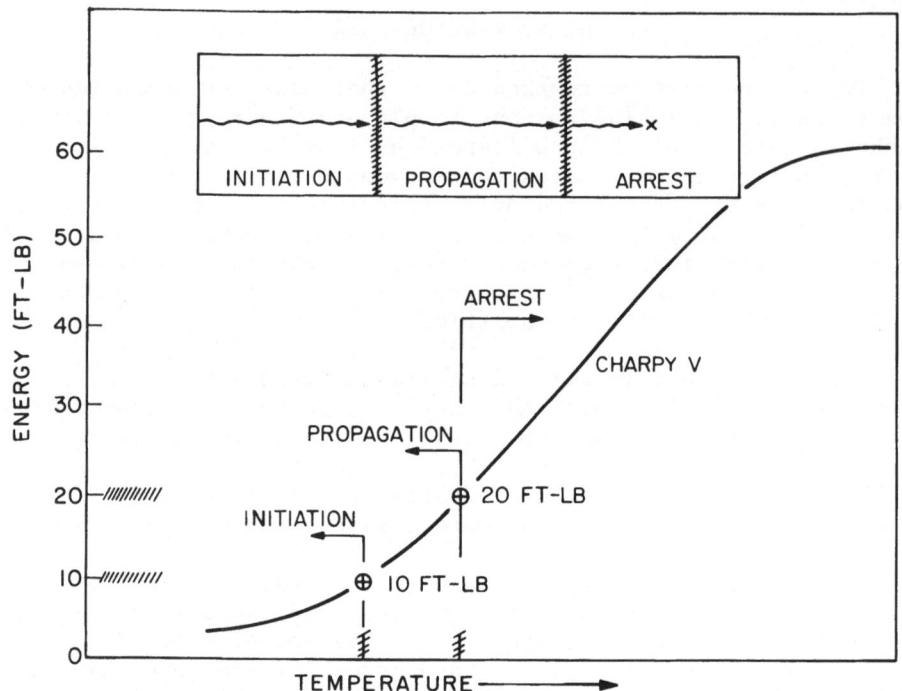

Figure 1. Correlation of WW II ship fractures with C_v energy values.

have been proposed are all proportional to each other. The dilemma of the C_v test is not due to the criterion of fracture resistance, but it is due to the design of the test piece.

The relatively dull notch in the standard C_v specimen is one reason the transition region of a C_v energy-temperature relationship is not in phase with structural performance. To improve this feature of the C_v specimen design, much recent research has been conducted on studies of the benefits of fatigue cracking the 10 mm C_v specimen. Fatigue cracking the specimen tends to steepen the transition region and shift it to a higher temperature, but the plane strain measuring capability of a 10 mm section remains at $0.4 = K_{Id}/\sigma_{yd}$. This limitation in plane strain measurements can be an important restriction to the characterization of the transition region of certain steels.

The dynamic plane strain fracture toughness relationship with temperature is nearly asymptotic to the $K_{Id}/\sigma_{yd} = 0.5$ level in some steels. For such materials there is a big shift in the temperature

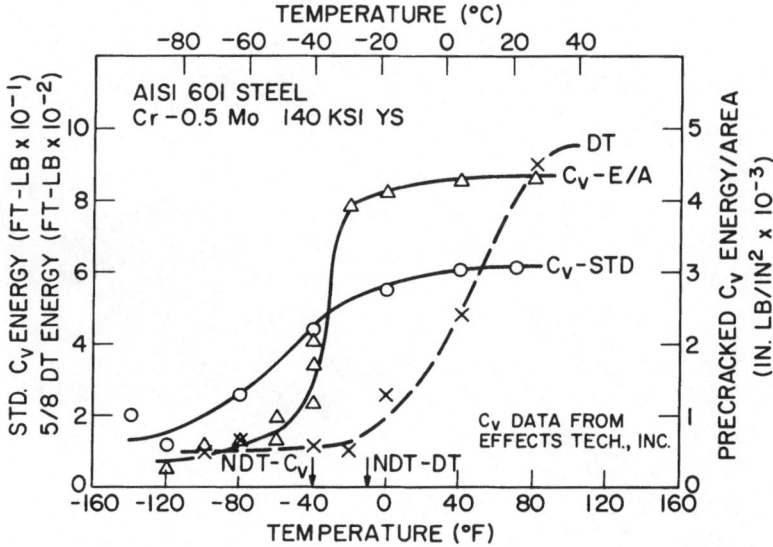

Figure 2. Comparison of the temperature transition characteristics of an SISI 601 steel using the standard C_V test, the fatigue precracked C_V test, and the 5/8-in. (16 mm) DT test. Note that the curves for the two sets of C_V test data are displaced to lower temperatures than that for the DT data.

of the transition region when the constraint conditions are slightly less than this critical level which is associated with the constraint level of a section 16 mm (5/8-in.) thick. This is illustrated by the data in Figure 2, where the influence of notch sharpness and section size on the temperature transition features of an AISI 601 steel heat treated to 140 ksi (960 MPa) yield strength can be seen. Note that the transition curve for the 10 mm of fatigue cracked Charpy specimen, the C_V-energy/area curve, is significantly steeper than the transition curve for the standard C_V energy values. However, the curve for the precracked Charpy specimen is displaced approximately 60°F (33°C) from that for the 16 mm DT specimen. Unfortunately, these displacements in the energy temperature relationship from 10 mm and 16 mm specimens are not predictable from steel to steel. Therefore, when the 10 mm C_V specimen is a subthickness specimen, care must be exercised in the interpretation of energy values in the transition (mixed-mode) region. This is also true for a 16 mm specimen, but with proper specimen design, analyses of energy values can be made more reliable.

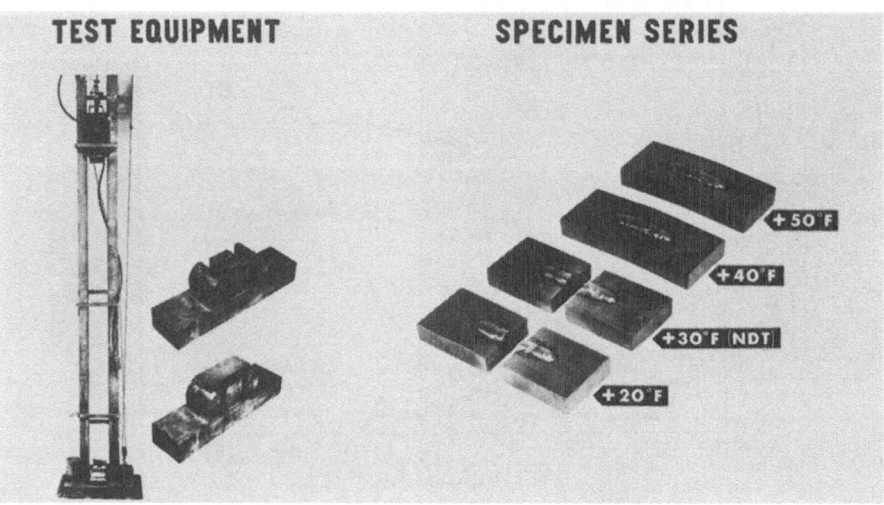

Figure 3.   NRL Drop-Weight Test. The drop-weight machine and a
             series of specimens illustrating the break – no-break
             performance at the Nil-Ductility-Transition (NDT)
             temperature.

DROP-WEIGHT, NIL-DUCTILITY TRANSITION (NDT)
TEMPERATURES TEST (ASTM E208)

Because of the difficulty and expense of developing correlations between C_v energy and the structural performance of every
ferritic steel, the Drop-Weight, Nil-Ductility Transition (NDT)
temperature test was devised at the Naval Research Laboratory in
the early 1950's [15]. This test is currently ASTM E208, and it was
the first test to be called a "Drop-Weight" test. This term was
used to emphasize the simplicity of a test that could determine the
temperature at which a steel sample can plastically deform under a
dynamic load in the presence of a small crack. It was the first
"pop-in" crack test.

The pop-in crack of the DWT-NDT specimen is initiated by a
brittle weld bead that fractures when the specimen is loaded in a
drop-weight machine, Figure 3. The crack from the weld bead is
either arrested or it continues to propagate across the top surface
of the specimen. If the crack does not propagate to one of the top
corners of the specimen before the surface strain approaches 2
percent, the deformation is stopped by an arrestor block on the

anvil and the test result is called a "no break". Specimens are
subsequently tested at lower temperatures until a "break" occurs.
The DWT-NDT test is, therefore, a "go" - "no-go" appearance test.

The interpretation of performance at an NDT temperature is
self explanatory, but the NDT temperature index can provide
broader meaning when certain generalities occur in the transition
features of a steel. These generalities in the transition features
of conventional steels were first seen from the results of the C_v
test and the explosion bulge test [4].

The Explosion Bulge Test

The intent of the explosion bulge test was to simulate a large
structural element being subjected to a dynamic, plastic overload.
Because of the size of the original specimen, 14x14x1-in. (356x
356x25-mm), the plate was loaded by means of an explosive charge.
The plates were conditioned to the desired test temperature, placed
upon a die with a circular hole and an explosive charge was then
detonated at a stand-off distance of 12 in. (305 mm) or more above
the plate surface. With this technique, the air transmits a shock
wave that causes an inertial loading of the plate and a strain rate
in the material that is comparable to that in the DWT-NDT test. In
Figure 4, a series of plates that were tested throughout the transition temperature range is shown and the three critical temperatures
are noted.

The Critical Temperature Concept

When a 1 in. (16 mm) thick plate of steel is tested in the
bulge test, the temperatures for three critical levels of performance are noted. The temperature below which a plate fractures in
a brittle (nil-ductile) manner is called the NDT temperature.
Approximately midway in the transition region, fractures are
arrested at the hold-down region. This performance is called
"Fracture Transition Elastic" (FTE), because plastic overloading
is required to propagate the fracture at temperatures above the FTE.
The "upper shelf" temperature of the transition region is called
"Fracture Transition Plastic" (FTP), because at temperatures above
the FTP, no cleavage fracture occurs and the plate is fully plastic.

The three critical temperatures, NDT, FTE, and FTP can occur
at approximately 60°F (33°C) increments for conventional, low-
strength structural steels. On the basis of this observation and
the analysis of many structural failures, a generalized Fracture
Analysis Diagram (FAD), as shown in Figure 5, was proposed in 1963
by Pellini. The FAD was indexed to the NDT temperature and the
stress level relative to yield strength [15]. This diagram was

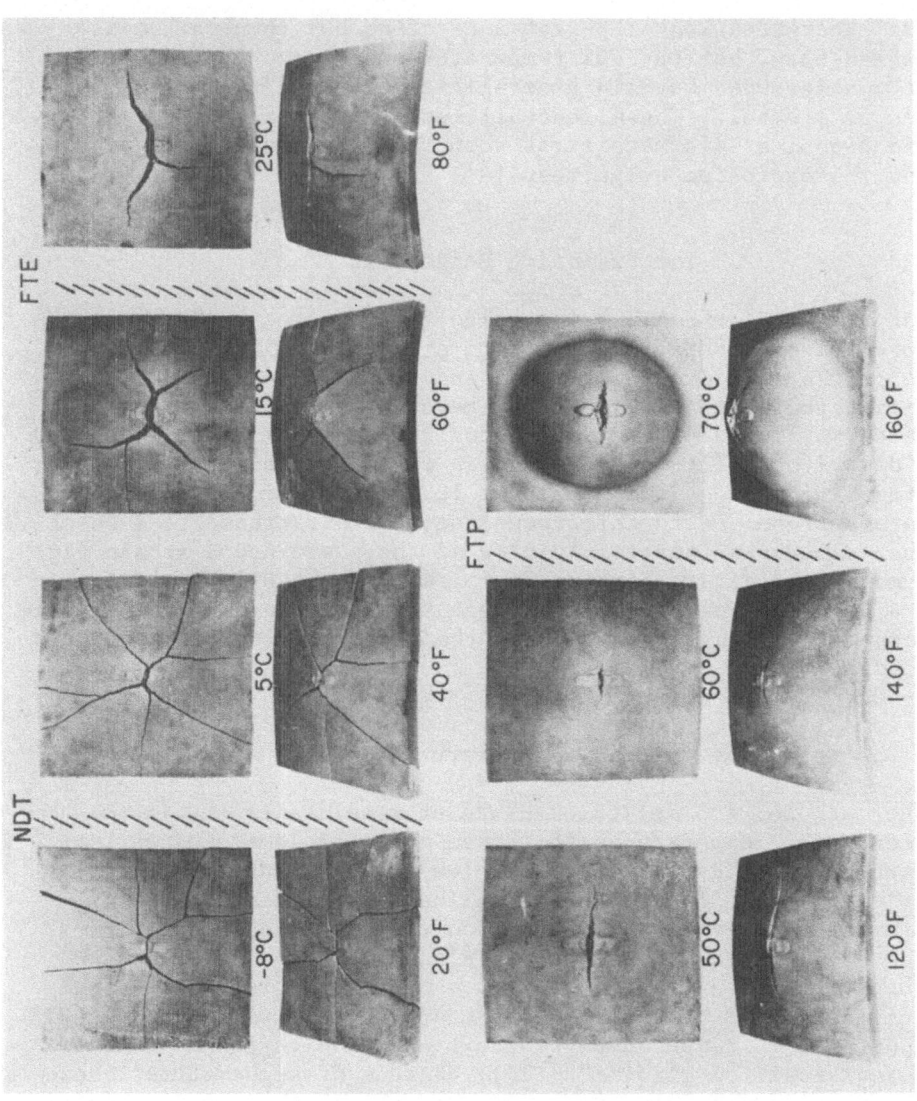

Figure 4. Performance of a ship plate steel in the Explosion Crack Starter test. The steel illustrated features a 15 ft-lb (20 J) C_v transition of approximately 30°F (0°C).

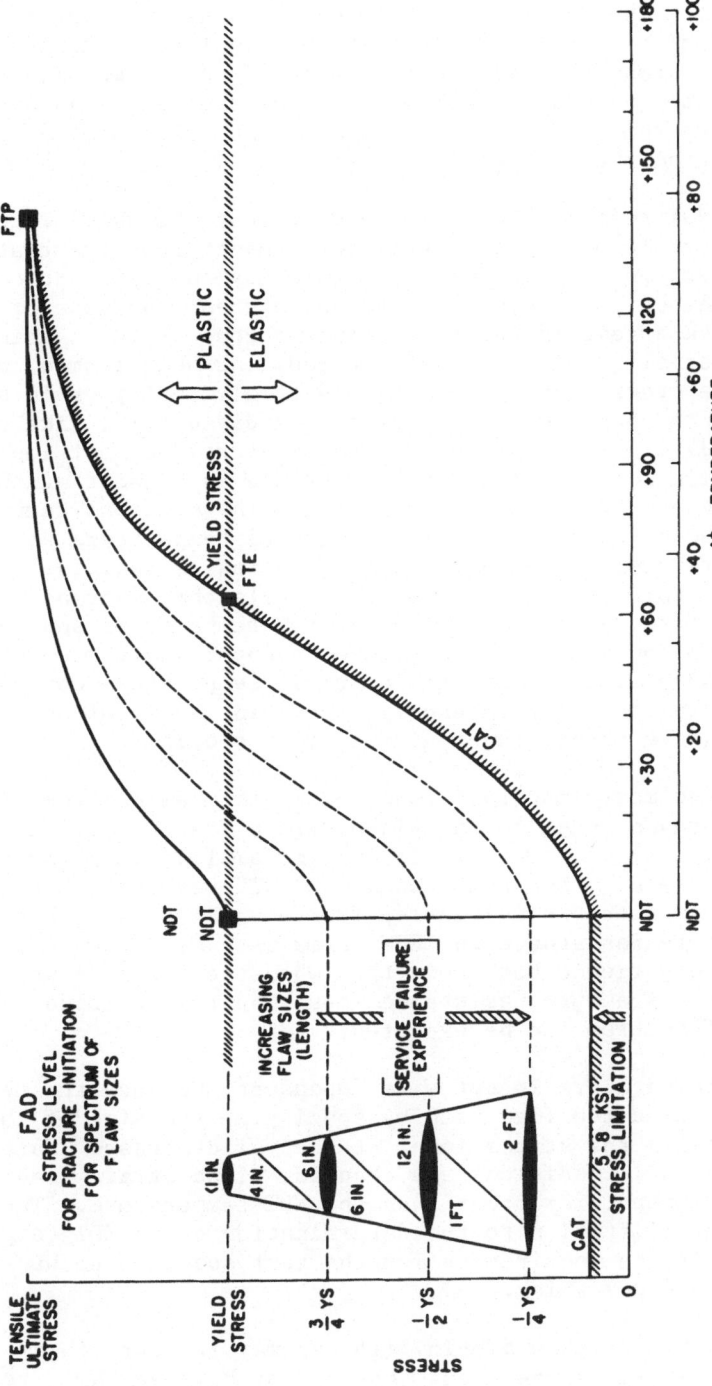

Figure 5.   Fracture Analysis Diagram (FAD).   The diagram illustrates a generalized
relationship between flaw size, effective stress, and temperature for steels
in their transition region.

later modified to apply to steel in thick sections, 3 to 12 in.
(76 to 305 mm). The FAD was the only engineering tool for translating the results of a laboratory fracture resistance test to
structural performance until linear elastic fracture mechanics
was developed to the stage where it could be applied as an analytical approach to the structural analysis of the lower half of the
transition region, below the FTE temperature.

The DWT-NDT test is simple to conduct, but the DWT-NDT test
and the interpretation of structural performance from the test
result have certain limitations. The test may not apply to cold-
worked material or to quenched-and-tempered steels because of
potential crack arrest in the heat-affected zone of the specimen.
Also, if the steel sample was heat treated, the heat from welding
on the brittle crack starter bead may adversely modify or in some
cases enhance the base metal properties. Additionally, like the
15 ft-1b (20 J) C_v criterion the NDT temperature does not imply
that a very high level of fracture resistance is present at the
temperature associated with that criteria. The critical temperature concept does imply that at increments of temperature above
the NDT temperature, fracture resistance increases rapidly.
Unfortunately, this is not true for the steels that increase very
slowly in fracture resistance above the NDT temperature and for
the steels with low fracture resistance at upper shelf temperatures.
Recognizing these limitations, the critical temperature concept can
still be a useful tool for referencing certain levels of performance
in the temperature transition region of many steels.

Perhaps the most important function of the DWT-NDT test is to
provide a reference index to the plane-strain fracture limit in
sections 5/8-in. (16 mm) thick. If one is familiar with the test
method, he can readily recognize that an "NDT" performance translates to brittle fracture initiating from a small flaw. This
level of fracture resistance in LEFM terms is: $K_{Id}/\sigma_{yd} = 0.5 \sqrt{in.}$
[9]. Service experience has shown that when the steel in welded
structures has a fracture resistance less than this critical level,
catastrophic fractures can be expected.

The NDT temperature is not size dependent, because the test
procedure has fixed the flaw size by specifying the size of the
weld bead and also the stress level at the yeild stress. Naturally,
if these mechanical conditions are changed, plane strain fracture
can occur at temperatures other than the NDT temperature. There is,
however, a size limitation to the determination of an NDT temperature and that is the requirement for the test specimen to be at
least 5/8-in. (16 mm) thick.

Sections thinner than 5/8-in. (16 mm) may fracture in a
brittle, plane-strain manner, but they do not have an "NDT" temperature.

The semantics of the term "NDT" temperature are sometimes mixed with a "Plane-Strain Limit" temperature, but the fracture resistance level implied by an NDT temperature performance should always refer to the mechanical conditions defined in ASTM E208. The restriction of the NDT temperature to relatively thick sections provided part of the incentive for the standardization of another test method that could locate the temperature transition region of steel in thin sections using a fracture appearance criterion. This test is the "Drop-Weight Tear Test".

DROP-WEIGHT TEAR TEST (ASTM E436)

The Drop-Weight Tear Test (DWTT) is used to define the temperature transition region for ferritic steels in sections from 0.125 to 0.75 in. (3.18 mm to 19.1 mm). Its primary application has been to determine the temperature at which shear type fractures occur in steel for linepipe. For certain pipes the propagation of cracks that are inadvertently initiated from a number of accident type sources can run for extended distances unless the appearance of the fracture is more than 80% shear. Although this criterion has limitations to linepipes of certain designs, the test does have general application for establishing the transition region for steels in thin sections that are not necessarily used in linepipes.

When a conventional steel with a yield strength under 120 ksi (827 MPa) is used in sections less than 5/8 in. (16 mm) thick, the transition in fracture resistance tends to be restricted to a narrower range of temperatures than when the same steel is in a thicker section. Therefore, for many structural applications, a fracture appearance criterion is all that is needed to predict a brittle or ductile performance because of the narrow range of temperatures involved, as illustrated in Figure 6. Fracture resistance rises so sharply in thin sections that a 50% shear temperature criterion, which is adequate to preclude fracture initiation except for plastic overload conditions around the flaw, is a very readily determined and unambiguous parameter for predicting service performance when a full-section specimen is used.

Because the DWTT has a fracture appearance criterion, it is not a test for generalized use. It cannot be used to quantitatively establish fracture resistance at upper shelf temperatures or to evaluate transitions for steels that do not undergo a sharp transition in fracture appearance. The test is mainly useful for indexing the temperature transition region for plain carbon and low-alloy steels in sections thinner than 3/4 in. (19 mm). Energy measurements from the DWTT are not meaningful because the small notch causes the ligament to plastically deform prior to fracture even when the fracture is brittle. These limitations in the design of the DWTT specimen that precluded the measurement of a meaningful energy

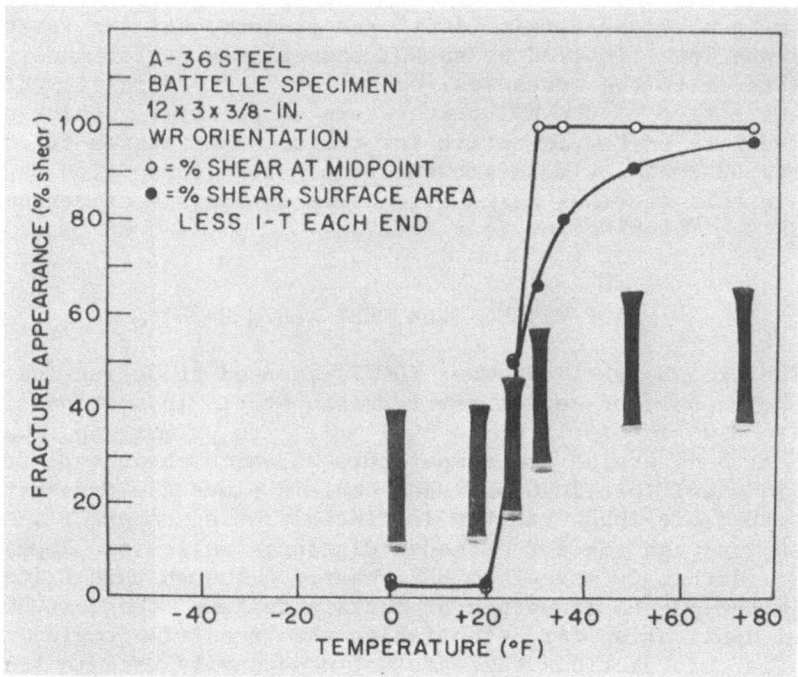

Figure 6. A typical temperature transition for fracture appearance in the Drop-Weight Tear Test (DWTT). Note the narrow transition region from a cleavage to a shear fracture mode for both the area and the linear method of measurements.

criteria were considered in the design of the DT test specimen.

DYNAMIC TEAR TEST (MIL-STD (SHIPS) 1601)

The Dynamic Tear (DT) test was developed to fill the need for a practical test that could precisely measure fracture resistance over a broad range such as that encountered in the transition region of the ferritic steels. For fractical reasons the specimen is not instrumented, and the total energy used to fracture a specimen is the criterion of fracture resistance. This empirical value of fracture resistance, DT energy (DTE), is translated to structural parameters by means of correlations with the analytical parameters or with specific structural performance.

The primary intent of developing the DT test was not to develop an inexpensive K_{Ic} or K_{Id} test but to provide a more

sensitive and reliable fracture resistance criterion for the
elastic-plastic and fully-plastic regimes. Although the DT energy
criterion is empirical, it has correlated very well with K_{Ic} values
for various steels, titanium, and aluminum alloys [3,16,17]. The
correlations cover a sufficiently broad range of alloys in each base
metal to justify the development of generalized diagrams for
structural analysis. The proposed diagrams are called "Ratio
Analysis Diagrams" where by simple graphical means a DT energy
value can be translated into a K_{Ic}/σ_{ys} value or a critical flaw
size as will be described later.

When practical, the DT test uses a full-thickness specimen so
that the constraint in the specimen is that in the structural element of interest. Obviously, this is not very practical when
thick sections are involved, so extensive studies have been made
on the effect of specimen size and the fracture characteristics of
all types of structural alloys, including tests on 12-in.-thick
sections. As a result of these studies, methods for extrapolating
DT energy values obtained with a subsize specimen to full-section
structural performance have been developed [4,9].

                        Specimen Design

The experiemence gained in early constraint effect studies on
the temperature transition features of steels dictated that for
standardization purposes a minimum subsize DT specimen should be
5/8-in. (16 mm) thick, Figure 7. The next critical dimension was

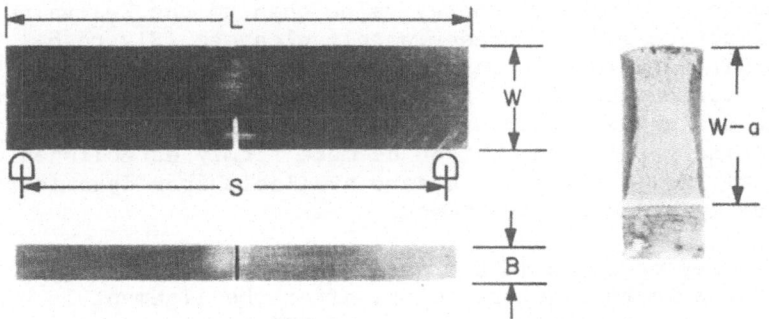

DIMENSIONS OF 5/8-in. (16 mm) DT SPECIMEN

	L	S	W	W-a	B
in.	7	6.5	1.6	1.125	0.625
mm	181	165	38	28.5	16

Figure 7.   The 5/8-in. (16 mm) Dynamic Tear (DT) specimen. The
            dimensions shown are in mm. The standard DT test is
            defined in MIL-STD 1601.

the notch depth, which was set at 0.6 times the thickness to preclude the plastic hinge from reaching the top surface. Another
important design feature is the ligament which was set at 2 times
the thickness. This ligament design provides for extensive stable
crack extension so that a normal Vee shape crack front can develop
in materials when fracture is preceded by extensive through-
thickness deformation. With these basic principles of specimen
design incorporated in the DT specimen, the DT test method has a
broad range measurement capability, and it is economical to conduct.

The 5/8-in. (16 mm) DT specimen, Shown in Figure 7, and DT
testing machines have been studied extensively, and in 1973 the test
method was standardized as MIL-STD 1601. The DT test is being
studied by an ASTM committee, and a modification of the MIL-STD test
method appears in Vol. 10 of the 1975 ASTM Book of Standards as a
"Recommended Practice".

Analysis of DT Energy

The empirical nature of an energy criterion for fracture resistance obtained from an impact test complicates a direct analytical
treatment of DT data. This is not a permanent limitation because
translations can be generated by means of correlations with the
more basic parameters, such as K_{Ic} or K_{Id} as previously stated.
For the iron base alloys, there appear to be two correlations
depending upon the dominating microfracture mode. In the plane-
strain and the elastic-plastic regions when the microfracture mode
is predominately microvoid coalescence, slightly higher K_{Ic} values
correspond to a certain DT energy value than to the K_{Id} value where
the microfracture mode is predominately cleavage, Figure 8. The
lower K_{Id}-DTE relationship in Figure 8 was calculated from DTE
values for various steels at their NDT temperature where fracture
resistance can be expressed as $K_{Id}/\sigma_{yd} = 0.5 \sqrt{in.}$ [18]. Hopefully,
this preliminary relationship can be more firmly established when
valid dynamic J_{Id} or K_{Id} data become available from dynamic tests
using large instrumented specimens.

The energy measured in a DT test can be related to specimen
size and shape where fracture occurs after the ligament is plastic.
The power-law relationship that equated DTE to the geometry of the
specimen is:

$$DTE = R_p \; (b)^2 \; (B)^{1/2} \tag{2}$$

where:

> DTE = dynamic tear energy
>
> R_p = plastic resistance factor

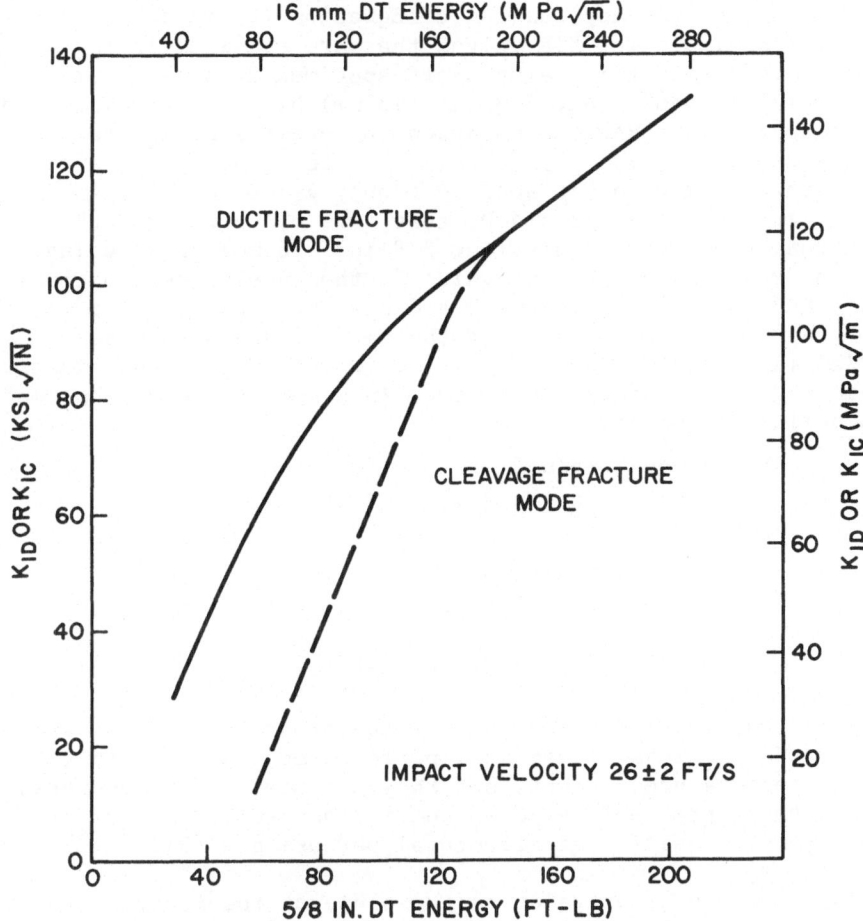

Figure 8.   The relationships between DT energy and the plane-strain
            stress-intensity factors, K_{Ic} and K_{Id}. For steels at
            their upper shelf temperatures, use the K_{Ic} relationship,
            and for steels at temperatures within the transition
            region, use the K_{Id} relationship.

        b = ligament
        B = thickness

    This relationship has been validated for four different
aluminum alloys, 5083, 5086, 6061, and 7005, and for a wide variety
of steels, including some 6-in. (152 mm) thick plates of A533
steel [19]. Unfortunately, the effect of specimen size on the DTE
of titanium alloys was consistent only for limited ranges within
the whole alloy system.

The most useful application of Equation (2) is to provide a
structural analysis for DTE values that are obtained from specimens
that are different from the standard specimen in size. Using
Equation (2) an equivalent 5/8-in. (16 mm) DTE can be calculated,
and the appropriate RAD can be drawn to obtain a structural performance analysis [17]. For example, if the geometry of a 1-in.
(25 mm) thick component is such that only a specimen 3/8-in.
(9.6 mm) can be removed for a DT test, the DT value from this specimen is converted to an equivalent 5/8-in. (16 mm) value using
Equation (2). The equivalent value is then positioned on the 1-in.
(25 mm) RAD as shown in Figure 9 and a structural analysis is
thereby derived. For projecting the performance of the same
material in sections other than 1-in. (25 mm) thick, the elastic-
plastic region is shifted up or down in accordance with the conditions defined as follows:

(L) Plane Strain Limit:

$$B = 2.5 \ (K_{Ic}/\sigma_{ys})^2 \eqno(3)$$

(YC) Yield Criterion:

$$B = 1.0 \ (K_{Ic}/\sigma_{ys})^2 \eqno(4)$$

where B = thickness of the section.

The mechanical conditions defined by Equations (3) and (4) are
arbitrary definitions of critical performance levels as in the
critical temperature concept, but they are useful for understanding
the results of trade-off studies and for assessing the effects of
improving metal quality on structural performance [9].

The effect of constraint on the slope of the fracture resist-
ance-temperature curve in the elastic-plastic regime of ferritic
alloys is complex. Certain generalities have been proposed for the
extrapolation of size effects using a generalized K_{Id}/σ_{yd} vs.
temperature curve, such as the upper bound curve in Figure 10.
There are exceptions to this generalized shape for defining the
upper bound of dynamic plane-strain fracture resistance, and the
temperature transition characteristics of a material of interest
should be generated. Unfortunately, the more flexible analytical
parameters, such as J_c, are also size dependent when the plastic
zone extends through the ligament of the specimen. Therefore, at
this point in time, elastic-plastic fracture mechanics is in the
research stage of development, and the interpretation of fracture
resistance data from subsize specimens to full-section performance
remains an empirical exercise.

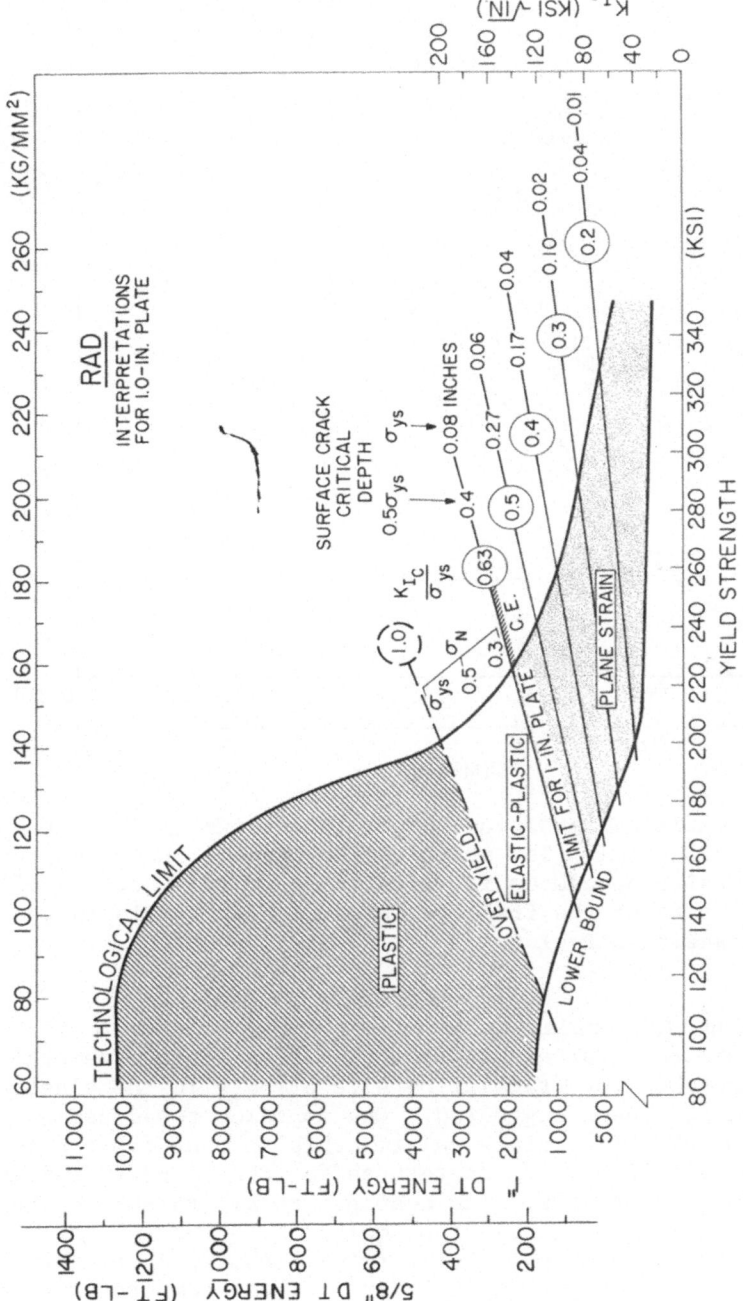

Figure 9.   The Ratio Analysis Daigram (RAD) for the steels.   The position of the
elastic-plastic regime is illustrated for a 1-in. (25 mm) thick
section.

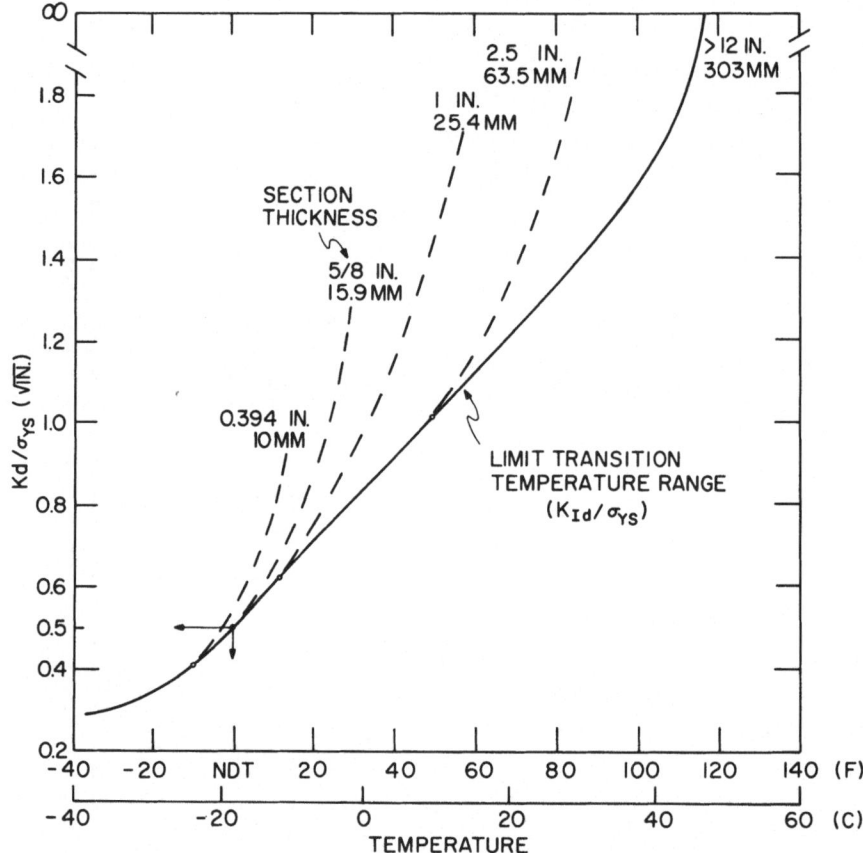

Figure 10.   The Limit-Transition-Temperature-Range (LTTR) for an
             A-533 steel. The start of the transition curves for
             subsize specimens is illustrated to indicate the effect
             of size on the fracture resistance of steels in the
             transition region.

     Recognizing the complexity of an analytical treatment of
fracture resistance of steels in the transition region, several
practical approaches for translating size effects in the elastic-
plastic region have been suggested. One approach to adjust a
5/8-in. (16 mm) DTE-temperature relationship to that for a thicker
section is to establish the temperatures for the L and YC indexes
for the section of interest. The appropriate DTE values for the L
and YC indexes can be obtained from Equations (3) or (4) and
Figure 10. For sections up to 3-in. (75 mm) thick, no temperature
adjustment is necessary for the L index, but the temperature for
the YC index of the thicker section on the 5/8-in. (16 mm) DTE-
temperature curve is shifted to a higher temperature by the
following increments:

Thickness, in. (mm)	1 (25)	2 (50)	3 (75)
°F	10	40	50
°C	6	22	28

If the dynamic yield strength of a steel is unknown, dynamic yield strength can be approximated with the following equation:

$$\sigma_{yd} = \sigma_{ys} + 30 \text{ ksi (207 MPa)}$$

Details on the use of this and other approaches to an analysis for size effects in the temperature transition region are presented in reference [9].

SUMMARY

The development of LEFM has helped to understand the inter-relationship between the mechanical and the metallurgical aspects of fracture. This includes the effects of constraint on the fracture mode and the transitions in fracture resistance of the structural metals. The elastic-plastic portion of a transition region can now be defined in terms of LEFM parameters, and by correlation the results obtained from the empirical impact tests can provide a much needed analysis in this important regime between brittle and ductile performance.

The DT test was designed to overcome some of the limitations of the C_v test and the DWT-NDT test. The DT specimen has an extended ligament in a C_v specimen or a K_{Ic} specimen. This feature increases the sensitivity for measurement of fracture resistance in the elastic-plastic regime by allowing a larger plastic zone to develop.

Impact testing of the DT specimen introduces a high enough strain rate to effectively simulate the high strain rate of a natural pop-in crack. Thus, the upper bound of the temperature transition characteristics of the ferritic alloys can be established for a structural reliability analysis. Although more research needs to be conducted on size effects, the economics of conducting a DT test and the availability of initial correlations with the more basic fracture mechanics parameters should help in the application of fracture mechanics to design.

ACKNOWLEDGMENT

The author gratefully acknowledges the help and contributions of his former coworkers, Mr. W. S. Pellini and Dr. P. P. Puzak, and coworkers, Mr. R. J. Goode, Mr. R. W. Judy, Jr., Dr. C. A. Griffis, and Mr. L. A. Cooley.

REFERENCES

1. "Standard Method for Notched Bar Impact Testing of Metallic
    Materials", Designation: E23-72 in 1973 Annual Book of ASTM
    Standards, Part 31. Philadelphia: American Society for
    Testing and Materials (1973), 277-93.

2. "Standard Method for Conducting Drop-Weight Test to Determine
    Nil-Ductility Transition Temperature of Ferritic Steels",
    Designation: E208-69 in 1973 Annual Book of ASTM Standards,
    Part 31. Philadelphia: American Society for Testing and
    Materials (1973), 597-616.

3. Impact Testing of Metals, Special Technical Publication 466.
    Philadelphia: American Society for Testing and Materials, 1970.

4. Pellini, W.S., "Evolution of Engineering Principles for Fracture-
    Safe Design of Steel Structures". Naval Research Laboratory,
    Washington, D.C., Report No. NRL-6957, September 1969. (AD
    697 631)

5. Nash, G.E. and Lange, E.A., "Mechanical Aspects of the Dynamic
    Tear Test", Trans. ASME, Ser. D, J. Basic Eng., 91 (1969), 535-
    43.

6. Shoemaker, A.K. and Rolfe, S.T., "The Static and Dynamic Low-
    Temperature Crack-Toughness Performance of Seven Structural
    Steels", Eng. Fract. Mech., 2 (1971), 319-39.

7. Loss, F.J., Hawthorne, J.R., Griffis, C.A. and Gray, R.A., Jr.,
    "Plane Strain Fracture Toughness at High Loading Rates", Report
    of NRL Progress (December 1975), 20-21.

8. Irwin, G.R. and Roberts, R., "Fracture Toughness of Bridge
    Steels", Phase I. Report, Lehigh University, Bethlehem, Pa.,
    1972.

9. Pellini, W.S., "Analytical Design Procedures for Metals of
    Elastic-Plastic and Plastic Fracture Properties", Weld. Res.
    Counc. Bull., No. 186 (1973), 17-38.

10. "Standard Methods and Definitions for Mechanical Testing of
    Steel Products", Designation: A370-74, in 1975 Annual Book
    of ASTM Standards, Part 10. Philadelphia: American Society
    for Testing and Materials (1975), 1-52.

11. Williams, M.L., "Analaysis of Brittle Behavior in Ship Plates",
    in Symposium on Effect of Temperature on the Brittle Behavior
    of Metals with Particular Reference to Low Temperatures, Special
    Technical Publication 158. Philadelphia: American Society for
    Testing and Materials (1954), 11-41.

12. Puzak, P.P. and Lange, E.A., "Significance of Charpy-V Test
    Parameters as Criteria for Quenched and Tempered Steels", Naval
    Research Laboratory, Washington, D.C., Report No. NRL-7483,
    October 1972. (AD 751 534)

13. Czyzewski, H., "Brittle Failure: The Story of a Bridge", <u>Metal Prog. (West)</u>, <u>1</u>, No. 1 (1975), W6-W12.

14. "Tentative Method for Drop-Weight Tear Tests of Ferritic Steels", Deisgnation: E436-71T, in <u>1973 Annual Book of ASTM Standards, Part 31</u>. Philadelphia: American Society for Testing and Materials (1973), 1049-54.

15. Puzak, P.P., Babecki, A.J. and Pellini, W.S., "Correlations of Brittle-Fracture Service Failures with Laboratory Notch-Ductility Tests", <u>Weld. J.</u>, <u>37</u> (1958), 391s-407s.

16. Judy, R.W., Jr., Goode, R.J. and Freed, C.N., "A Character-ization of the Fracture Resistance of Thick-Section Titanium Alloys", Naval Research Laboratory, Washington, D.C., Report No. NRL-7427, July 1972. (AD 747 230)

17. Pellini, W.S., "Criteria for Fracture Control Plans", Naval Research Laboratory, Washington, D.C., Report No. NRL-7406, May 1972. (AD 743 058)

18. Lange, E.A. and Cooley, L.A., "Factors Determining the Per-formance of High-Strength Structural Metals (Nil-Ductility-Transition (NDT) Temperature in 5/8-in. (16mm) Dynamic Tear (DT) Energy for Steels)", <u>Report of NRL Progress</u> (May 1971), 33-34.

19. Judy, R.W., Jr., and Goode, R.J., "Ductile Fracture Equation for High-Strength Structural Metals", Naval Research Labora-tory, Washington, D.C., Report No. NRL-7557, April 1973. (AD 759 351)

THE ANALYSIS OF FATIGUE CRACK GROWTH RATE DATA

W. G. Clark, Jr. and S. J. Hudak, Jr.

Mechanics Department
Westinhouse Research Laboratories
Pittsburgh, Pennsylvania

ABSTRACT

The analytical variables associated with the characterization
of fatigue crack growth rate performance are discussed and the
influence of these variables on predicted cyclic life is illustrated.
It is shown that the analytical techniques chosen to process laboratory test results and to develop upper-bound design information can
have a significant effect on subsequent life predictions. The
development of an optimum methodology for evaluating and utilizing
fatigue crack growth data is also considered.

INTRODUCTION

The successful use of fracture mechanics concepts in predicting
the cyclic life of structures requires fatigue crack growth rate
data which accurately represents the service conditions likely to
be encountered in the intended application [1,2]. Consequently,
prior to generating fatigue crack growth rate data for a given
application, it is necessary to identify the pertinent variables
associated with the expected service conditions so that meaningful
data can be developed. Table 1 presents some of the variables
which can have a significant effect on the fatigue crack growth
rate properties of structural alloys and ultimately, component life.
Obviously, such variables must be considered in a fatigue crack
growth rate testing program conducted to provide useful design data.

The material, environment, and loading variables noted in
Table 1 are generally well recognized as important factors in

TABLE 1

FACTORS THAT CAN INFLUENCE FATIGUE CRACK
GROWTH BEHAVIOR

Material Variables

    Composition
    Inclusion Content
    Microstructure
    Processing

Environmental Variables

Environmental Chemistry
Temperature
Pressure

Loading Variables

    Load Spectrum

    Load Range
    Stress Ratio
    Frequency

    State of Stress
            Thickness
            Constraint
            Type of Loading
                    Uniaxial
                    Complex

---

predicting fatigue life. As a result, considerable effort is
usually expended in identifying and subsequently, controlling these
variables in fatigue crack growth rate testing. However, additional
and often neglected aspects of fatigue crack growth rate testing
which must be considered in life predictions are the variables
associated with the analysis of test data. These variables include
the data processing methods used to convert the raw test data
(crack length versus elapsed cycles) to crack growth rate data
(growth rate versus stress intensity range) and the techniques used
to interpret and present the test results in a form suitable for
use in design. Often, these analytical variables can have as much
or more influence on the measured rate of crack growth and predicted
cyclic life than the testing variables summarized in Table 1.
Consequently, it is important to recognize the significance of
these parameters in fatigue crack growth rate testing and formulate
an optimum method to analyze test results.

This paper presents a review and detailed discussion of the
pertinent variables associated with the analysis and interpretation

of fatigue crack growth rate data. The influence of these parameters on predicted cyclic life is illustrated and the consideration of an optimum analytical methodology is discussed.

ANALYTICAL VARIABLES

Figure 1 presents a schematic representation of the analytical
aspects associated with the development of fatigue crack growth

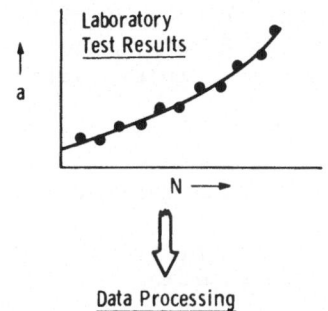

Data Processing

Converting a vs. N Data to da/dN vs. ΔK_I Data

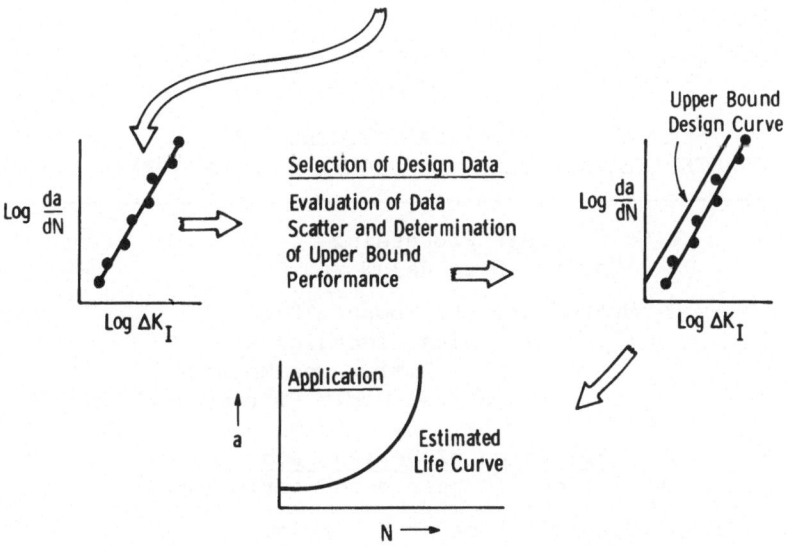

Figure 1.   Schematic of the analysis of fatigue crack growth rate
data.

rate data suitable for use in design. Two pertinent areas of consideration are involved. These include the data processing methods used to convert the crack length versus elapsed cycles test data ("a" vs. N) to crack growth rate data expressed as a function of the stress intensity factor range (da/dN vs. ΔK_I) and the analytical methods used to evaluate data scatter and subsequently, establish upper-bound design data. Table 2 presents a summary of the analytical methods currently being used to characterize fatigue crack growth rate performance. The influence of the above factors on the prediction of cyclic life are evaluated and discussed in subsequent portions of this paper. This evaluation of the analytical methods used to characterize fatigue crack growth rate performance is limited to the consideration of linear elastic conditions and crack growth rate behavior which can be characterized by a simple power relationship (da/dN = $C_o \Delta K_I^n$). However, the observations are equally applicable to other more complex fatigue crack growth rate behavior.

Data Processing Variables

The problem encountered in converting "a" vs. N data to crack growth rate data is that of determining the derivative (slope) of a function, a = f(N), which is known only at certain pivotal points

TABLE 2

SUMMARY OF ANALYTICAL METHODS USED TO
CHARACTERIZE FATIGUE CRACK GROWTH RATE PERFORMANCE

Data Processing
("a" vs. N to da/dN vs. ΔK_I)

Method used to Compute Rate
Graphical Techniques
Finite Difference Methods
Numerical Curve Fitting Techniques

Selection of Design Data
(determination of upper-bound performance)

Visual Determination of Upper-Bound
Upper-Bound Regression Line
Confidence Bands Based on Standard Deviation

---

corresponding to a set of "a" and N measurements. Presently, at least six different data processing techniques are being used to analyze fatigue crack growth data [3]. In general, these techniques can be classified into three basic categories: graphical techniques, finite difference methods and numerical curve fitting techniques. Examples of each of these data processing techniques are illustrated in Figure 2 and described below.

The graphical method of determining crack growth rate from "a" vs. N data simply involves visually fitting a smooth curve through the plotted "a" vs. N data and estimating the rate of growth at a given crack length, a_i, from the slope of a straight line drawn tangent to the curve at a_i. The corresponding ΔK_I parameter is computed using the crack length at the point of tangency, a_i.

The most commonly used finite difference method of analyzing fatigue crack growth data is the secant or point-to-point technique. This method involves calculating the crack growth rate from the slope of the straight line connecting two adjacent points on the "a" vs. N curve. Thus, the rate is simply

$$da/dN = \frac{a_{i+1} - a_i}{N_{i+1} - N_i} \tag{1}$$

Since da/dN is the average crack growth rate over the $a_{i+1} - a_i$ increment, the average crack length, $1/2 (a_{i+1} - a_i)$, is normally used to calculate ΔK_I.

The numerical curve fitting methods currently in use in the analysis of fatigue crack growth data include several sophisticated techniques of fitting data with complex polynomial expressions. Of these, a relatively simple method which is currently gaining wide acceptance is the so-called incremental polynomial technique. This method involves fitting a 2nd order polynomial (parabola) to sets of successive data points (usually 5 to 7 data points). The form of the equation for this local fit is

$$a = b_0 + b_1 \left(\frac{N - C_1}{C_2}\right) + b_2 \left(\frac{N - C_1}{C_2}\right)^2 \tag{2}$$

where b_0, b_1 and b_2 are the regression parameters which are determined by the least squares criterion (to minimize the square of the deviations between observed and fitted values of crack length) over the local range of crack lengths. The parameters C_1 and C_2 are used to scale the input data; thus avoiding numerical difficulties in determining the regression parameters. For the case of a seven point subgroup; $C_1 = 1/2 (N_{i-3} + N_{i+3})$, $C_2 = 1/2 (N_{i+3} - N_{i-3})$ and

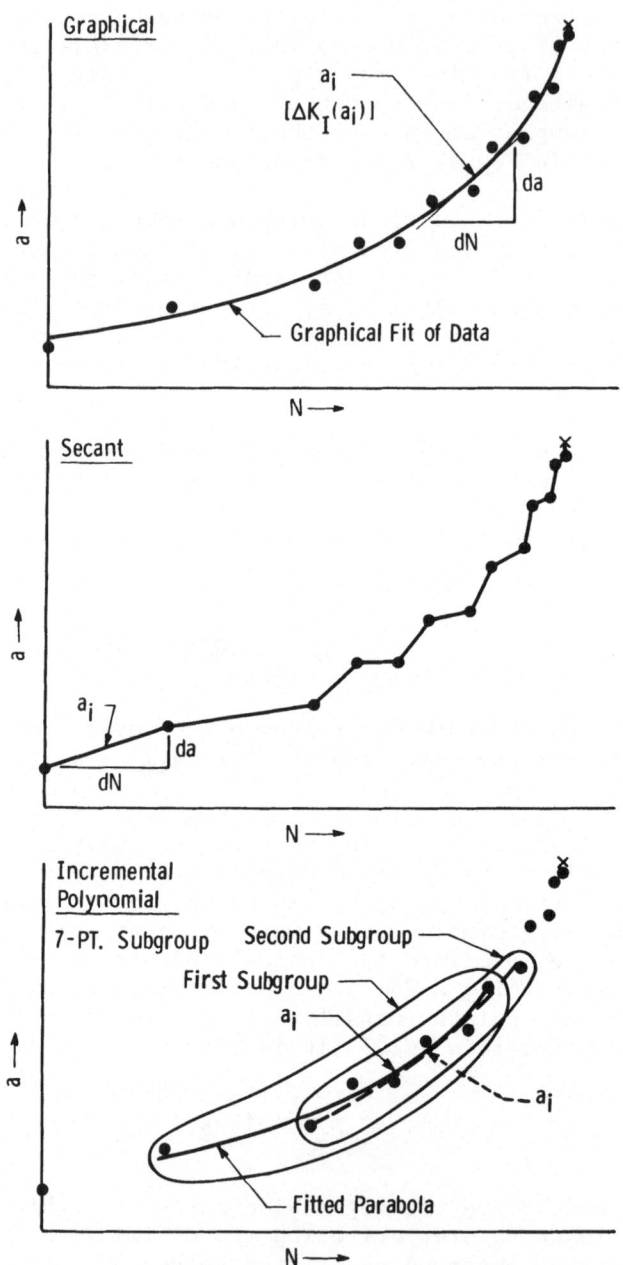

Figure 2.   Schematic illustration of various data processing
            techniques.

the local range of crack lengths is $(a_{i-3} \leq a \leq a_{i+3})$. The rate of crack growth at (a_i, N_i) is obtained from the derivative of the above parabola which is given by the following expression:

$$da/dN = \frac{b_1}{C_2} + 2b_2 \ (N_i - C_1)/C_2^{\ 2} \tag{3}$$

Generally, the ΔK_I value corresponding to this growth rate is computed for the fitted crack length, $\hat{a}_i$, corresponding to the median number of cycles, N_i, over the local elapsed cycle range. The value N_i is used in the above expression to calculate da/dN. The rationale for using these particular values of crack length and elapsed cycles is as follows. The recorded N_i value is used since this variable is essentially error-free, having been simply read from a digital devise. On the other hand, the recorded a_i always contains measurement error. In using a regression technique to determine $\hat{a}_i$, one recognizes this error and attempts to establish a best estimate of the true crack length.

From the brief descriptions presented above, it is obvious that the data processing techniques currently being used to evaluate fatigue crack growth data vary greatly in terms of their approach and degree of sophistication. It is also apparent that these techniques include inherent limitations which are likely to have a significant effect on the characterization of fatigue crack growth behavior. For example, the graphical procedure suffers from subjectivity, thus precluding consistent, reproducible results. The finite difference techniques which are based on point-to-point differences tend to magnify variations in "a" vs. N test data and consequently, may yield excessive variability in the processed data. In addition, this variability is sensitive to crack length measurement interval. The numerical curve fitting techniques generally involve a least squares regression analysis which may induce excessive "smoothing" of the data that can mask important behavior. In view of the basic differences in the data processing techniques and the associated limitations, it is obvious that the potential effect of data processing variables must be considered in the development of meaningful fatigue data.

In order to further illustrate the potential effect of data processing techniques on the characterization of fatigue crack growth rate data, let us consider the analysis of the same "a" vs. N test results with both the secant and incremental polynomial (7 point subgroup) methods. Due to the subjective nature of the graphical method this procedure is not considered in our analysis. In addition, in order to simplify the comparison of results, let us consider a set of hypothetical test data which represent idealized bheavior to which we have added a uniform error, $\pm \varepsilon$, to the crack length measurements. Figure 3 presents the hypothetical test

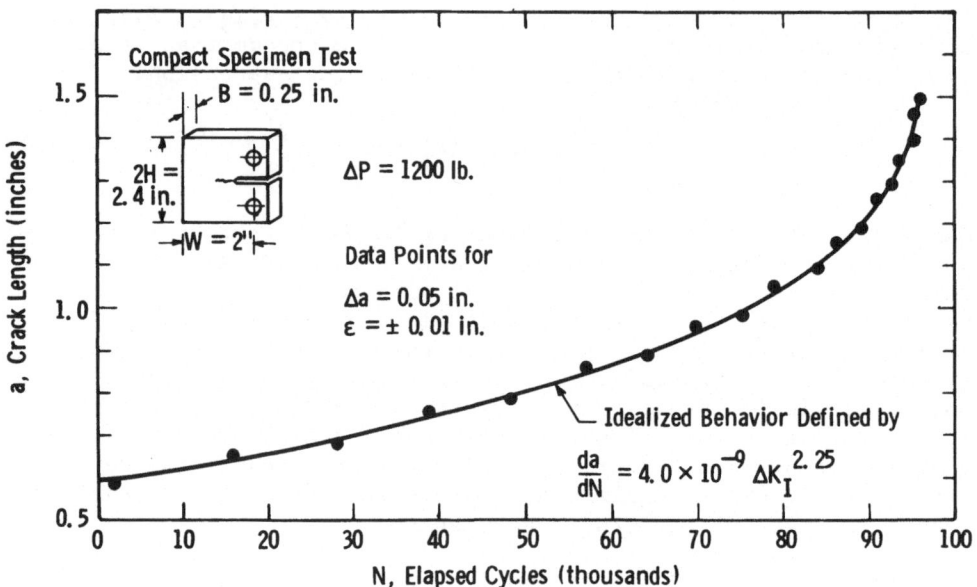

Figure 3.   Hypothetical crack growth test data.

results.   These data represent "a" vs. N behavior for a 0.25 in.
thick compact specimen (H/W = 0.6, W = 2 in.) computed from the
growth rate behavior given by the equation:

$$da/dN = 4 \times 10^{-9} \Delta K_I^{2.25} \qquad (4)$$

(da/dN in inches/cycle, ΔK_I in ksi$\sqrt{in.}$)

Note that the solid curve in Figure 3 presents the idealized "true"
behavior and the data points represent data collected at a crack
length interval, Δa, of 0.050 in. with a uniform error of $\pm$ 0.010
in. (The data points alternate about the "true" value of "a" in a
regular pattern of a_1 - 0.01 in. followed by a_2 + 0.01 in., etc.)
These data are fairly typical of a "good" set of laboratory test
results.   Note that although it is usually possible to make crack
length measurements with an error of less than $\pm$ 0.01 in., the
crack length as measured on each side of the specimen often varies
by at least 0.010 in. [3].   Thus, we have chosen an error of
$\pm$ 0.01 in. for our example.

Figure 4 presents the results of the analysis of the hypothetical crack growth data by means of both the secant and incremental
polynomial data processing methods.   Included in Figure 4 are the
crack growth rate data points generated from the secant and

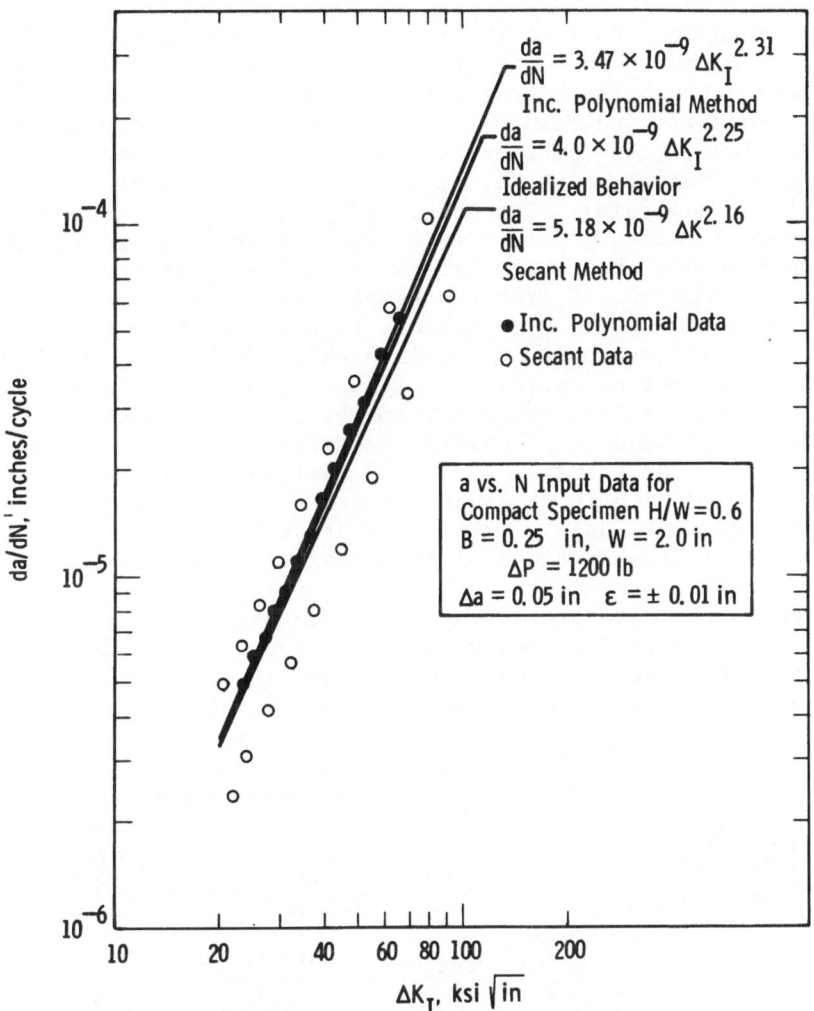

Figure 4.   Results of the crack growth rate analysis.

incremental polynomial analyses, the respective "best fit" linear
regression lines and the idealized curve used to develop the "a"
vs. N data in Figure 3.   These results clearly illustrate that
the data processing technique has a small but significant effect
on the "central tendency" behavior as defined by the regression
lines and a very significant effect on the extent of data scatter.
Note that for the ΔK_I region being considered, the secant regression

line tends to underestimate the growth rate as compared to the
idealized behavior, whereas the incremental polynomial line over-
estimates the growth rate. Figure 4 shows that the difference in
growth rate between the secant and polynomial methods as estimated
from the respective regression lines is dependent on the ΔK_I level
being considered (i.e., the regression lines are not parallel) and
reaches a maximum of about 20 percent at a ΔK_I level of 100 ksi$\sqrt{\text{in}}$.
(limit of our analysis). The effect of data processing techniques
on the central tendency growth rate behavior is further illustrated
in Figure 5, where the cyclic life of a 0.25 in. thick compact
specimen, as determined from the respective regression lines, is
compared to the original hypothetical test data. Note that for our
example, the difference in cyclic life as estimated from either
data processing method varies by only about 10 percent and these
estimates vary by only 5 percent from the idealized behavior. Thus,
it appears that the data processing technique has a relatively
small effect on the "central tendency" fatigue crack growth rate
behavior defined by a least squares regression line established for
the da/dN versus ΔK_I data. This observation is consistent with the
results of a previous evaluation of actual test data [3].

Although the data processing technique has a relatively small
effect on the "central tendency" fatigue crack growth rate behavior
this is not the case with regard to the extent of data scatter.
The difference in data scatter encountered in the secant and

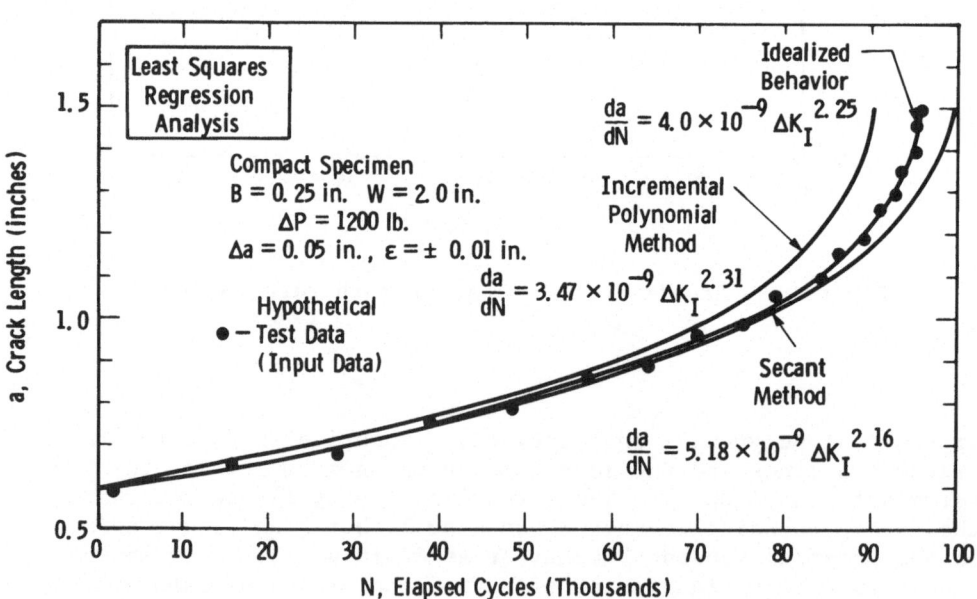

Figure 5.   Estimated life based on regression analysis.

incremental polynomial analyses (Figure 4), as characterized in
terms of a variability factor based on $\pm$ 2 standard deviations
(da/dN + 2 std. dev. $\div$ da/dN - 2 std. dev.), is 6 versus 1.1,
respectively. In order to illustrate the potential influence of
the data scatter arising from the combination of raw data variability and variability introduced by data processing techniques, it is
necessary to consider the various methods used to establish design
information from fatigue crack growth rate data.

Selection of Design Data

As noted in Table 2 there are essentially three methods
currently in use for dealing with scatter and ultimately, determining upper-bound fatigue crack growth behavior suitable for use
in design considerations. One technique simply involves visually
fitting a straight line through the upper data points to establish
an upper-bound. The other techniques involve the determination of
the least squares regression line for a group of data which is then
used to establish an upper-bound. In one method, the upper-bound
is established by drawing a line parallel to the regression line
at a location which encompasses all of the data and in another
method, the standard deviation about the regression line is established and an upper-bound determined on the basis of two or three
standard deviations. (The standard deviation, or more precisely
the estimated standard deviation, can also be used to establish any
one of several types of upper confidence bands to a level of
confidence throught appropriate for a given application [4].) The
subsequent discussion is limited to the consideration of design data
developed by means of a regression analysis. Like the graphical
method of data processing, the graphical determination of design
data is considered too subjective for use in the analysis of fatigue
crack growth rate data.

Figure 6 presents a comparison of the various methods which
can be used to establish design data from the results of both the
secant and incremental polynomial analyses. The upper-bound determined by "sliding" the regression line to encompass all of the test
data as well as the upper-bound defined by +2 standard deviations
are included. The upper-bound line for the incremental polynomial
data is not shown since it essentially falls on top of the respective
+2 standard deviation line. The standard deviation about the regression line was computed in terms of log da/dN as described in
Reference [3], where it is shown that the error in da/dN follows a
lognormal distribution. It should be noted that the standard deviation bands are not confidence statements and cannot be used to make
inferences concerning the probability that all data fall below
these bands. Additional statistical analyses are required to
establish confidence bands more suitable for use in actual design
considerations [4]. Such analyses are beyond the scope of this

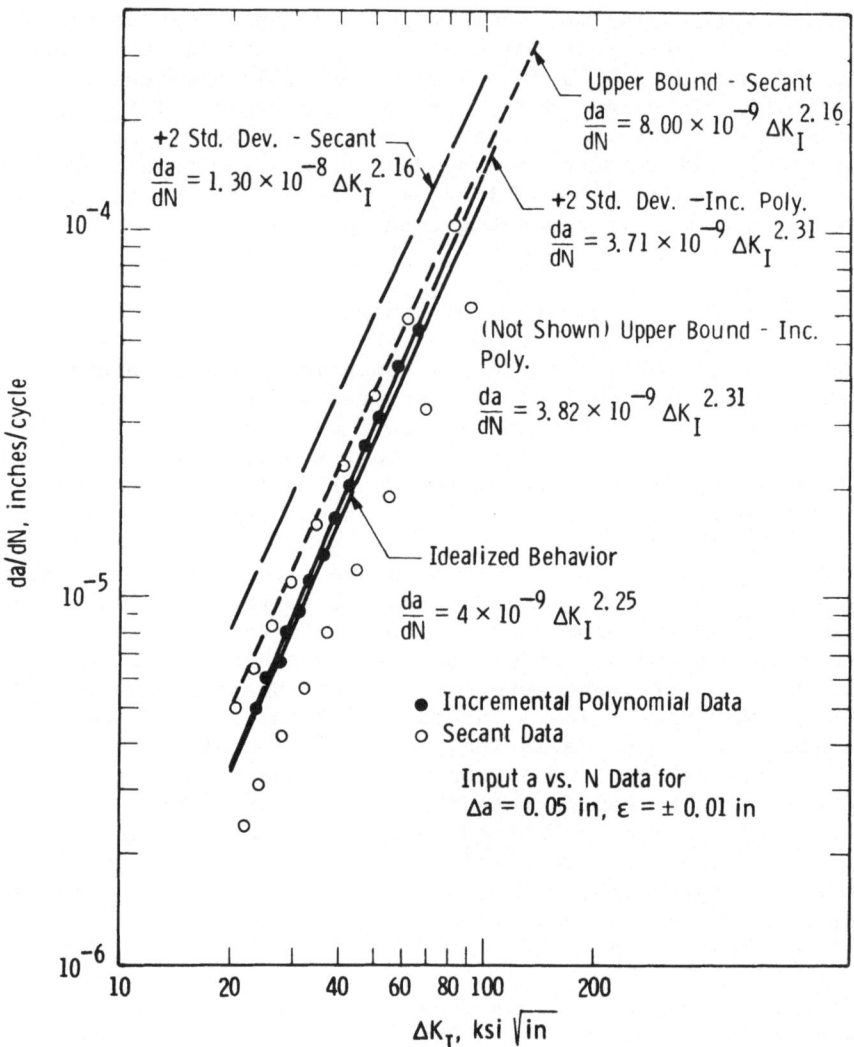

Figure 6.   Comparison of methods used to establish design data.

paper, however, the consideration of the standard deviation is adequate for illustration purposes.

Figure 6 clearly illustrates the potential influence of data scatter and ultimately, data processing technique on the characterization of fatigue crack growth rate data for use in design. Note that for the input data being considered (Figure 3), the incremental polynomial method of data processing yields little scatter and

consequently, essentially the same upper-bound performance is
obtained from either a standard deviation or "sliding" regression
line analysis. However, for the secant data, there is a factor of
2.5 difference in growth rate between the idealized behavior and
the +2 standard deviation upper-bound. In addition, the crack
growth rate associated with the upper-bound defined by the
"sliding" regression line is on the order of 1.5 times faster than
the idealized behavior. The influence of data scatter and the
effect of the method used to establish design information on crack
growth performance is further illustrated in Figure 7, where the
cyclic life computed from the various upper-bound curves is compared to the original test data. Note that depending on the
methods used to process the test results and to establish upper-
bound design information, the predicted cyclic life of the original
test specimen can range from 39,000 cycles to 84,200 cycles compared
to the "idealized" life of 95,500 cycles. Both Figures 6 and 7
clearly demonstrate the potential influence of data analysis methods
on the characterization of fatigue crack growth performance. It is
obvious from these results that the analytical aspects associated
with fatigue crack growth rate testing can be as important as the
actual test variables. It is also apparent that a realistic
approach to the characterization of fatigue crack growth rate
behavior must include an optimum method of data processing.

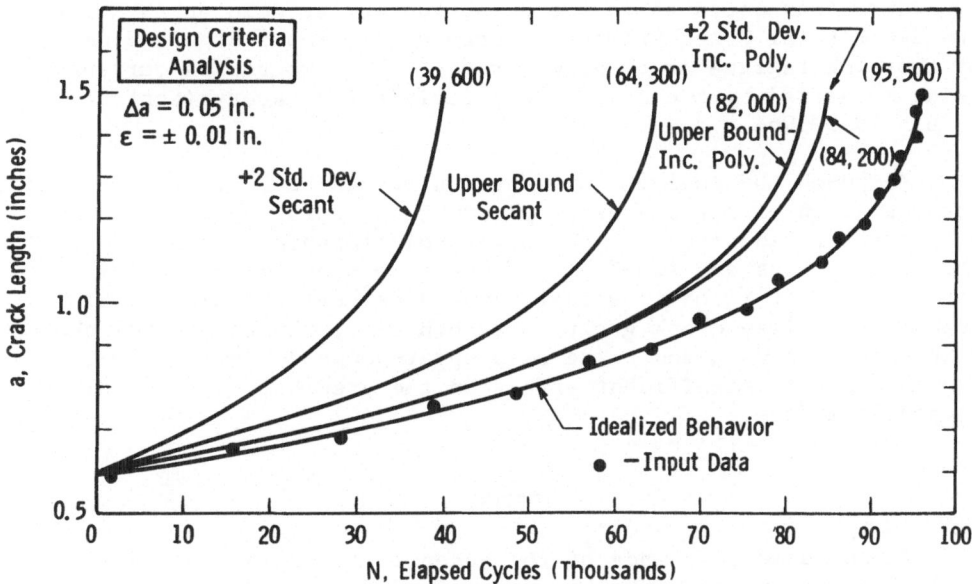

Figure 7. Life estimates based on various design criteria.

DISCUSSION

Based on the preceding analysis, it appears that a numerical curve fitting method of data processing combined with the use of a statistical method of developing upper-bound limits on da/dN represents an optimum method for evaluating fatigue crack growth rate data. However, additional aspects of the problem must be considered before an optimum technique can be established. For example, it is apparent from the preceding analysis that both the crack length measurement interval, Δa, and the crack length measurement error, ε, can influence the effect of data processing methods on predicted cyclic life. Specifically, for the case of the secant method, as the ratio of ε to Δa decreases the slope between successive data points exhibits decreased oscillations and the inherent scatter in the method decreases. In addition, the shape of the "a" vs. N curve, as determined by the specific test specimen geometry, may influence the effect of data processing methods on the characterization of fatigue crack growth. Other practical aspects of fatigue crack growth rate testing must also be considered in the selection of an optimum data processing method. For example, in the region of very low crack growth rate or fatigue threshold testing where it generally takes an extremely long time to develop a single data point, it may be impractical to use an incremental polynomial method of analysis which requires five to seven data points to develop a single growth rate value. In addition, since the incremental polynomial technique requires a computer for analysis it is not convenient to compute growth rate data during a test. The considerations noted above are being investigated in an ASTM E24.04 (On Subcritical Crack Growth) program to develop a standard method of test for fatigue crack growth rate. This program is currently being sponsored by the Air Force Materials Laboratory (Contract F33615-75-C-5064).

Although the analysis presented here has been limited to the consideration of fatigue crack growth rate performance which can be characterized with a simple power relationship (da/dN = $C_0 \Delta K^n$), the observations are equally applicable to other more complex situations. It is obvious that, regardless of the relationship used to formalize crack growth rate behavior, the techniques chosen to process the data and to develop upper-bound design information can have a very significant effect on the prediction of useful component life.

SUMMARY

An assessment is made of the potential influence of analytical variables associated with the processing of raw fatigue crack growth data, as well as with the techniques used to establish

upper-bound crack growth rate curves for use in design. Both secant
and incremental polynomial data processing methods were considered
in conjunction with two popular methods for establishing upper-
bound design curves. Examples are given which show that these
analytical variables can have a significant influence on subsequent
cyclic life predictions. Due to the scatter introduced into the
processed data by the secant method, subsequent life estimates were
found to be excessively conservative (by as much as a factor of
about 2.5) and were also dependent on the technique employed to
establish the upper-bound design curve. The incremental polynomial
processing method resulted in slightly conservation predictions (by
about a factor of 1.2) which were essentially independent of the
methods used herein to establish the design curve. These results
indicate the need for the development of optimum methods for processing and utilization of fatigue crack growth rate data. It is
concluded that data processing methods based on numerical curve
fitting techniques over a local region, such as the incremental
polynomial method, represent near-optimum methods for situations
where ample raw data are available. Statistically based confidence
bands, established for a level of confidence and probability
thought appropriate for a given application, need to be given more
consideration in establishing upper-bound design curves.

ACKNOWLEDGMENTS

The work described in this paper was sponsored in part by the
Air Force Materials Laboratory under Contract F33615-75-C-5064.
Mr. A. Gunderson is the Air Force Project Engineer.

REFERENCES

1. Clark, W.G., Jr., "How Fatigue Crack Initiation and Growth
    Properties Affect Material Selection and Design Criteria",
    Metals Eng. Quart., 14, No. 3 (1974), 16-22.

2. Hoeppner, D.W. and Krupp, W.E., "Prediction of Component Life
    by Application of Fatigue Crack Growth Knowledge", Eng. Fract.
    Mech., 6 (1974), 47-70.

3. Clark, W.G., Jr., and Hudak, S.J., Jr., "Variability in Fatigue
    Crack Growth Rate Testing", J. Test. Eval., 3 (1975), 454-76.

4. Draper, N.R. and Smith H., Applied Regression Analysis. New
    York: John Wiley & Sons, Inc., 1966.

PROOF TEST CRITERIA FOR THIN WALLED PRESSURE VESSELS

R. W. Finger

Boeing Aerospace Company

Seattle Washington

INTRODUCTION

Service failures of pressure vessels are generally a consequence of inherent part-through surface of sub-surface defects.
The catastrophic pressure vessel failures of the mid 50's and 60's
necessitated the development of a technology for the understanding
and thereby eventual elimination of these failures. Fracture
mechanics evolved as the technology base for the analysis of
failures and for the development of methodologies for the prevention
of these failure occurrences. Paramount in the prevention of
service failures has been the development of fracture resistant
materials and proof testing procedures. Reference [1] ("Fracture
Control of Metallic Pressure Vessels") defines the necessary procedures for determining proof testing parameters for pressure
vessels in which the critical flaw depth at proof pressure is less
than the wall thickness. The increased usage of fracture resistant
materials has resulted in an increase in the critical flaw size at
proof pressure. Presently a significant percentage of aerospace
pressure vessels have critical flaw sizes at proof that are in
excess of the wall thickness. For most aerospace pressure vessels
fabricated from 2219 aluminum the failure made at proof pressure
is leakage. It is the purpose of this chapter to present a proof
test criterion for assuring minimum service life requirements for
these pressure vessels. The criteria proposed are based on the
results of several experimental programs conducted on surface flaw
specimens fabricated from 2219 aluminum base and weld metal. A
discussion of the stable crack growth behavior of surface flaws
during loading is presented followed by a discussion of using proof
testing to assure minimum service life requirements.

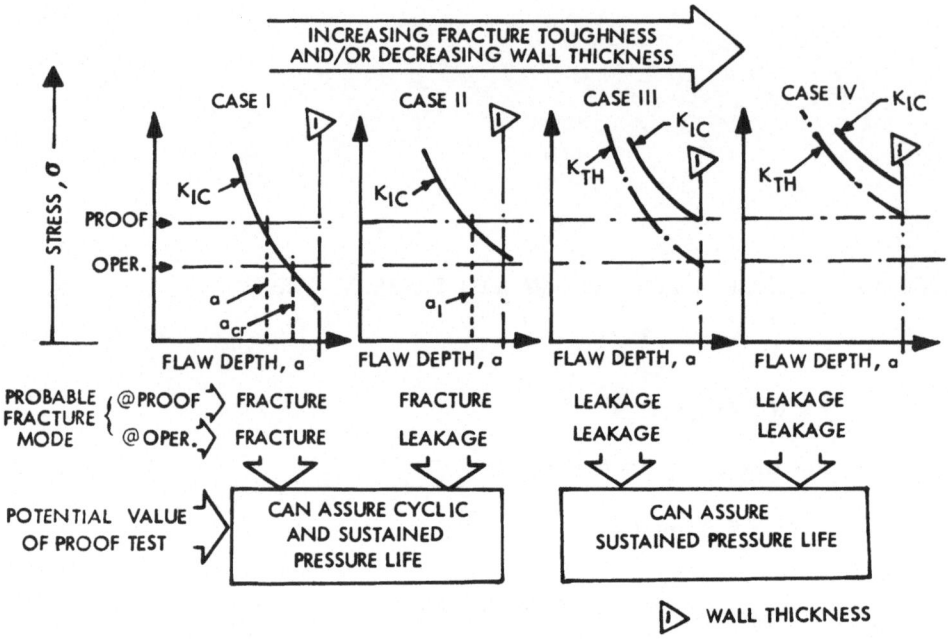

Figure 1. The effect of wall thickness on the value of proof test.

BACKGROUND

Reference [1] has defined the methodology for determining
proof testing procedures for pressure vessels in which the critical
flaw size at proof is less than the wall thickness. These procedures
can be employed in determining the proof testing parameters which
will ensure a minimum service life consisting of both cyclic and
sustain loading. For vessels in which the failure mode is leakage
at proof pressure Reference [1] concluded that the proof test could
only be designed to assure a subsequent sustain load capability.
The potential value of proof testing as effected by increased
fracture resistant material or decreasing wall thickness is presented in Figure 1 from Reference [1]. The increased usage of
fracture resistant materials in aerospace pressure vessels and the
typical gage thicknesses of these vessels have resulted in a significant portion of the present aerospace vessels having a leakage
failure mode at proof pressure. Pressure vessels fabricated from
2219 aluminum (an alloy used extensively in aerospace pressure
vessels) generally are leakage critical at proof pressure. Therefore it was essential to develop a proof testing criterion which
could be used on leakage critical vessels for assuring minimum post
proof test cyclic lives. The first experimental program directed

at developing a better understanding of the behavior of deep flaws
in 2219 aluminum was initiated in 1967 [2]. This program was
directed towards developing deep flaw magnification factors which
could be applied to Irwin's surface flaw stress intensity solution
[3]. Although instrumentation for determining if the flaw penetrated the rear surface prior to failure was not available, the
presence of this phenomenon was suspected. Instrumentation for
determining breakthrough was added in follow-on programs [4,5].
These programs were directed towards providing additional data on
the flaw growth during loading and the effects of a simulated
proof pressure cycle on subsequent cyclic life. A subsequent
study [6] was directed at better defining the stable crack growth
during loading characterisitics of the 2219 aluminum alloy and the
minimum assured cyclic life of a vessel which survived a proof
test which inflicted the maximum stable crack growth possible without causing leakage.

MATERIALS AND PROCEDURES

The 2219 aluminum alloy material, 0.50 inch thick plate and
0.250 inch thick sheet, was purchased in the T87 condition per
Boeing BMS 7-105C (equivalent to MIL-A-8920-ASG). Both parent
metal and as-welded weld metal specimens were tested. The weld
metal specimens were prepared from 0.50 and 0.25 inch thick weldments. The weld panel halves were prepared with a square butt
edge prep and were welded using the direct current straight polarity
(DCSP) gas tungsten arc (GTA) process. Two passes were required to
produce the welds and filler wire (2319) used on the 0.250 inch
thick panels. All of the weld panels were of typical aerospace
quality.

Specimens having gage thickness of either 0.375, 0.250 or
0.125 inch thick were fabricated from both the parent metal and
weld panels. The 0.375 inch thick specimens were prepared from
the 0.50 inch thick weld panels and plate material. The other
specimens were prepared from the sheet stock and the 0.250 inch
thick weld panels. All of the specimens were uniaxial surface
flawed specimens. An Electric Discharge Machine was used to
introduce starter slots into the specimens from which fatigue
cracks were grown. The fatigue crack procedure was monitored
visually with the aid of a 30-power microscope. The fatigue cracking was terminated when a crack existed over the entire periphery
of the EDM starter slot. In general, 10,000 cycles at a maximum
stress of either 12 or 19 ksi for the base and weld metal specimens,
respectively, was suggicient to produce the desired fatigue crack.
The surface flaws were orientated parallel to the rolling direction
for the base metal specimens and in the center of and parallel to
the weld nugget for the weld metal specimens.

All of the testing was accomplished at either R.T., -320°F or -423°F. The cryogenic temperatures were obtained by surrounding the test section with either liquid nitrogen (-320°F) or liquid hydrogen (-423°F). Specimen instrumentation consisted of EDI clip gage, to provide a continuous record of crack opening displacement, and pressure cups to determine when the flaw penetrated through the rear surface of the specimen. The pressure cups were applied directly over and directly behind the surface flaw. The front cup was pressurized to approximately 10 psig with helium, while the pressure in the rear cup was monitored continuously. Breakthrough (i.e., when the flaw penetrated through the rear surface) was denoted by a sharp increase in pressure in the rear cup. Immediately prior to testing, the rear cup was vented in order to relieve any vacuum which might have occurred as a result of the cryogenic cool-down.

During the course of the experimental program three different types of tests were conducted, a) static fracture, b) load-unload plus cyclic, and c) load-unload plus marking. The static fracture tests were conducted at a loading rate which would result in failure within approximately one minute. For the load-unload plus cyclic tests the initial loading cycle was applied in approximately one minute, followed by either 60 or 1 cpm cyclic loading. A sinusoidal loading profile was used for the 60 cpm loading and an equally segmented trapezoidal profile (15 second and fall combined with 15 second holds at maximum and minimum load) was used for the 1 cpm tests. The minimum cyclic load in all of the tests was approximately zero. Generally, the 60 cpm tests were continued until failure had occurred; whereas, the 1 cpm tests were terminated after 100 cycles if failure had not occurred. In all of the testing, failure was considered to be either the flaw penetrating through the rear surface or the specimen fracturing. For the load-unload plus marking tests the initial loading cycle was applied in approximately one minute, followed by low stress fatigue marking. The fatigue marking stresses used were generally the same as the initial precracking stress.

After testing, the fracture faces of the test specimens were examined visually with the aid of a 30-power microscope. The load sequencing testing procedures previously described delineated the flaw sizes at each change in loading. Therefore, all of the flaw dimensions were determined directly from the fracture faces of the specimens. The crack opening displacement records provide further substantiation of the visual measurements.

RESULTS AND DISCUSSION

The objective of the subject program was the characterization of the stable crack growth behavior of surface flaws during proof

TABLE 1

2219-T87 ALUMINUM TEST PROGRAM

MATERIAL CONDITION	TEST TEMPERATURE	a/2c		
		0.15	0.30	0.45
BASE METAL	R.T.	X	X	X
	-320°F	X	X	
	-423°F	X		
WELD METAL	R.T.	X	X	
	-320°F	X		
	-423°F	X		

IDENTICAL TEST PROGRAMS WERE CONDUCTED FOR EACH OF THREE
THICKNESSES 0.125, 0.250 AND 0.375 INCH.

X ~ DENOTES CONDITIONS UNDER WHICH BOTH GROWTH -ON-
LOADING AND POST PROOF CYCLIC TESTS WERE CONDUCTED

testing and the development of a criteria for the determination of
the minimum assured service life subsequent to a proof test for a
thin-walled pressure vessel fabricated from 2219-T87 aluminum.
Prior studies [4,5] have shown that significant stable crack growth
can be encountered during the loading of test specimens. Because
the failure mode for the majority of the conditions tested was
leak-before-break, the possibility existed that a proof test cycle
could obliterate any usable service life. The test program conducted has been summarized and is presented in Table 1. Initially,
the results of the growth-on-loading and cyclic crack growth tests
will be discussed separately and then their implication on the
proof test procedures for thin walled 2219-T87 aluminum pressure
vessels will be considered.

The initial flaw sizes were selected such that the failure
stresses would be 45, 50 and 59 ksi for the R.T., -320°F and -423°F
base metal specimens and 22.5, 25.0 and 29.5 ksi for the R.T., -320°F
and -423°F weld metal specimens. The base metal failure stress
levels represent 90% of the material's minimum yield strength at
the corresponding temperature and are typical of proof test stress
levels. The weld metal failure stress levels were selected to be

one half the base metal value because weld lands twice the nominal
base metal thickness are common in 2219-T87 aluminum pressure
vessels. The flaw sizes were selected after a review of available
data had been made. Deviations from the selected flaw sizes were
not made after testing had been initiated, even though the target
failure stresses were not always obtained. In all cases either
the flaw penetrated the rear surface of the specimen or the specimen fracturing was considered to constitute a failure. The failure
stresses were generally within 10 percent of the targeted value,
therefore it is believed that the minor variation in failure stresses
did not adversely affect the data trends established.

The growth on loading behavior of surface flaws was determined
from specimens which were either loaded directly to failure or to
a load level somewhat less than that required to cause failure.
The specimens were unloaded at a variety of different stress levels
in order that the crack growth behavior could be obtained from
initiation to failure. In some instances the loading was terminated immediately prior to failure. The crack opening displacement
records were used in these instances to determine when loading
should be terminated. Figure 2 shows the COD records for two
similar specimens; one of which was loaded to failure (3BN21-2) and
the other (2BN21-2) which was unloaded immediately prior to failure.
It is obvious from Figure 2 that failure was imminent at the time
specimen 2BN21-2 was unloaded.

The results of the growth on loading tests are presented in
Figures 3 through 14. The figures are plots of stress versus
flaw depth in which the initial and final flaw depths are shown at
the stress level corresponding to the maximum gross section stress
applied to the specimen. Throughout the course of the experimental
program particular attention was paid to insuring all similar
specimens were prepared and tested under identical conditions.
Particular emphasis was placed on the precracking and testing procedures because it was felt that differences in these areas would
influence the test results. Surface flawed specimens fabricated
from 2219-T87 aluminum oocasionally experience delamination at
the crack tip during testing. This phenomena would definitely
have a very significant influence on the growth-on-loading behavior. Delaminations of sufficient size to be visible under a
30 power microscope were present in only three of the specimens.
These three specimens appear in Figure 5 and have been identified
as having delaminated. The other variations in the growth-on-
loading behavior of seemingly identical specimens is not related
to test technique or to delamination. The differences are possibly
a consequence of minor variations in the material properties,
micro delamination not visible under a 30 power microscope, or the
position of the crack front in relation to the grain boundaries.
Since it is impossible to exercise any control over these parameters, the data scatter encountered must be considered inherent

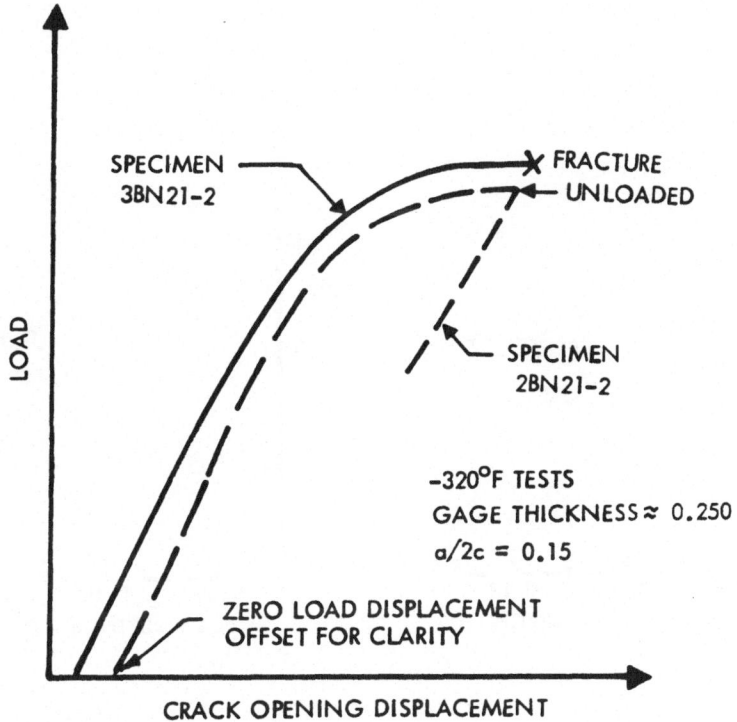

Figure 2.   Load versus crack opening displacement.

to this type of testing. Regardless of the data scatter several things are apparent from the figures. First, the extent of the crack growth encountered during loading is dependent upon the initial flaw shape and the proximity of failure when the specimen was unloaded. Additionally, significant crack growth occurs only within a very narrow band of stress level.

Summary plots of the data have been prepared and are presented in Figures 15 through 19. In these figures the percent increase in flaw depth is compared to K_{Ii}/K_{cr}. The K_{Ii} and K_{cr} values were calculated using the following formula:

$$K_{Ii} = 1.1 \; \sigma \; M_K \; (a\pi/Q)^{1/2}$$

$$K_{cr} = 1.1 \; \sigma_{cr} \; M_K \; (a\pi/Q)^{1/2}$$

σ = Maximum Gross Section Stress

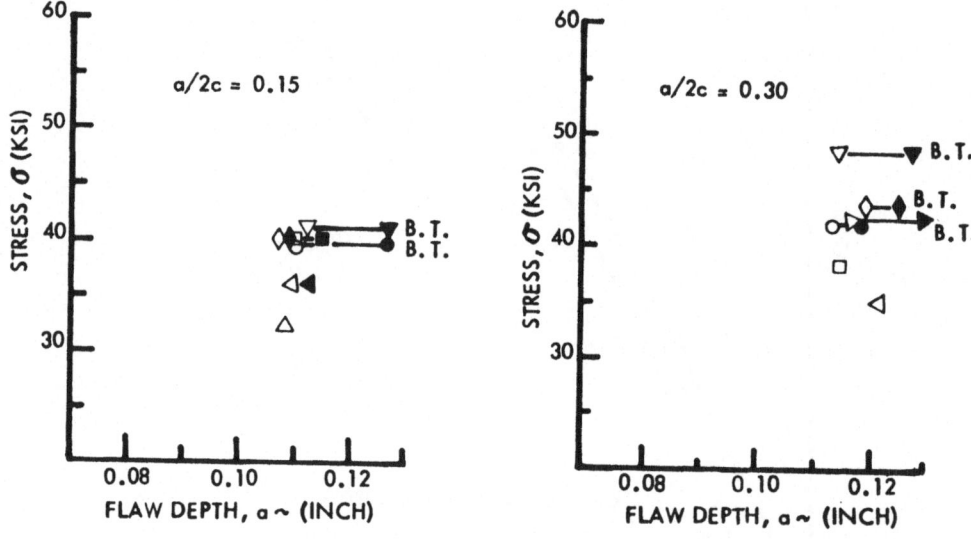

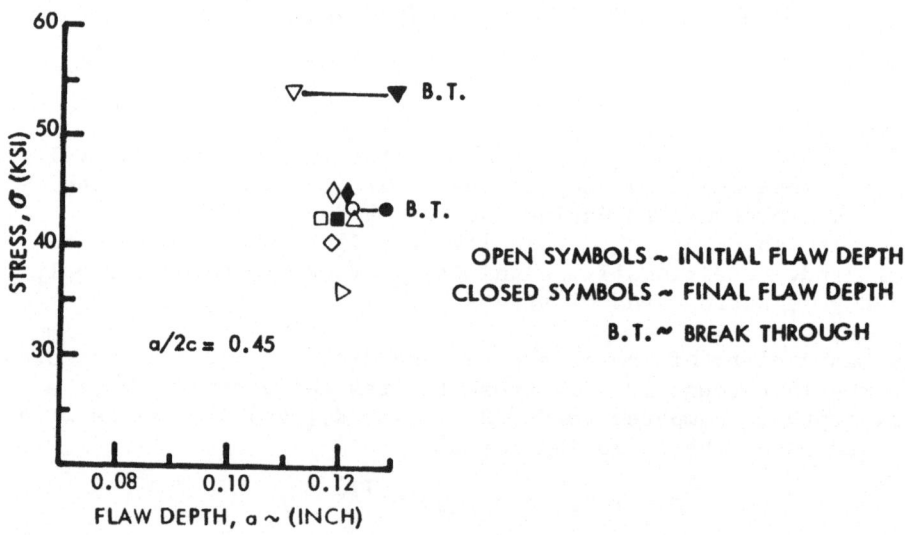

Figure 3.   Growth-on-loading test results for 0.125 inch thick
             2219-T87 aluminum base metal at room temperature.

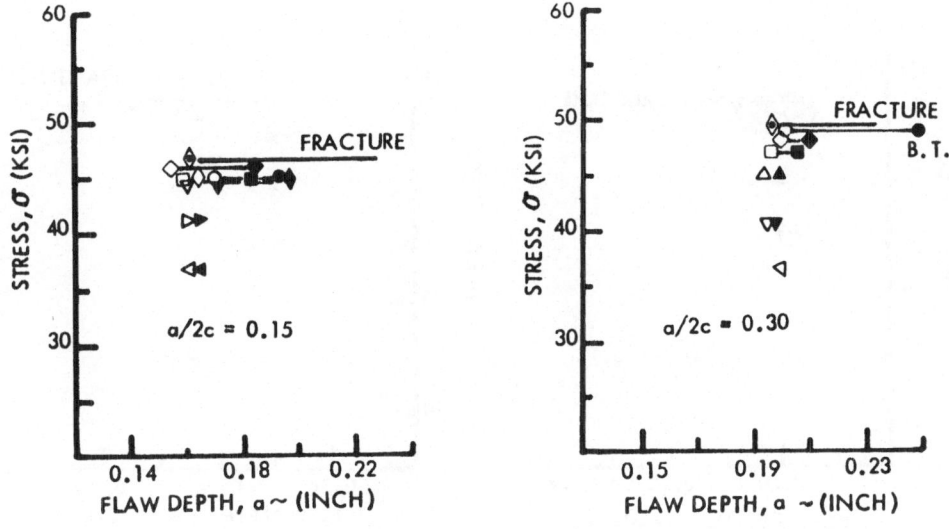

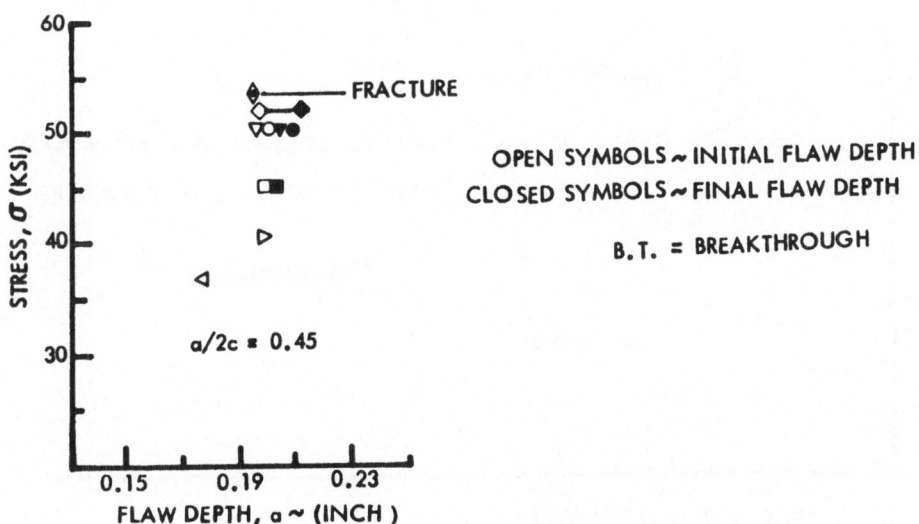

Figure 4. Growth-on-loading test results for 0.250 inch thick 2219-T87 aluminum base metal at room temperature.

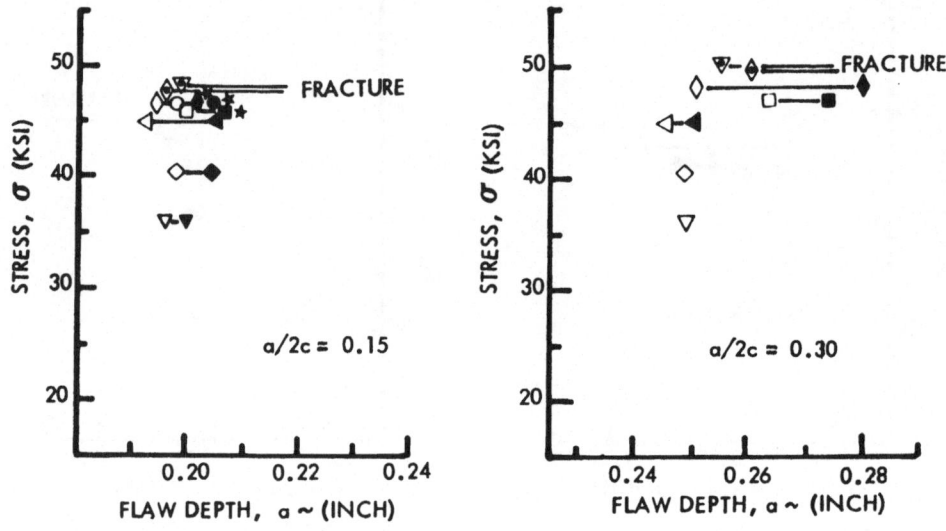

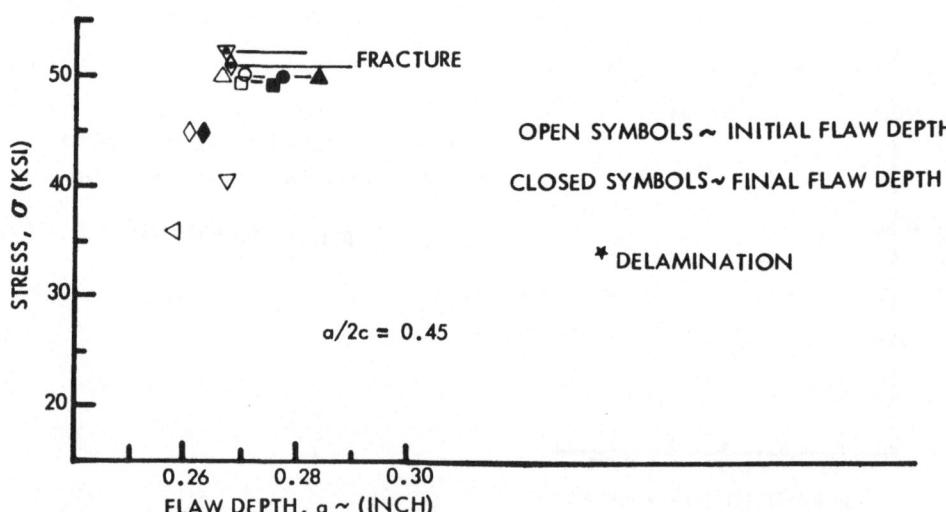

Figure 5.   Growth-on-loading test results for 0.375 inch thick
            2219-T87 aluminum base metal at room temperature.

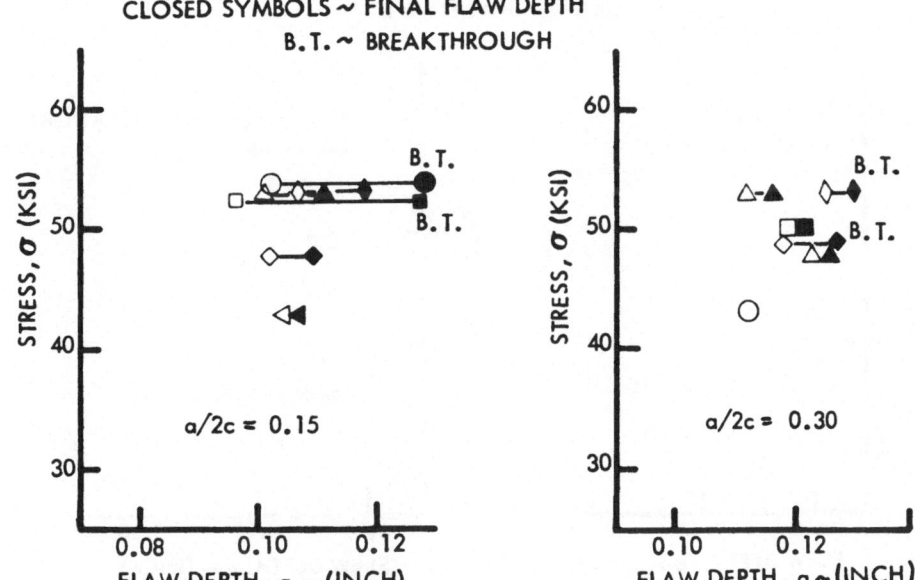

Figure 6.   Growth–on–loading test results for 0.125 inch thick
            2219–T87 aluminum base metal at –320°F.

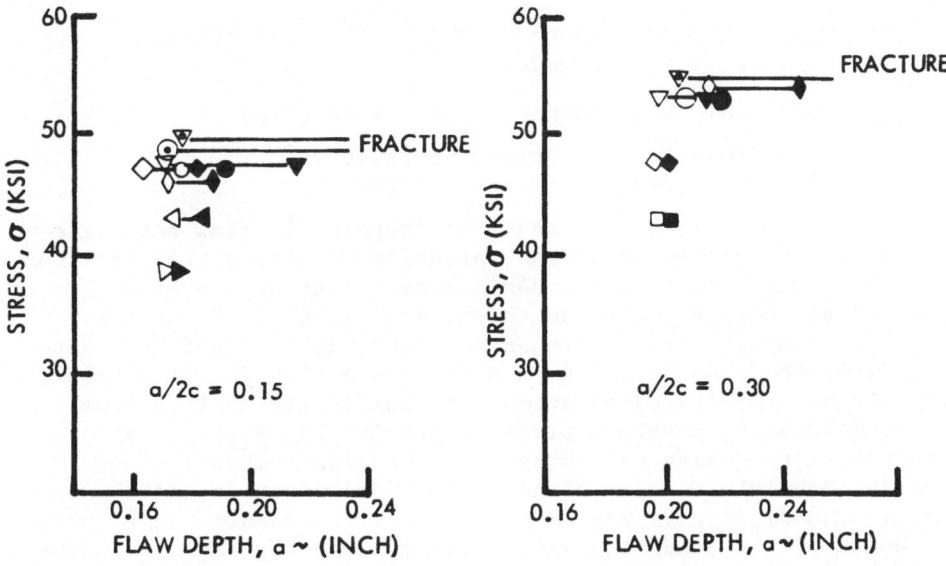

Figure 7.   Growth–on–loading test results for 0.250 inch thick
            2219–T87 aluminum base metal at –320°F.

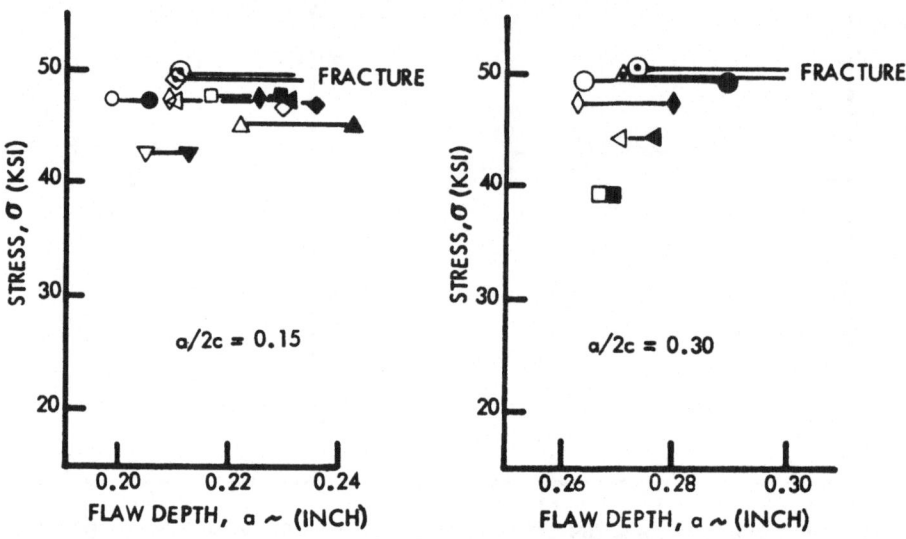

Figure 8.   Growth-on-loading test results for 0.375 inch thick
            2219-T87 aluminum base metal at -320°F.

M_K = Deep Flaw Magnification Factor (see Figure 20)

a = Initial Flaw Depth

Q = Flaw Shape Parameter (see Figure 21)

σ_{cr} = Gross Section Stress at Failure

The parameters K_{Ii}/K_{cr} and % of increase in flaw depth are
related to the change in stress intensity resulting from the change
in flaw depth. Although the basic constraints of the linear
elasticity fracture mechanics theory are violated the stress
intensity concept still represents a reasonable method for characterizing the results. The deep flaw magnification term used in
the stress intensity calculation was empirically derived from
tests of 2219-T87 aluminum plate material. The K_{Ii}/K_{cr} ratios
were calculated using an average K_{cr} for each combination of
material condition, gage thickness, test temperature and flaw
shape. The K_{Ii}/K_{cr} ratios in excess of 1.0 presented in Figures
15 though 19 are a consequence of using an average K_{cr} calculated
from the failure points and the data scatter.

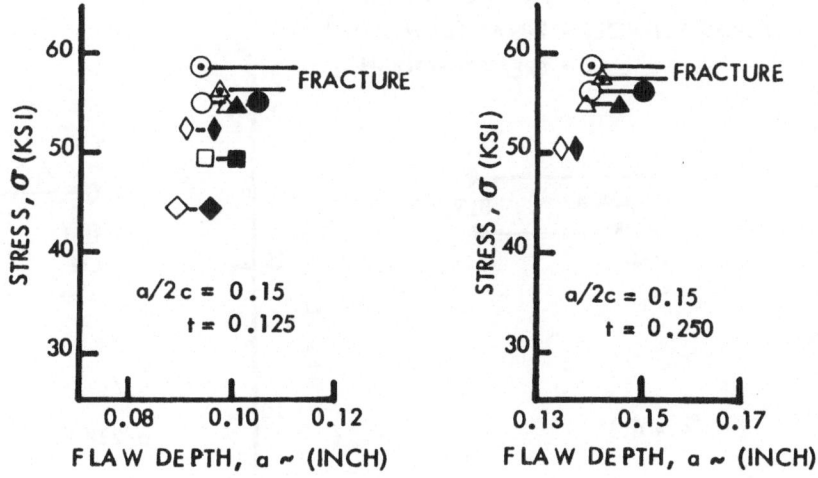

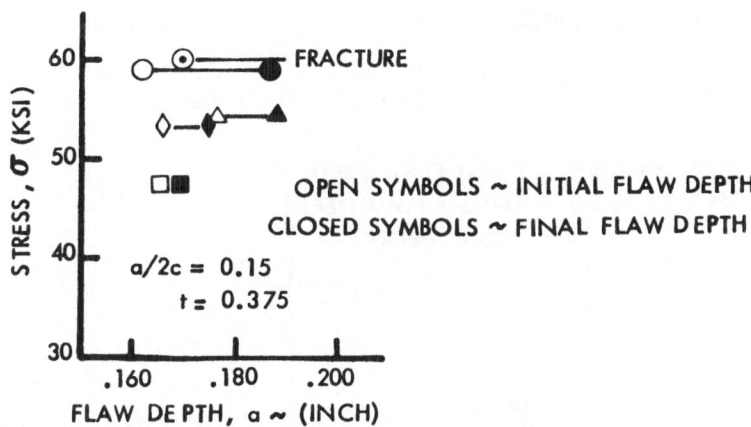

Figure 9. Growth-on-loading test results for 2219–T87 aluminum
base metal at –423°F.

From the summary figures several additional observations pertinent to the growth-on-loading behavior of surface flaws in 2219-
T87 aluminum can be made. It is apparent that the increase in
crack depth, on a percentage basis, is independent of test temperature or gage thickness. Since both fracture toughness and yield

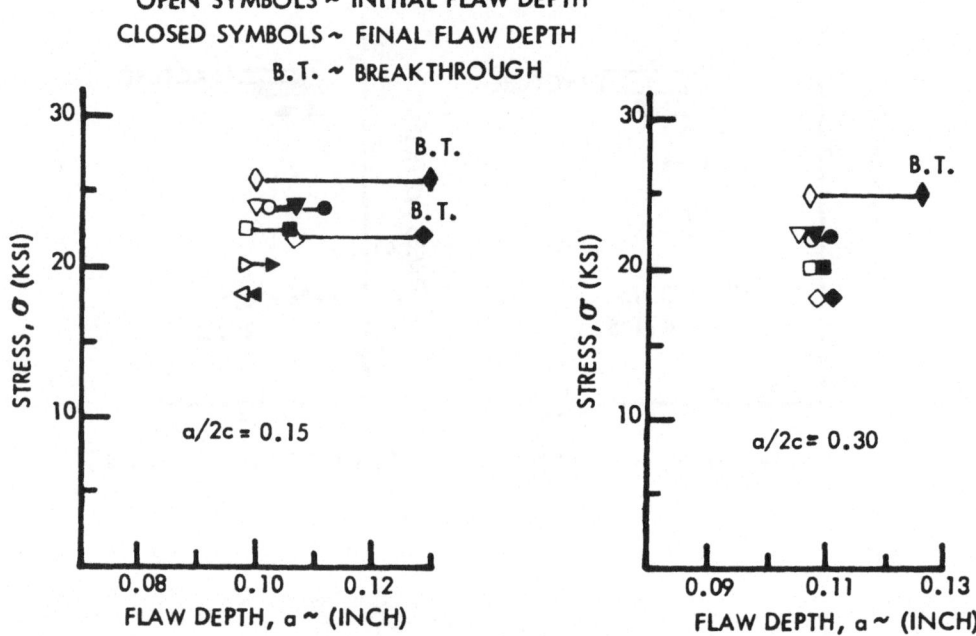

Figure 10.   Growth-on-loading test results for 0.125 inch thick
             2219 aluminum weldments at room temperature.

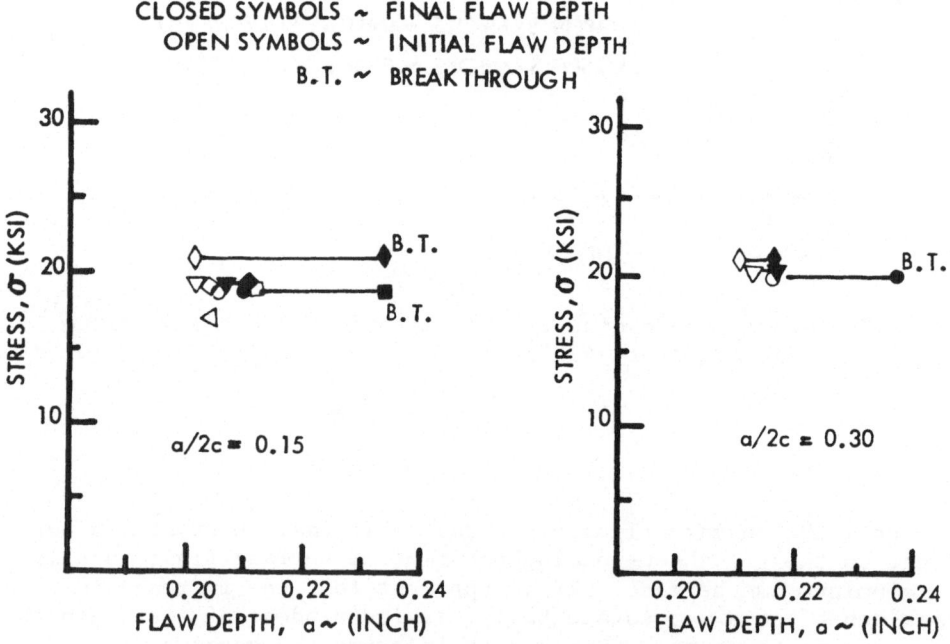

Figure 11.   Growth-on-loading test results for 0.250 inch thick
             2210 aluminum weldments at room temperature.

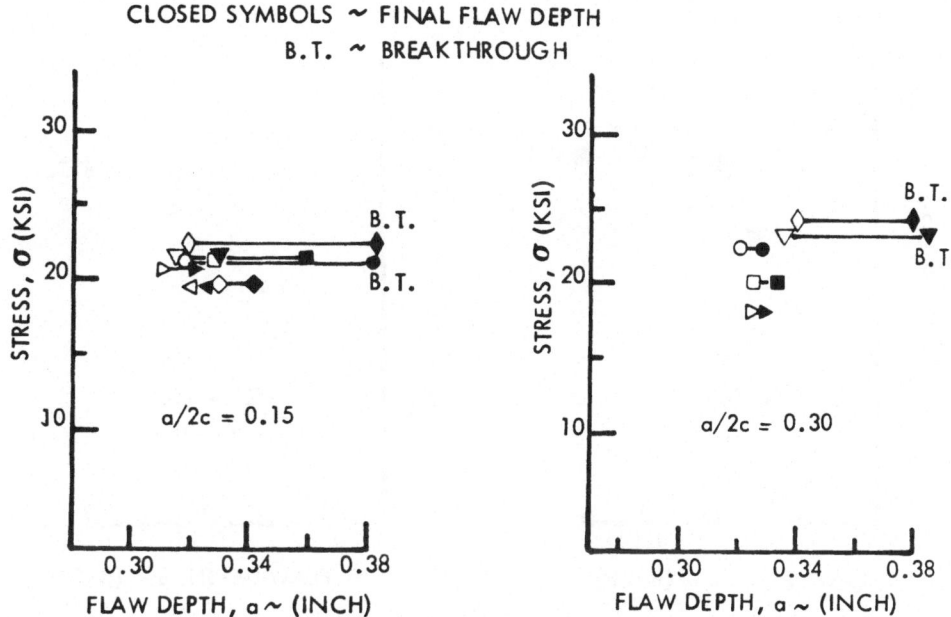

Figure 12. Growth-on-loading test results for 0.375 inch thick 2219 aluminum weldments at room temperature.

strength increase similarly with decreasing temperature, the lack of a temperature influence on crack growth during loading is not surprising. Although the percent increase in crack depth was not influenced by gage thickness, the absolute values were, since the thicker specimen had larger initial flaws. Since the percent increase in flaw depth was unaffected by temperature gage thickness, the ratio of final to initial stress intensity would also be independent of these parameters. The percent increase in crack depth was influenced by the initial flaw shape and material condition. The higher aspect ratio flaws (a/2c = 0.30 and 0.45) experienced less growth in the depthwise direction than the low aspect ratio flaws. The weld metal specimens experienced more stable depthwise growth than the base metal specimens and the growth initiated at lower K_{Ii}/K_{cr} ratio for the weld metal than the base metal. From the summary figures it is apparent that stable depthwise crack growth can be anticipated at K_{Ii}/K_{cr} values in excess of 0.60 and 0.70 for 2219-T87 aluminum weld metal and base metal, respectively. However, even for the worse case tested (a/2c - 0.15) K_{Ii}/K_{cr} ratios of 0.80 for the weldments and 0.90 for the base metal are required to produce a 10 percent increase in

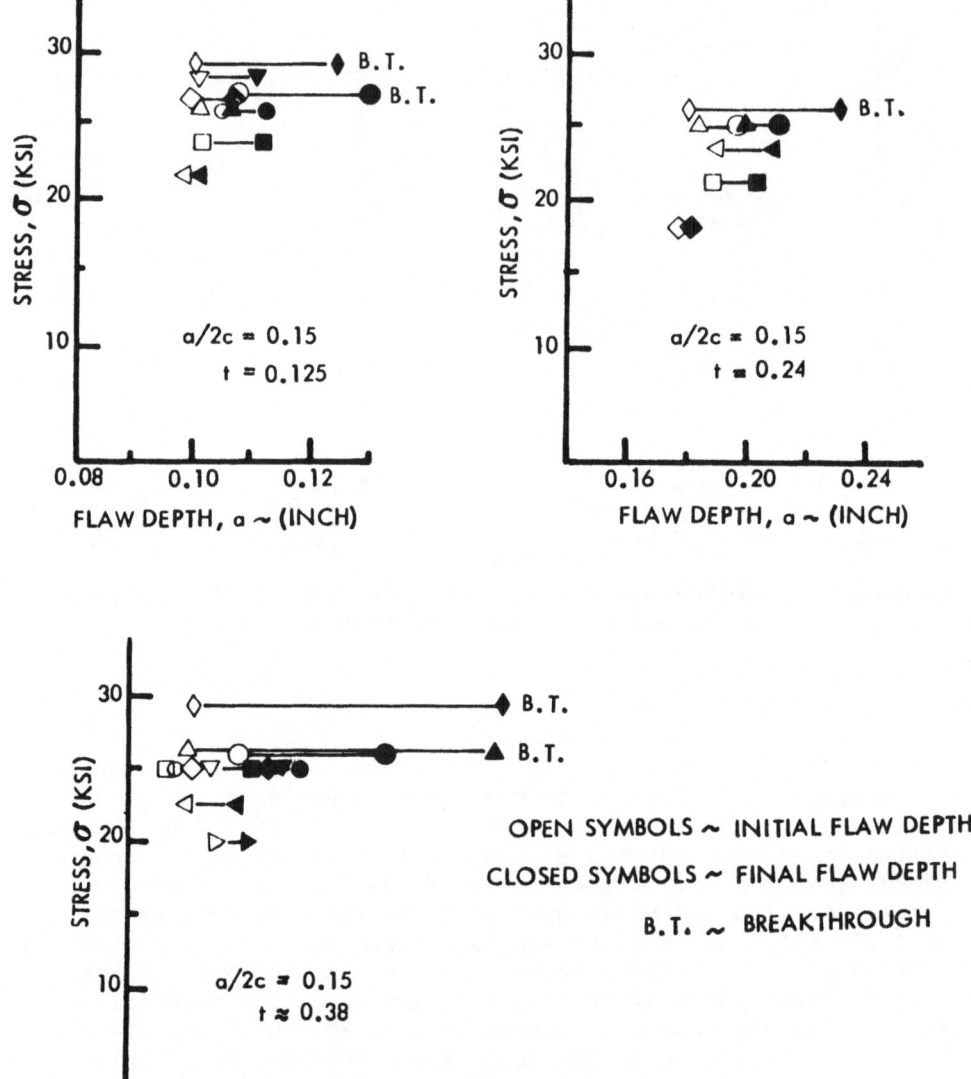

Figure 13.   Growth-on-loading test results for 2219 aluminum
weldments at −320°F.

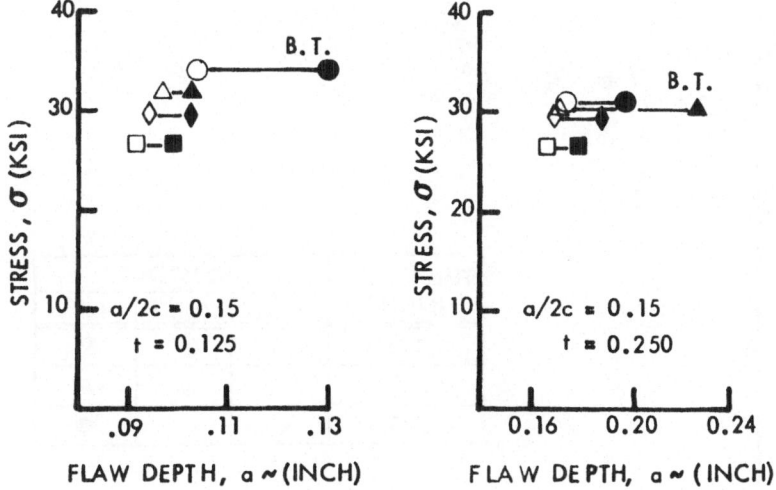

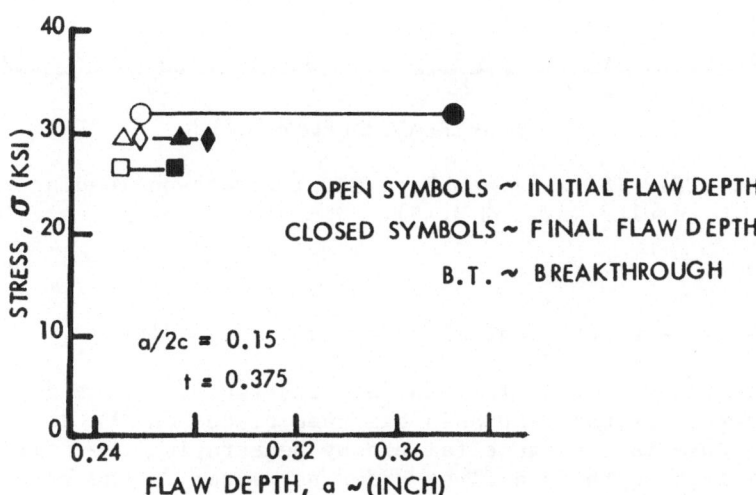

Figure 14. Growth-on-loading test results for 2219 aluminum
weldments at -423°F.

crack depth. Prior to the subject program a limited amount of
work aimed at determining the effect of loading rate on growth on
loading had been conducted at The Boeing Company. The work was
conducted on the 2219-T87 aluminum alloy and the conclusion of
the study was that the loading rate did not influence the extent

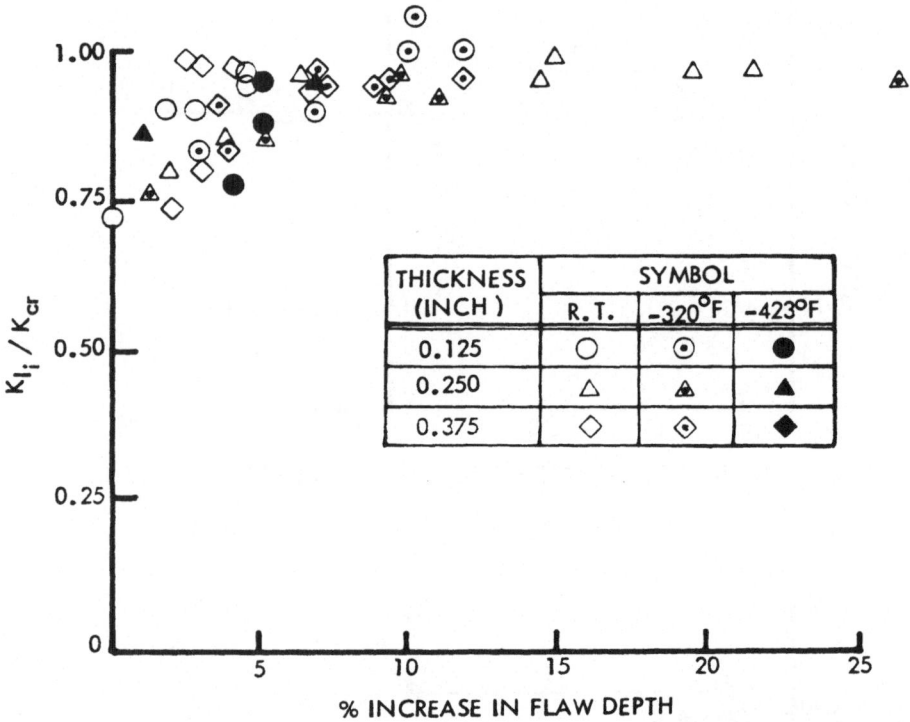

Figure 15.   2219-T87 aluminum base metal growth-on-loading test
results (a/2c ∿ 0.15).

of stable crack growth that can occur during loading.

The discussion this far has been restricted to the depthwise
crack growth. Primary emphasis has been placed on the depthwise
growth because is can cause failure by penetrating the rear surface and crack depth is a first order parameter in the stress
intensity equation. In a limited number of cases there was an
increase in crack length, as well as crack depth. These cases
were almost exclusively limited to the base metal specimens which
had initial a/2c values of 0.30 and 0.45. The manner in which
the different flaw shapes grew is illustrated in Figure 22. The
maximum laterial dimension was considered to be the final crack
length. This was done because none of the specimens experienced
any lateral growth on the front surface. Approximaterly 30% of
the base metal specimens having an initial a/2c = 0.30 and 0.45
experienced lateral crack growth during loading. Only a very few
of the other specimens experienced any growth in the lateral
direction and in all the cases it was less than a 10 percent

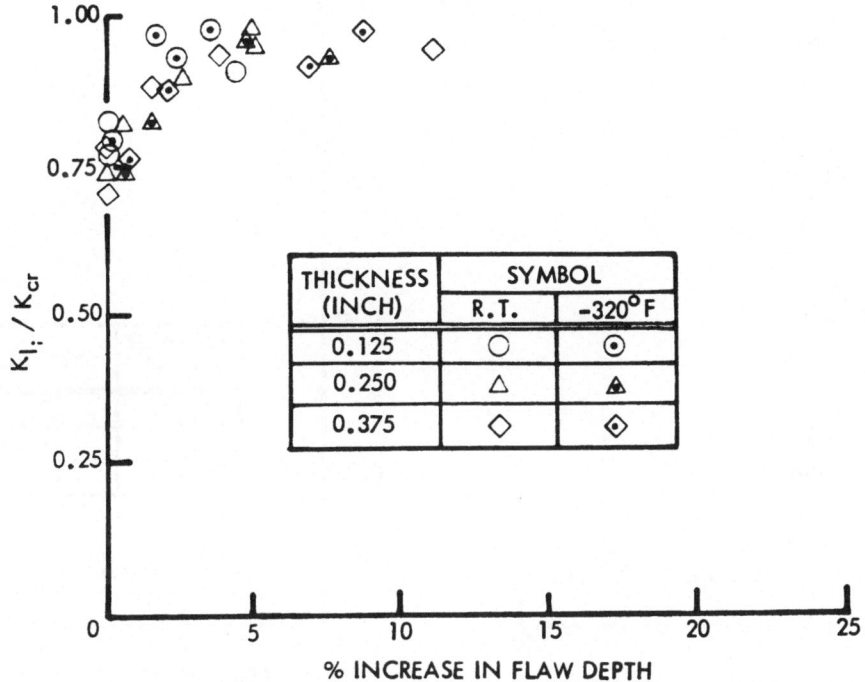

Figure 16.

increase. A summary of the lateral crack growth results in presented in Figure 23. The maximum percent increase in flaw length
was greater than the corresponding increase for flaw depth, however,
increases in flaw length were not experienced below a K_{Ii}/K_{cr}
ratio of 0.90, whereas depthwise extension was.

The results of the cyclic tests are presented in Figures 24
through 27. K_{cr} was calculated in the same manner as previously
described and K_{Ii} was calculated using the prior to proof flaw
size and the cyclic stress level. Tests which were terminated
prior to failure have been noted on the figures. All other specimens were cycled until failure or until failure was imminent.
Although the tests were conducted at both 60 and 1 cpm the cyclic
frequencies have not been distinguished because they had no influence on the results. The cycles to failure curves presented in
Reference [7] for thick specimens whose failure mode was fracture
have been drawn on Figures 24 through 27. Reference [7] did not
present any cycles to failure curves for 2219 aluminum weldments
so the base metal curves have been drawn on the weld metal figures.

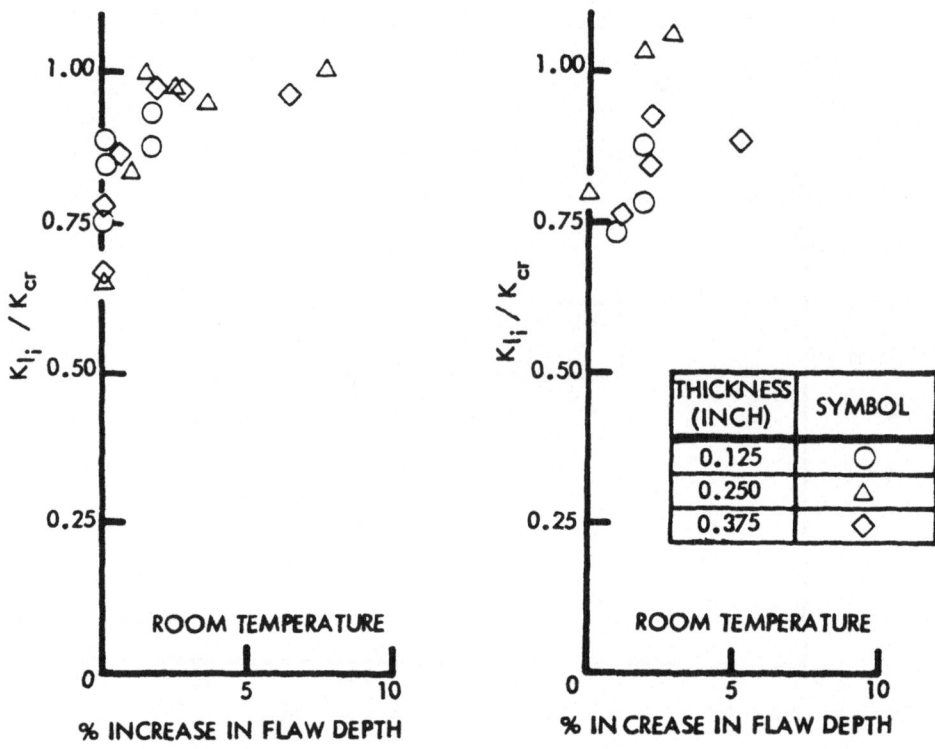

Figure 17.   2219-T87 aluminum   Figure 18.   2219 aluminum weld-
             base metal growth-   ments growth-on-
             on loading test re-   loading test results
             sults (a/2c ≈ 0.45)   (a/2c ≈ 0.30)

The Reference [7] curves were "best fit" not lower bound curves,
therefore the dispersion of the data about the curves are indicative of agreement between the K_{Ii}/K_{cr} vs. cycles to failure relationship for both failure by leakage and by fracture. All of the
data, with the exception of the room temperature base metal results,
are evenly dispersed about the reference curves. The room temperature base metal results (Figure 24) tend to be to the right of the
reference curve indicating a slight increase in cycles to failure
for a given K_{Ii}/K_{cr} ratio. In total 107 cyclic tests were conducted, all of which were subjected to a simulated proof test
prior to failure. The proof tests in 91 of the tests were designed
to produce the maximum possible flaw growth without causing leakage
during the proof loading, thereby minimizing subsequent cyclic life.
In all but three incidences the cyclic lives of proof tested

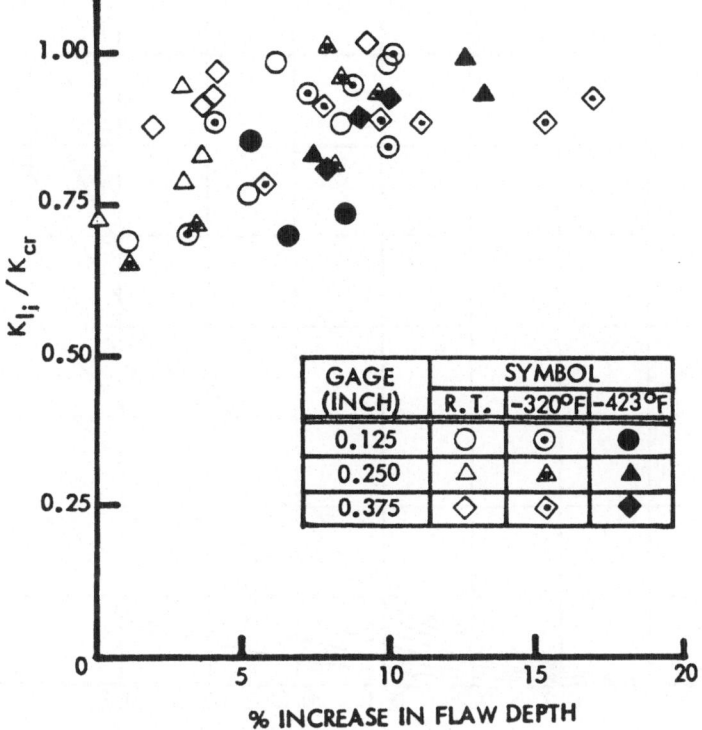

Figure 19. 2219 aluminum weldments growth-on-loading test results
(a/2c $\approx$ 0.15).

specimens correspond very well to the cyclic life of non-proofed
specimens. The three specimens (R-1, R-2 and N-1) which have not
been presented in Figures 24 through 27 all failed, by leakage, on
the first cycle after proof testing. All three of these specimens
were 0.125 in. thick base metal, two were room temperature tests
and one a -320°F test. Two of the specimens, R1 and N-1, were
cycled at 1 cpm and breakthrough was noted after the peak cyclic
load had been obtained. Specimen R-1 was tested at room temperature,
and had an initial a/2c of 0.45. The proof stress was 42.5 ksi and
the cyclic stress was 38.2 ksi. Specimen N-1 was tested at -320°F
and had an initial a/2c of 0.15; the proof stress was 47.0 ksi and
the cyclic stress was 37.5 ksi. Both of these specimens were subjected to the trapezoidal cyclic loading profile. The leakage
rate of the helium was light, but detectable, on the first cycle.
Because there was a hold time at peak load for the cyclic test and
there was not any during the proof cycle, the possibility does

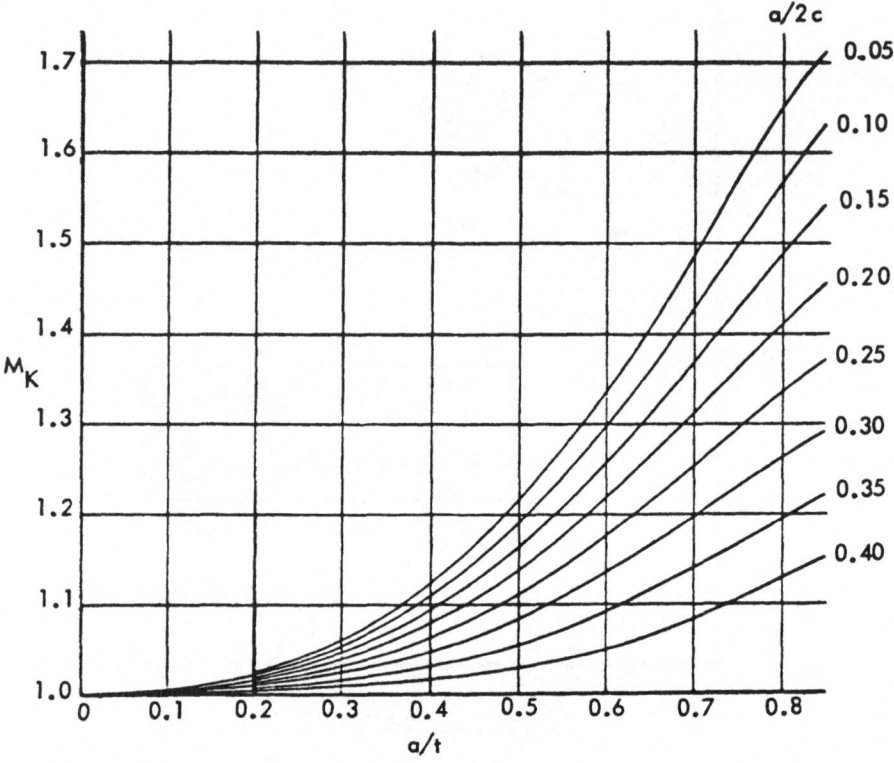

Figure 20.   Deep flaw magnification curves (Reference [2]).

exist that breakthrough occurred during the proof cycle and was
not detected. There was, however, no indication on the pressure
traces that this had occurred. Specimen R-2 had an initial a/2c
of 0.15, and was subjected to a room temperature proof cycle to
40.0 ksi. The cyclic test was to be at 60 cpm with a peak stress
of 32 ksi. All of the test machines are equipped with a shutdown
system which is activated by an increase in pressure in the rear
cup. When the cyclic loading was initiated the shutdown switch
was actuated at 18.8 ksi. Since the machines was programmed to
run at 60 cpm and the shutdown was roughly half the programmed
load, the shutdown was activated approximately 1/4 second after
the test had been initiated. The unloading time from the proof
overload level to 18.8 ksi was at least two seconds. Although it
is possible, it is extremely unlikely that breakthrough occurred
undetected on the proof overload cycle.

     The objective of the cyclic test program was to determine
if a proof test could cause a pre-existing flaw to grow sufficiently

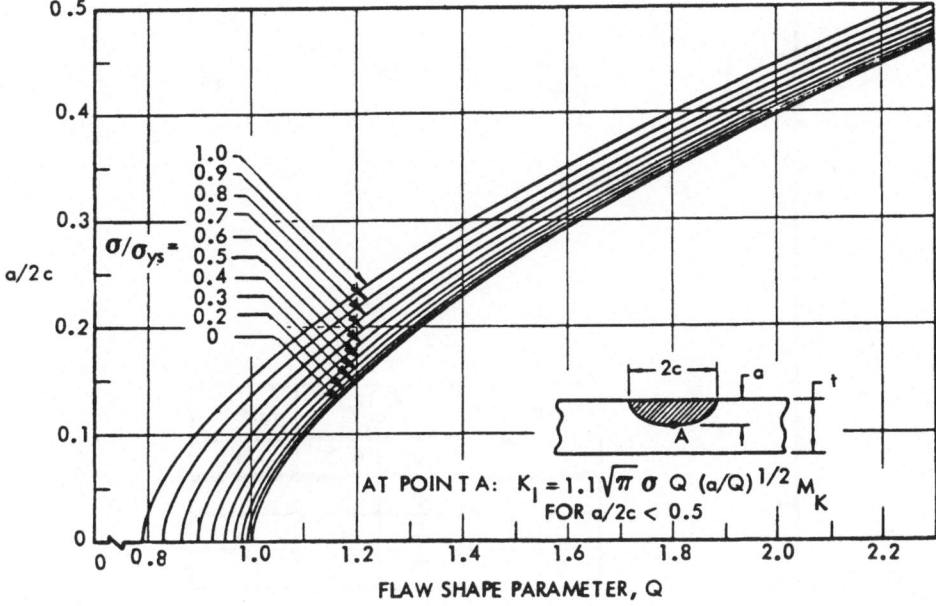

Figure 21.   Shape parameter curves for surface and internal flaws.

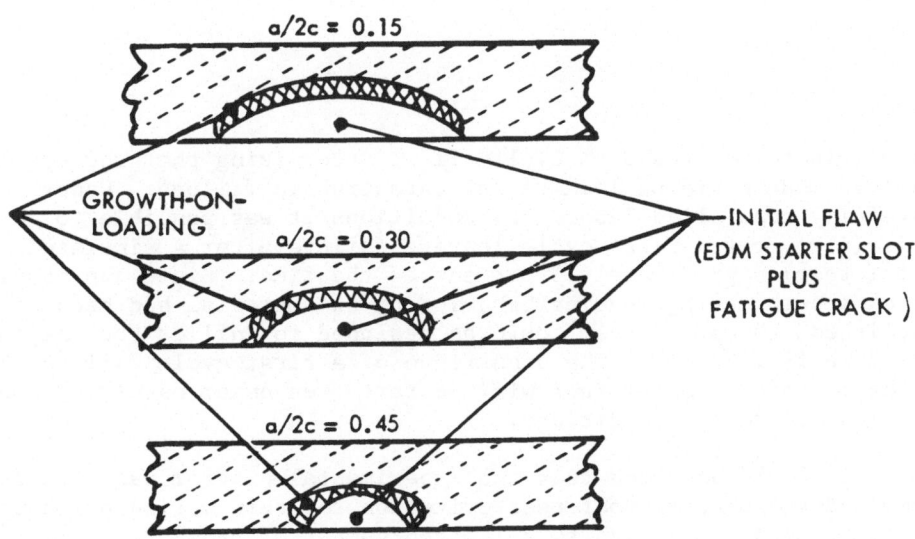

Figure 22.   Illustration of growth-on-loading for various flaw
             shapes.

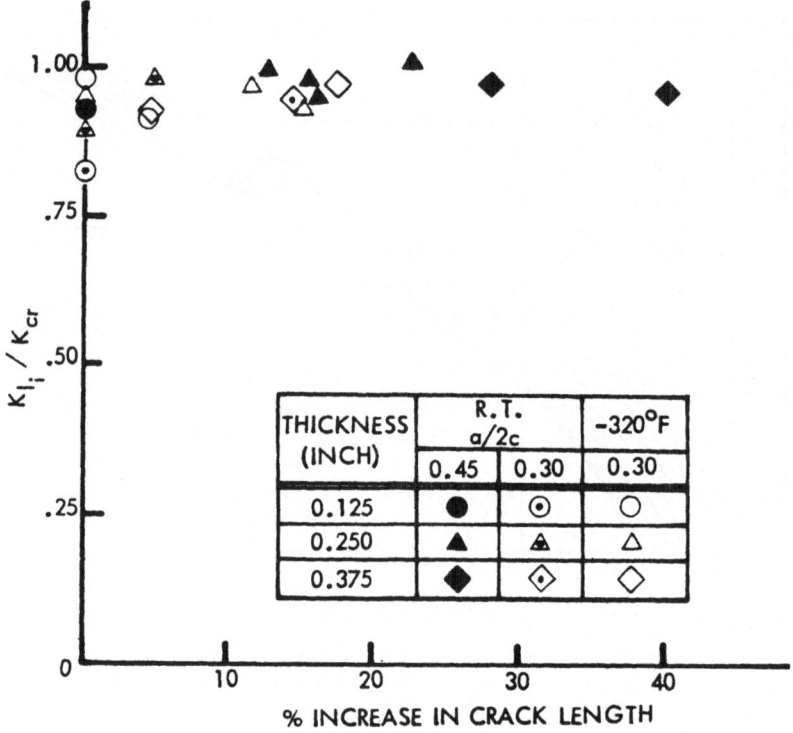

Figure 23.   2219-T87 aluminum base metal lengthwise growth-on-
              loading test results.

to eliminate any residual cyclic life, recognizing that the cyclic
failure mode would be leakage not catastrophic failure. Under
carefully controlled laboratory conditions it was possible to do
this (i.e., get a first cycle leakage failure after a simulated
proof test) approximately 3 percent of the time. Since the stable
crack growth during loading behavior of the material had been established, the proof test could be designed to inflict the maximum
possible flaw growth. The occurrence of a first cycle failure, by
leakage, after a proof test will be rare even under carefully controlled laboratory conditions.

    It should be noted that the experimental program was directed
toward developing a proof test criterion for thin walled pressure
which would be subjected to a limited number of pressure cycles.
The 1 cpm tests were terminated at 100 cycles, if failure had not
occurred, because the program was directed at pressure vessels

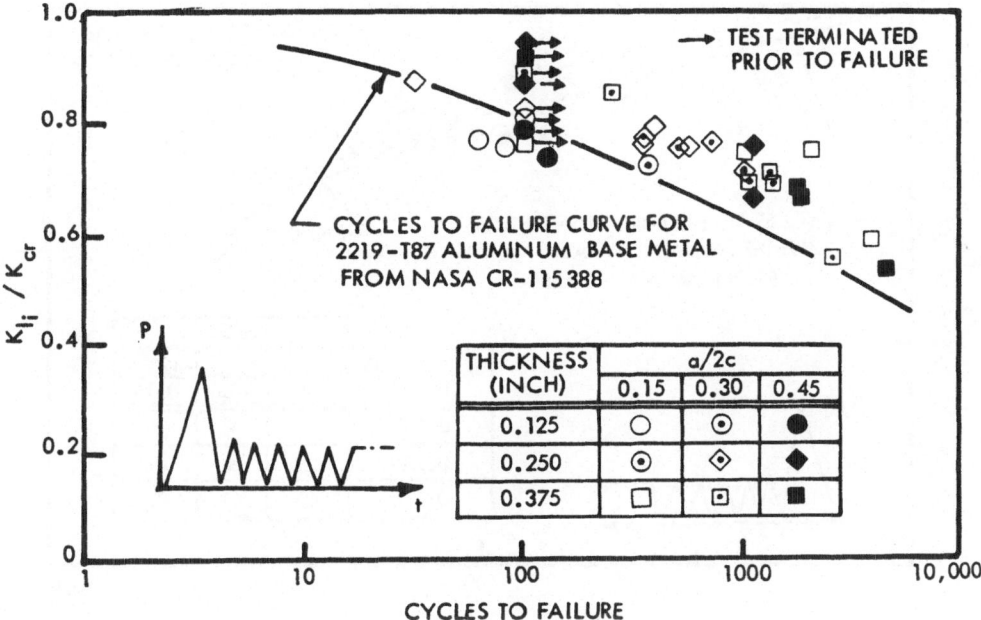

Figure 24. K_{Ii}/K_{cr} versus cycles to failure for proof loaded 2219-T87 aluminum base metal at room temperature.

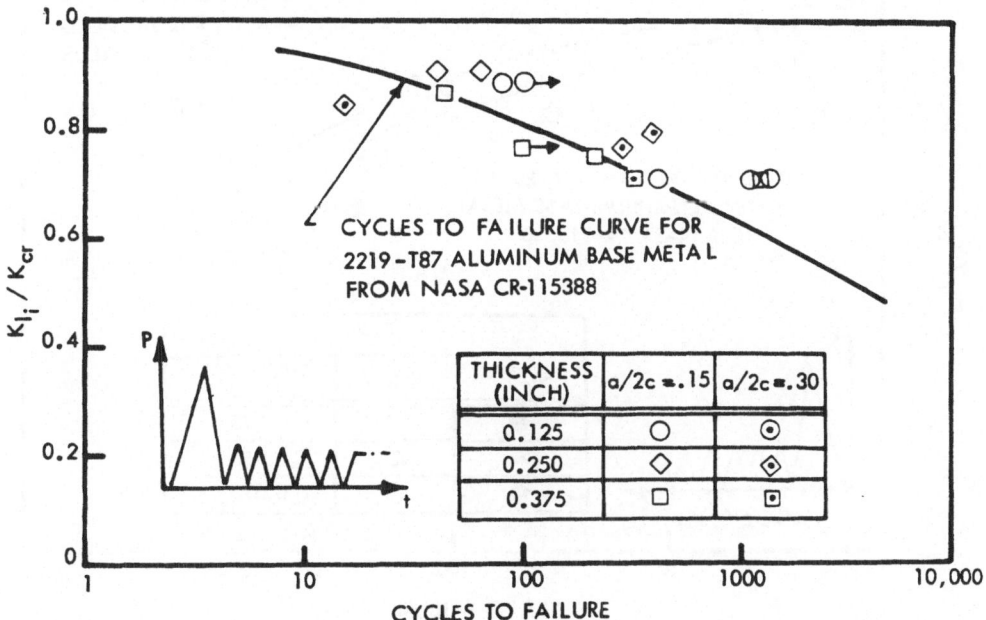

Figure 25. K_{Ii}/K_{cr} versus cycles to failure for proof loaded 2219 aluminum weldments at room temperature.

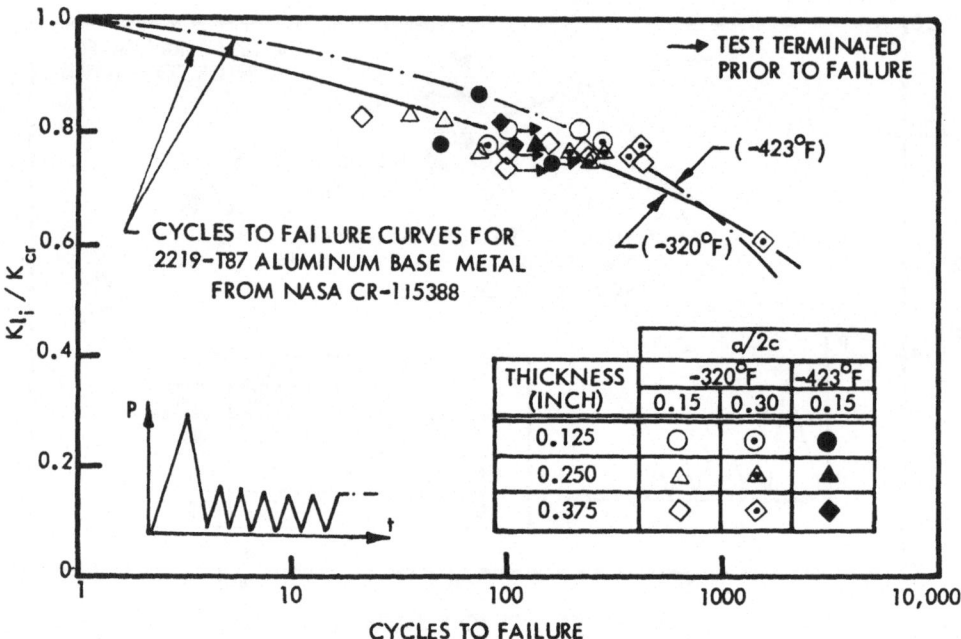

Figure 26. K_{Ii}/K_{cr} versus cycles to failure for proof loaded 2219-
T87 aluminum base metal at −320°F and −423°F.

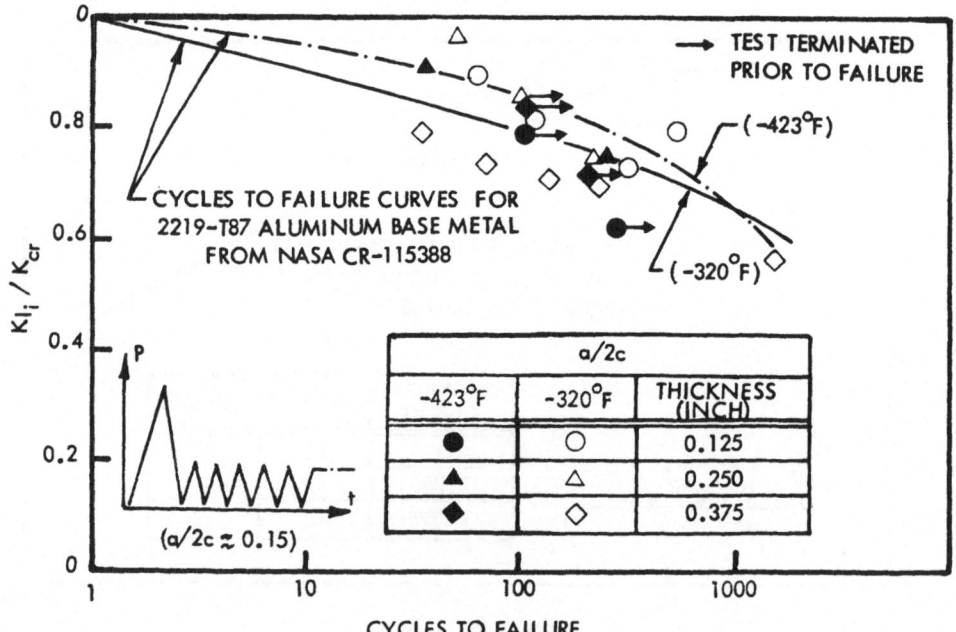

Figure 27. K_{Ii}/K_{cr} versus cycles to failure for proof loaded 2219
aluminum weldments at −320°F and −423°F.

which characteristically see less than a couple hundred operational cycles after proof testing. Attempts to extrapolate the K_{Ii}/K_{cr} vs. cycles to failure curves should not be made.

CONCLUSIONS

The following conclusions were derived from an experimental program conducted on surface flaw specimens of 2219-T87 aluminum base metal and weld metal. Three thicknesses of material, 0.125, 0.250 and 0.375 in. were tested at each of three different temperatures, 72°F, -320°F and -423°F. All of the tests were conducted using uniaxial specimens. The following conclusions apply to the aofrementioned conditions and attempts to extrapolate them beyond the bounds of the conditions tested or to other alloy systems is not advised.

1. Significant stable crack growth under increasing load can occur prior to failure. However, significant variability in results can be anticipated even when carefully controlled laboratory procedures are employed.

2. Initial flaw shapes and material conditions have a significant influence on the extent of growth occurring during the loading cycle.

3. Stable crack growth initiates at a lower K_{Ii}/K_{cr} ratio and is more severe in weld metal specimens than in base metal specimens. The ratios of K_{Ii}/K_{cr} required to initiate stable crack growth are approximately 0.70 for base metal and 0.60 for weld metal.

4. Low aspect ratio flaws (a/2c - 0.15) experience more growth in the depthwise direction than higher aspect ratio flaws (a/2c = 0.30 and 0.45). However, crack growth in the lengthwise direction is more prevalent in the rounder flaws, but only at K_{Ii}/K_{cr} ratios in excess of 0.90.

5. Proof testing assures that any first cycle failure will be by leakage not a catastrophic failure. Although possible, first cycle leakage failures after proof testing will be rare, even under carefully controlled laboratory conditions. Therefore, it is possible to design a proof test that will eliminate the possibility of a first cycle catastrophic failure and also provide a high degree of assurance that the vessel will meet its service requirements.

ACKNOWLEDGMENT

The author wishes to acknowledge the financial support of the National Aeronautics and Space Administration's Lewis Research Center through Contract NAS3-18906 under the administration of Mr. Gordon T. Smith.

REFERENCES

1. Tiffany, C.F., "Fracture Control of Metallic Pressure Vessels. NASA Space Vehicle Design Criteria, Structures", Boeing Company, Seattle, Wash. National Aeronautics and Space Administration Contract Report No. NASA-SP-8040, May 1970. (N71-14130)

2. Masters, J.N., Haese, W.P. and Finger, R.W., "Investigation of Deep Flaws in Thin Walled Tanks", Boeing Company, Seattle, Wash. National Aeronautics and Space Administration Contract Report No. NASA-CR-72606, December 1969. (N70-16465)

3. Irwin, G.R., "Crack Extension Force for a Part-Through Crack in a Plate", Trans. ASME, Ser. E, J. Appl. Mech., 29 (1962), 651-54.

4. Masters, J.N., Bixler, W.D. and Finger, R.W., "Fracture Characteristics of Structural Aerospace Alloys Containing Deep Surface Flaws", Boeing Aerospace Company, Seattle, Wash. National Aeronautics and Space Administration Contract Report No. NASA-CR-134587, December 1973. (N74-19542)

5. Masters, J.N., Engstrom, W.L. and Bixler, W.D., "Deep Flaws in Weldments of Aluminum and Titanium", Boeing Aerospace Company, Seattle, Wash. National Aeronautics and Space Administration Contract Report No. NSAS-CR-134649, April 1974. (N74-32932)

6. Finger, R.W., "Proof Test Criteria for Thin-Walled 2219 Aluminum Pressure Vessels", Boeing Aerospace Corporation, Seattle, Wash. National Aeronautics and Space Administration Contract Report No. NASA-CR-135036, August 1976.

PRACTICAL FRACTURE MECHANICS APPLICATIONS TO DESIGN OF HIGH

PRESSURE VESSELS

T. E. Davidson and J. F. Throop

Watervliet Arsenal, Watervliet, New York

ABSTRACT

Fracture mechanics provides the rationale by which thick-walled pressure vessel designs may be assessed for adequacy of material toughness, for length of fatigue life, for necessary frequency of inspection and for the fatigue life remaining after the detection of a given size crack.

Examples of the fatigue performance and fracture characteristics of thick-walled cylinders of high-strength steels are illustrated, and interpreted with pertinent fracture mechanics calculations. These provide insight for the designer into the various aspects of metallurgical control of strength and toughness, optimization of stress state through the use of residual stresses and importance of monitoring fatigue crack propagation.

INTRODUCTION

The expanding use of high pressure technology is producing the need for thick-walled pressure vessels that are capable of sustaining internal pressures of 5,000 to 100,000 psi or more. Such vessels are generally cylinders. As a matter of definition in this chapter, a cylinder with a wall thickness greater than one-tenth of the internal diameter is considered to be thick walled. With a ratio of outside to inside diameter of 1.2 or more, this means that the stresses in the wall have a gradient that is properly expressed by the Lamé equations of elasticity theory.

The increasing use of such vessels, combined with increasing

severity of service conditions, requires a critical re-examination
of the philosophy of pressure vessel design in order to reduce the
probability of unsatisfactory service life and/or catastropic failure.
The concept of designing for strength alone and using the highest
strength material available without regard to toughness is likely
to give dangerous results, since the attainment of high strength
frequently results in low ductility and low fracture toughness. Low
fracture toughness leads to fracture at small critical defect size,
and under fatigue loading conditions the inherent defects in the
material may be enlarged to this critical size quickly by sub-critical crack propagation. This results in short fatigue life and
brittle fracture.

    The nature of such behavior is illustrated by the results of
tests of cannon tubes, which are essentially high-performance thick-
walled cylinders. Figure 1 shows the fragments from a brittle fracture that occurred during the test firing of a developmental 175mm
cannon tube which had inadequate properties. Figure 2 shows fracture
surfaces from laboratory fatigue tests of other 175mm tubes which
were cycled at the firing pressure. These show the shallow critical
crack depth corresponding to a fracture toughness of 111 ksi $\sqrt{in}$ and
yield strength of 170-190 ksi, and the much greater critical depth
for a fracture toughness of 130 ksi $\sqrt{in}$ and yield strength of 140-
160 ksi in steel cylinders.

    It is generally recognized that fatigue is a three stage process
which involves initiation, stable crack propagation and, finally,
unstable fast fracture. The latter two stages can be analyzed by
fracture mechanics concepts. The portion of fatigue life endured
during the initiation stage, however, must be considered at present
as a bonus that depends upon favorable conditions. Crack initiation,
being controlled by the shear strain range, is sensitive to yield
strength. A high yield strength may provide a long initiation stage
if the bore surface is smooth and undamaged in service, but any type
of surface damage or stress discontinuity may cancel this advantage.
When high strength metals are used the sensitivity to defects, surface conditions, environment and temperature makes the duration of
the initiation stage difficult to estimate. Hence, in many applications, it may be best to neglect it and consider the fatigue life
as that from some measurable size of flaw until failure.

    The concepts of fracture mechanics provide a powerful tool
which, when incorporated into the design process, directly address
the questions of adequate strength and toughness, fatigue life expectancy, necessary frequency of inspection and the fatigue life
remaining after the detection of a fatigue crack of given size.
They show that it is best to optimize the combination of yield
strength and fracture toughness so that at the operating pressure
the toughness is sufficient to sustain a critical crack depth
approaching or equal to the wall thickness. This permits a "leak-

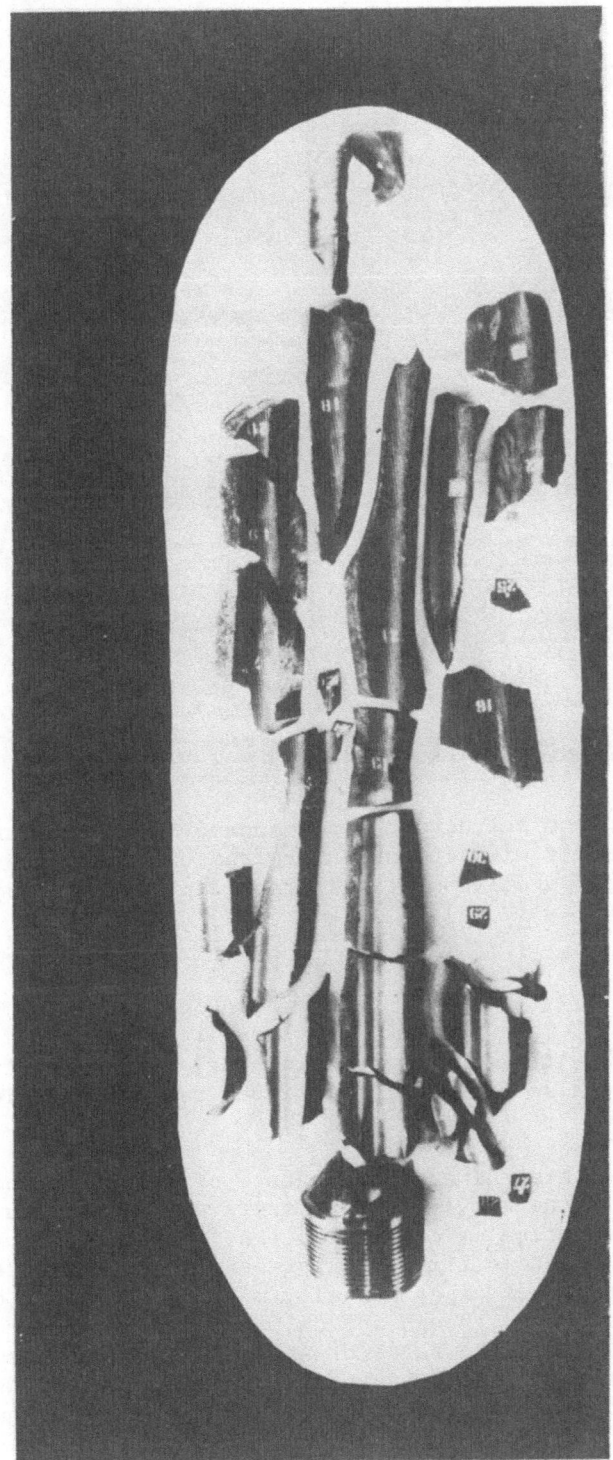

Figure 1. Fatigue in gun tube. Fragments from brittle fracture produced by fatigue during firing tests of a developmental 175mm cannon tube.

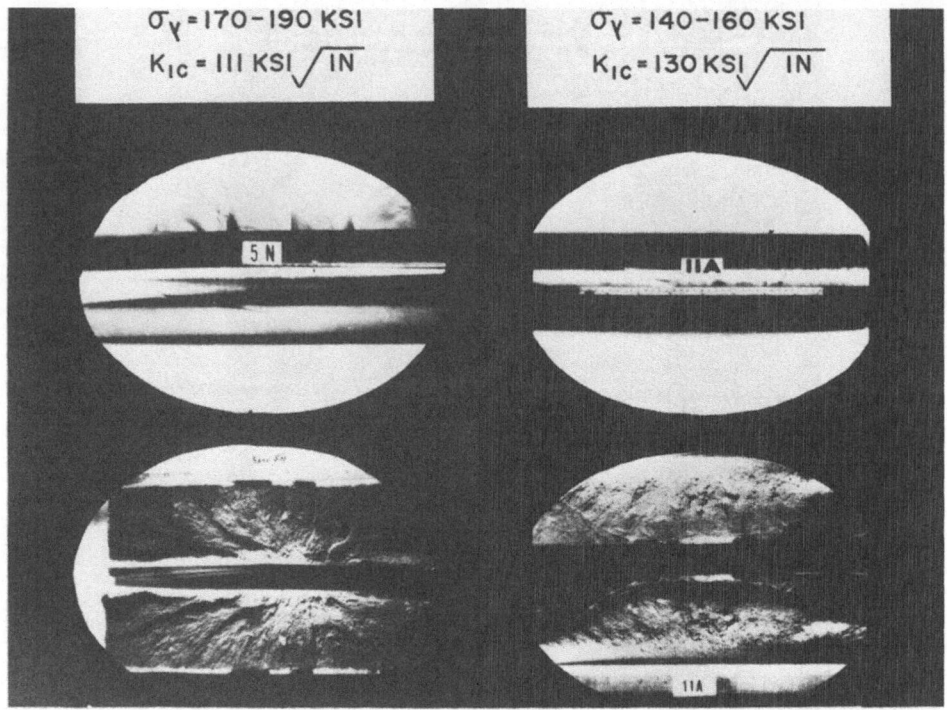

Figure 2.   Macroscopic fracture appearance of fracture surface of
            cylinders with low and high fracture toughness showing
            the comparative critical crack depth at failure.

before-fracture" with its attendant warning, making brittle fracture
unlikely, and it also extends the fatigue crack propagation as far
as possible.

    Because the three-dimensional nature of the crack and stress
configurations in thick cylinders presents severe difficulties in
the analytical solution of crack problems, the practical application
of fracture mechanics is limited at present to approximate solutions
and to idealizations of experimentally measured behavior of cylinders.
Such applications, even though approximate, may provide the designer
with much insight into the roles of strength and toughness, the optimization of stress state by autofrettage or prestressing and the
usefulness of monitoring of fatigue crack growth. The concepts,

approximations and idealizations that may be applied will be illustrated with examples from fatigue tests of high strength steel
cylinders.

Fracture Mechanics

The basic fracture mechanics equations applicable to the problem are the Paris [1] equation for fatigue crack propagation rate

$$\frac{db}{dN} = \frac{1}{M}(\Delta K)^m \qquad (1)$$

and the expression for the stress intensity factor for the opening
mode [2]:

$$K = Y \, S\sqrt{\pi b} \qquad (2)$$

where, in the cylinder, b is the radial depth of the crack into the
wall from the bore, N is the number of fatigue crack cycles, M is a
material constant and the exponent m also depends upon the material.
The ratio db/dN is the fatigue crack propagation rate at the deepest
point of the crack. In cylinders pressurized from zero to maximum
pressure, $\Delta K = (K_{max} - 0)$. Crack instability and fast fracture will
occur when the stress intensity factor, from Equation (2), equals
the fracture toughness K_{Ic} of the material. The relationship for
the stress intensity factor is, then, basic to the problem of prediction of the fatigue and fracture of a pressurized cylinder.

In Equation (2) the value of S is that of the tangential bore
stress given by the equation [3]:

$$S = (\frac{w^2 + 1}{w^2 - 1})p \qquad (3)$$

where p is the internal pressure and w is the ratio of outside diameter to inside diameter. The term Y is a function of w and of the
ratio (b/B) of crack depth b to the wall thickness B. It takes into
account the variation of stress through the wall thickness. Stresses
other than those caused by the pressure can be present, which will
modify the stress intensity factor and, in turn, the crack growth
rate. One important example is the residual stress state induced
by autofrettage, jacketing, etc. These processes result in a compressive stress near the bore. This counteracts the tensile tangential stress expressed in Equation (3), resulting in a lower
stress intensity factor for any given crack shape and a correspond-
ingly lower crack growth rate.

An upper bound of stress intensity factors for wall cracks in

TABLE I

The Normalized Correction Factor, H
Bowie and Freese, Single Crack

(b/B)	w = 1.25	w = 1.50	w = 1.75	w = 2.00	w = 2.25	w = 2.50
0.0	1.00	1.00	1.00	1.00	1.00	1.00
0.1	--	0.99	0.96	0.94	0.91	0.88
0.2	--	1.03	0.98	0.93	0.88	0.84
0.3	1.15	1.14	1.03	0.96	0.89	0.83
0.4	1.40	1.27	1.11	1.00	0.91	0.84
0.5	1.66	1.42	1.20	1.05	0.94	0.86
0.6	1.90	1.56	1.28	1.11	0.99	0.90
0.7	--	1.70	1.39	1.19	1.06	0.97
0.8	--	1.83	1.51	1.31	1.18	1.08
0.9	--	2.09	1.75	1.56	1.42	1.32

$$K = H \left[\frac{2.24 \, w^2}{w^2 - 1} \right] T \sqrt{\pi b}$$

cylinders is provided by the Bowie and Freese [4] solution for a
straight fronted crack in a circular ring acted upon by an external
traction T. In a tube this applies to a machined notch of depth b
with its leading edge parallel to the axis of the tube. The tangential bore stress caused by an external traction T is [3]:

$$S = (\frac{2w^2}{w^2 - 1}) \; T \tag{4}$$

The Bowie and Freese expression for K_I is

$$K_I = H \; (\frac{1.12 \; (2w^2)}{w^2 - 1} \; T) \; \sqrt{\pi b} \tag{5}$$

in which the Y of Equation (2) is seen to be 1.12 H. Values for H
are given in Table I as a function of w and (b/B). For the diameter
ratio w = 2 the value of H is seen to be close to unity for b/B
between 0 and 0.5.

For cylinders of diameter ratio w = 2, if T = p the value of
K expressed in terms of the internal pressure is

$$K_I = H \; (2.987) \; p \; \sqrt{\pi b} \tag{6}$$

This value of K is approximated for crack depths up to half the
wall thickness by the expression

$$K = (1.12 \; \frac{w^2 + 1}{w^2 - 1} \; p + 1.13 \; p) \; \sqrt{\pi b} \tag{7}$$

which gives K = 3 p $\sqrt{\pi b}$ for cylinders of w = 2, as developed by
Underwood et al. [5] from the superposition of solutions for a side-
notched plate under axial tension and pressure in the notch. The
two terms within the bracket reveal the relative contributions to K
for shallow cracks from the bore stress and the pressure in the
crack. Their experimental strain and compliance data showed that
this expression applied for a machined frontal notch in a 1-inch
diameter steel cylinder. This is confirmed by the photoelastic
measurements of C. W. Smith et al. [6] using the stress freezing
technique on similar notches in 1-inch diameter cylinders of this
diameter ratio.

A graph of K/p for w = 2 from Equation (5) is shown as Curve
(a) in Figure 3. Curve (b) is a plot of Equation (7) for the
machined frontal notch. This corresponds to Equation (2) with a
fixed value of α = 1.792 for the term Y. The stress intensity
factor for this frontal notch is so large that it exceeds the
fracture toughness K_{Ic} of even the toughest steels at very shallow
crack depths under usual operating pressures. Fortunately the

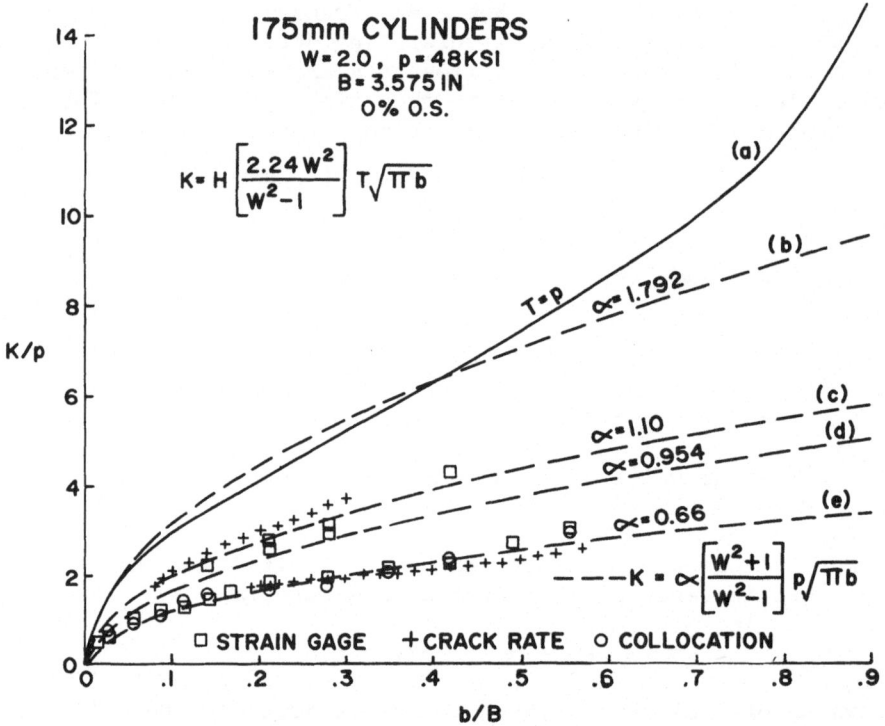

Figure 3.   K/p versus b/B, 175mm cylinders. The graph shows the
           variation for a machined frontal notch, a and b; a long
           curved crack, c; a semi-elliptical crack, d; and a
           semicircular crack e.

natural crack growth does not maintain this shape, but tends to
change toward a semi-elliptical shape which has a lower stress
intensity factor. Otherwise many more brittle fractures of pressure
vessels would have been experienced in service.

     The values for K for curved fronted cracks are much smaller,
as illustrated by Curves (c), (d) and (e) in Fig. 3. These were
obtained experimentally in 175mm cylinders of diameter ratio w = 2
from initial E.D.M. notches. The notches were 1/4 inch deep in a
wall thickness of 3.575 inches. The stress intensity factors decrease with increasing curvature of the crack front. Curve (c)
with α = 1.10 is for a long curved crack grown from a 20-inch long
notch. Curve (d) with α = 0.954 is for a semi-elliptical crack
grown from a 4-inch long notch and Curve (e) with α = 0.66 is for
a semicircular crack grown from a 1/2-inch notch. The latter is
confirmed to a depth of one-half the wall thickness by the results

of collocation analysis by Hussain et al. [7], as indicated by the circles on the graph. This lower curve represents the lower bound of K for wall cracks in non-autofrettaged cylinders of w = 2 diameter ratio.

Experimental data from compliance measurements are represented by squares on Fig. 3. Data from crack rate comparison with a C-shape specimen of known K-calibration and of the same material are shown by crosses. The curves and data are in agreement for crack depths up to 4/10 of the wall thickness. Since over 80% of the fatigue life is endured before this depth is reached the integration of Equation (1) using the constant values of α for Y in Equation (2) appears to be reasonable.

Fatigue Crack Growth

Expressing K by using a constant α for the term Y in Equation (2), as plotted in Fig. 3, permits the simple integration of Equation (1) in the form

$$\int_{N_i}^{N_b} dN = M \int_{b_i}^{b} (\Delta K)^{-m} db \qquad (8)$$

This gives the expression for remaining fatigue crack propagation life, $(N_f - N_i)$, from a crack depth b_i to any final crack depth b_f at a given cyclic pressure as:

$$(N_f - N_i) = M \pi^{\frac{-m}{2}} (\alpha S)^{-m} \int_{b_i}^{b_f} b^{\frac{-m}{2}} db \qquad (9)$$

Analysis of the fatigue crack growth monitored by ultrasonics during tests of 175mm cylinder specimens indicates that the integer exponent which represents the data most satisfactorily is m = 3.

With the value m = 3 the remaining crack life becomes

$$(N_f - N_i) = 2M\pi^{-3/2} (\alpha \frac{w^2 + 1}{w^2 - 1} p)^{-3} [\frac{1}{\sqrt{b_i}} - \frac{1}{\sqrt{b_f}}] \qquad (10)$$

This shows a strong influence of crack shape on fatigue life because it is expressed by α raised to the power of three. This influence is illustrated in Figure 4 with plots of crack depth versus cycles for (a) the long curved crack, (b) the semi-elliptical crack and (c) the semicircular crack. The data points shown are from crack depths measured by ultrasonic techniques during the fatigue tests of 175mm cylinder cycled at 48 ksi internal pressure. The crack sizes were monitored from an initial depth of 1/2-inch until failure.

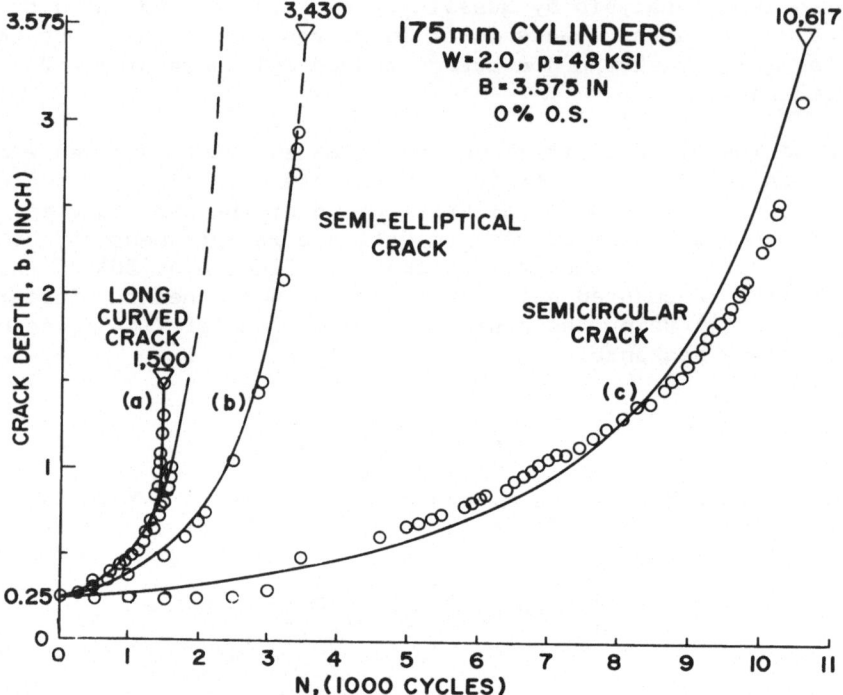

Figure 4.   Crack depth versus N, cycles, in 175mm cylinders. The
             graph shows fatigue crack propagation curves for three
             crack shapes in cylinders of 2.0 diameter ratio under
             cyclic pressurization to 48 ksi, showing data and
             calculated curves.

     The smooth curves on the graph were calculated from Equation
(10), using the idealized expressions

$$\frac{db}{dN} = 3.379 \times 10^{-10} \ (\Delta K)^3 \tag{11}$$

and

$$K = \alpha \left(\frac{w^2 + 1}{w^2 - 1}\right) p \ \sqrt{\pi b} \tag{12}$$

with α = 1.10, 0.954 and 0.66 for the three shapes, respectively.
Thus, with these values of α, and with M = 2.959 x 10^9, w = 2, p =
48, and b_i = 0.25 in Equation (10), the calculated curves approximate the measured crack growth data reasonably well from the 1/4-

TABLE II

Representative Composition, Properties, and Crack Rates

C	0.34	E	=	3×10^4 ksi
Mn	0.50	S_u	=	190 ksi
P	0.012	S_y	=	170 ksi
S	0.011	ELONG	=	18 %
Si	0.22	RA	=	50 %
Ni	3.08	υ	=	0.3 Poisson's Ratio
Cr	1.15	C_v	=	25 Ft.Lb. (-40°F)
Mo	0.58	K_{Ic}	=	140 ksi(in)$^{1/2}$
V	0.13	R_c	=	40 Rockwell Hardness

$$da/dN = 1.51 \times 10^{-8} \ (\Delta K)^2$$

for $K < 94$ ksi(in)$^{1/2}$

after Barsom [8]

$$da/dN = 4.5 \times 10^{-12} \ (\Delta K)^4$$

for $K > 94$ ksi(in)$^{1/2}$

after Throop and Miller [9]

$$db/dN = 3.379 \times 10^{-10} \ (\Delta K)^3$$

$$20 < K < 200 \ \text{ksi(in)}^{1/2}$$

inch initial crack depth to failure. The fatigue lives for the
three crack shapes were 1,500, 3,430 and 10,617 cycles, respectively.

Figure 5 shows the corresponding crack propagation rates on a
log-log plot of db/dN versus b. The data points show the crack
rates obtained from the measured depth – cycles data. The ideali-
zation of these rates are shown by the straight lines on the graph,
calculated using the coefficient

$$\frac{1}{M} = \frac{C}{(E\ S_Y\ K_{Ic})} \tag{13}$$

in Equation (1), as proposed by Throop and Miller [9] to relate the
fatigue crack rate to the mechanical properties in tempered martensitic steels. With the exponent m = 3 and the material properties
listed in Table II for these cylinders, Equation (11) corresponds
to the expression:

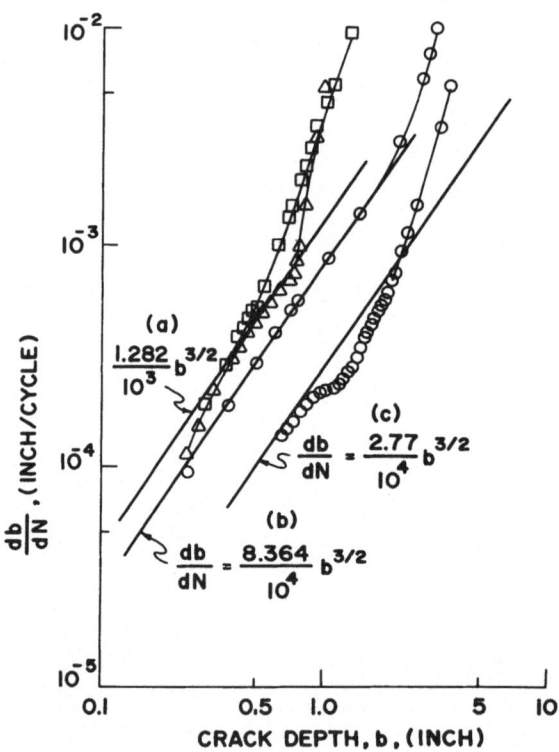

Figure 5.   Crack rate versus crack depth, 175mm cylinders. The log-
log graph shows the fatigue crack propagation rate for
three crack shapes, showing data and calculated lines.

$$\frac{db}{dN} = \frac{0.24}{(E \, S_Y \, K_{Ic})} \, (\Delta K)^3 \qquad\qquad (14)$$

and results in the lines of constant slope on Fig. 5. Since ΔK is
a function of the square root of the crack depth b and the crack
rate is proportional to $(\Delta K)^3$, the idealized crack propagation rate
is a function of $b^{3/2}$ as represented by these lines. Other crack
rate relationships for higher strength steels, dependent on $(\Delta K)^2$
as suggested by Barsom [8] and on $(\Delta K)^4$ as proposed by Throop and
Miller [9] did not approximate this data as well when used with ΔK
as expressed in Equation (12).

   Figure 5 shows that, although the measured crack propagation
rates are not matched exactly, approximating them in the range
between 10^{-4} and 10^{-3} inch/cycle is sufficient to approximate the
depth versus cycles data shown on Fig. 4. Closer approximation
will require the use of variable coefficients and exponents and
recourse to numerical integration. This will require more precise
knowledge of the effects of material properties, crack shape changes,
stress gradients, etc. on the crack propagation rate.

Critical Depth

The critical crack depth can be estimated by the expression

$$b_c = \frac{1}{\pi} \left(\frac{K_{Ic}}{\alpha \, S}\right)^2 \qquad\qquad (15)$$

Our experimental results indicate that for curved crack shapes in
non-autofrettaged cylinders this is a conservative estimate. The
depths, b_f, at fatigue failure shown in Table III are at least 60%
larger than b_c calculated using K_{Ic} = 140 ksi for this material.
This corresponds to a value of $K_{If} = \alpha \, S\sqrt{\pi b_f}$ at fatigue failure
which is at least 1.26 K_{Ic}. The measured K_{Ic} value was obtained
using a C-shaped specimen of known K-calibration taken from one of
the cylinders. It appears that because the rest of the curved crack
front is at lesser K than the deepest point the fracture may be
delayed by triaxial constraint even though the crack rate at the
deepest point is related to the value of ΔK at that point.

Crack Shape

   The Bowie and Freese solution for K of the frontal notch takes
into account the effect of size, stress gradient and diameter ratio
through the quantities H and w. A useful empericism, as yet unveri-
fied, is to assume that the K for other crack shapes may be found
by taking a fraction of the K for the frontal crack. This is

TABLE III

Cylinder No.	Initial Notch	Crack Shape	α	S ksi	b_c (inch)	b_f (in)	$(\frac{b_f}{b_c})$	K_{If} (ksi $\sqrt{in}$)	$(\frac{K_F}{K_{Ic}})$
AM2448B	20"	Long curved	1.10	100	0.516	1.0	1.94	195	1.39
AM2437B	20"	Long curved	1.10	80	0.806	1.5	1.86	191	1.36
CM2437B	4"	Semi-ellipse	0.954	80	1.071	2.95	2.75	232	1.66
B2448B	1/2"	Semicircle	0.660	80	2.238	3.575	1.60	176	1.26

Note: $K_{Ic} = 140$ ksi $\sqrt{in}$; $b_c = \frac{1}{\pi}(\frac{K_{Ic}}{\alpha S})^2$; $K_{If} = \alpha S \sqrt{\pi b_f}$

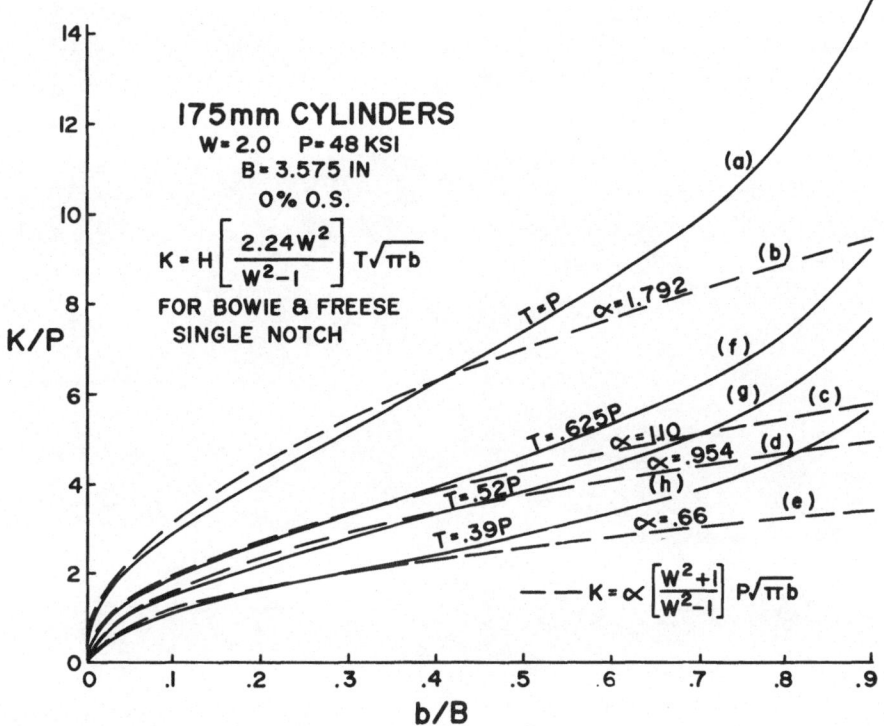

Figure 6.   K/p versus b/B at fractional traction. The graph shows
curves calculated from the Bowie and Freese solution by
use of a "reduced traction" corresponding to each crack
shape in 175mm cylinders of 2.0 diameter ratio.

equivalent to using a reduced traction T in Equation (5); for instance
T = 0.39 p for the semicircle, T = 0.52 p for the semi-ellipse and
T = 0.625 p for the long curved crack. The latter value corresponds
to the external traction T that, acting along with w = 2, would produce the same tangential bore stress as would the internal pressure
acting alone, since this requires that T = $[(w^2 + 1)/(2w^2)]$ p =
5/8 p. Curves for these reduced equivalent values of T are shown in
Figure 6 along with the curves previously shown in Figure 3. Curves
(f), (g) and (h) are for the long curved crack, the semi-elliptical
crack and the semicircular crack, respectively. The use of appropriate "reduced tractions" for similar crack shapes in tubes of other
bore sizes and wall ratios is yet to be experimentally evaluated.

In Fig. 6 Curves (a), (f), (g) and (h) represent the K for
cracks of constant crack shape, whereas curves (c), (d) and (e) represent K for cracks that are changing shape as they deepened by
fatigue. It is apparent that the latter cut across curves of lower

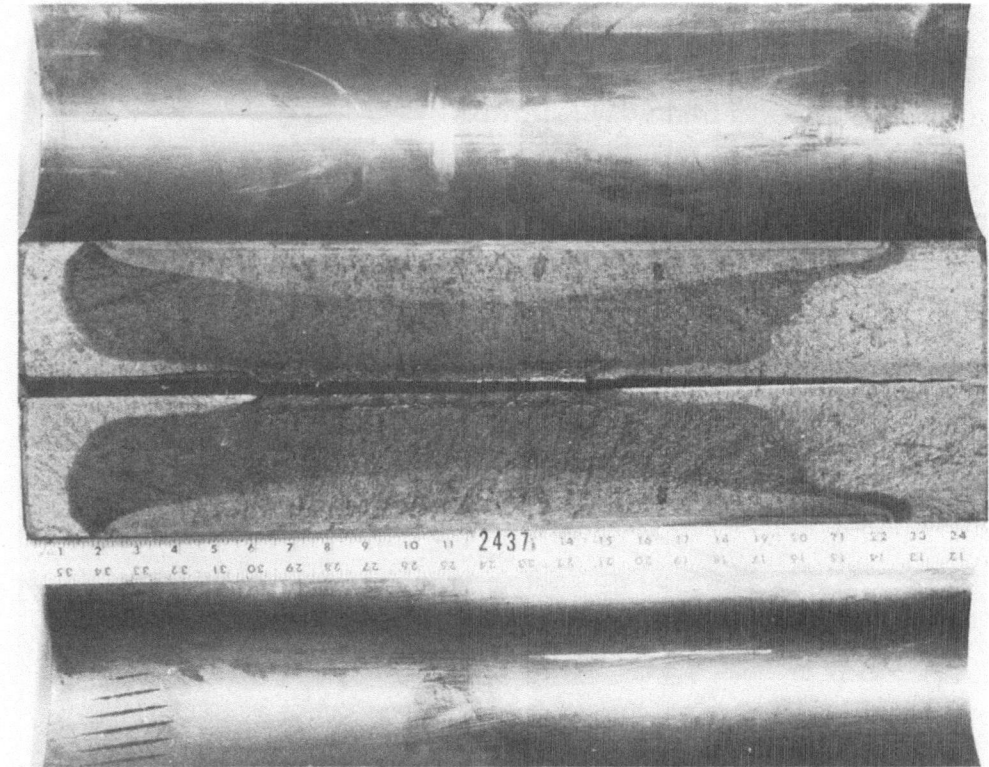

Figure 7.   Fracture surface, long curved crack. The photo shows
            the 20-inch long notch, 1/4 inch deep in 175mm cylinder
            AM2437B and shows the long curved crack developed by
            fatigue to N = 1,500 cycles and failure from 1.5 inch
            crack depth.

and lower K as the crack grows. The changes in crack shape from
the three initial notches are shown in Figures 7, 8 and 9.

Similitude

For wall cracks in cylinders where $K = \alpha \ S \ \sqrt{\pi b}$ is a reasonable
approximation for the stress intensity factor and which have sufficient toughness to sustain a "leak-before-fracture", the failure
crack depth, b_f, can be taken as the wall thickness B, and Equation
(10) can be written as:

$$(N_f - N_i) = \mathcal{D} \ [\sqrt{B/b} - 1] \tag{16}$$

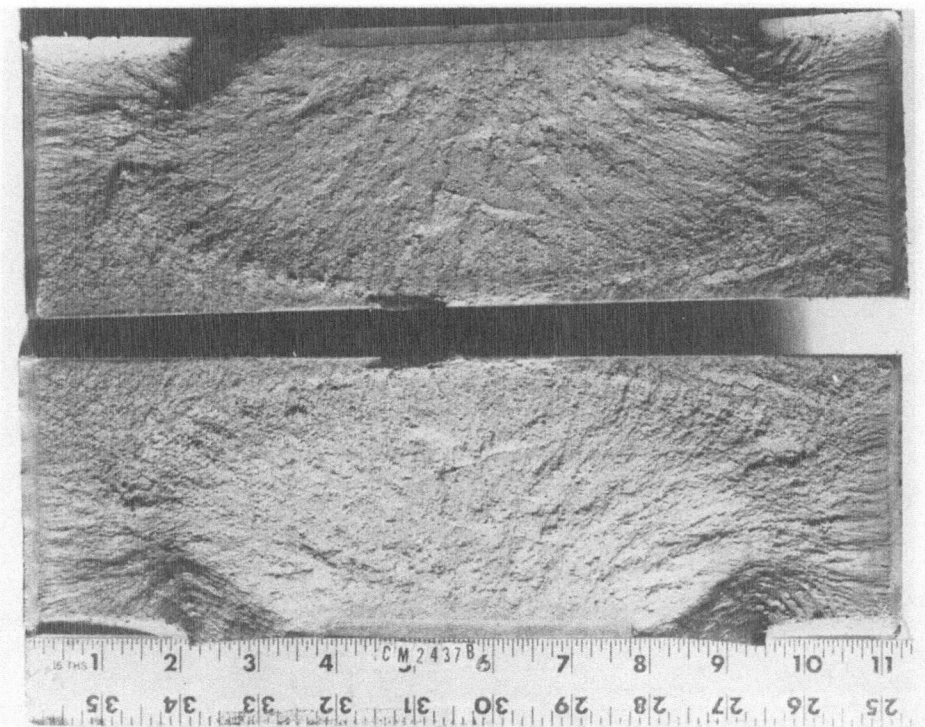

Figure 8.   Fracture surface, semi-elliptical crack. The photo shows
             the 4-inch long notch, 1/4 inch deep in 175mm cylinder
             CM2437B and shows the semi-elliptical crack shape devel-
             oped by fatigue to N = 3,430 cycles and failure through
             the 3.575 inch thick wall.

where

$$\mathcal{D} = [2\ \pi^{-3/2}\ M\ \alpha^{-3}\ (\frac{w^2 + 1}{w^2 - 1}\ p)^{-3}\ \frac{1}{\sqrt{B}}] \tag{17}$$

Thus, two cylinders will have the same fatigue crack propagation
lives for the same ratio of initial crack depth to wall thickness,
(b_i/B), if their values of $\mathcal{D}$ are equal. If they are unequal, the
one with the larger value of $\mathcal{D}$ will have the longer life.

    The constant M depends on the choice of material. The
quantities p, w and B can be adjusted by design geometry, while the
value of α for a given crack shape can be changed by autofrettage

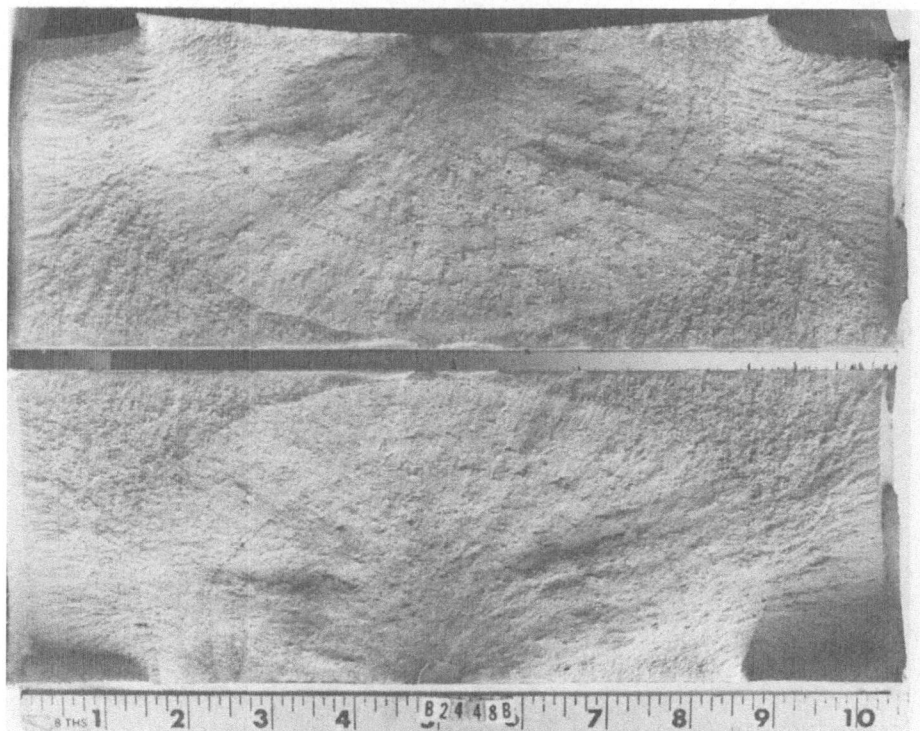

Figure 9.   Fracture surface, semicircular crack. The photo shows
the 1/4 inch radius initial notch and the semicircular
crack growth which finally became elliptical by N =
10,617 cycles and fatigue failure through the 3.575 inch
thick wall.

or prestressing to produce residual compressive stress at the bore.
The quantity $\mathcal{D}$, then, expresses the effect of these design variables
upon the remaining fatigue life ($N_f - N_i$).

Practical Application

Since the remaining fatigue life ($N_f - N_i$) is linearly proportional to the value of $\mathcal{D}$, this quantity enables the comparison
of lives of different cylinder designs, with different sizes,
operating pressures and amounts of overstrain on a basis which
accounts for the differences.

The major difficulty in establishing the proper value of $\mathcal{D}$

lies in the evaluation of α, the remainder of the variables being
materials and configuration characteristics which can be selected.
In the absence of residual stresses the value of α depends upon
the crack shape and ranges from 1.792 for the frontal notch to
0.66 for the semicircular crack. Thus one is faced with a wide
range of choice, for which some general guidance can be given. If
in the application there are not discontinuities occuring along
the bore, so that crack initiation can be more or less random, and
if toughness is sufficient to sustain considerable crack depth, one
can safely assume that the crack of concern will be of semi-elliptical or semicircular shape. Thus the value of α will be between
0.945 and 0.66. Use of the higher value is the more conservative
estimate. If toughness is low or if a continuous longitudinal
discontinuity exists or may be introduced in operation, one must
assume that a long crack will form and value of α from 1.792 to
1.10 should be chosen. It is likely that the lower value (1.10)
will safely apply to the majority of cases of this type, except
where K_{Ic} is very low.

As stated previously, residual stresses will modify the value
of the stress intensity factor for a given crack shape. Autofrettage overstrain produces compressive residual stresses at the
bore and alters the stress distribution through the wall considerably. This serves to reduce α in the stress intensity factor expression, particularly at shallow crack depths. As the crack
deepens into the region of tensile residual stress the stress intensity factor may continue to be expressed in terms of the reduced
α, presumably because of the gradual relaxation and equilibration
of the residual stresses near the advancing crack front. Although
quantitative definition of the effects of autofrettage and residual
stresses on α is the subject of current study, some insight into
the effect can be obtained from examining the values of α associated
with some cases of observed and measured crack growth in autofrettaged cylinders. Moreover, this will serve to demonstrate the
important role that residual stresses can play in enhancing fatigue
behavior.

One example is a sample of four 175mm tubes [10] which were
partially overstrained (40% of the wall thickness) by hydraulic
autofrettage and fatigue tested to failure. The graphs of crack
depth versus cycles, measured with ultrasonics, are shown on Figure
10 along with a theoretical curve shown as a dashed line. This was
calculated using the values $\alpha = 0.77$, $b_i = 0.1$ inch, $B = 3.575$
inch in Equations (10), (11) and (12). This gives 6033 cycles to
go from 0.10 inch to 0.25 inch depth where the first cracks were
detected, and a remaining life of 7,640 cycles from there to failure,
a total of 13,673 cycles. The average life of the four tubes is
13,673 cycles. These tubes were tested at the pressure of 46 ksi.
The spread in life from 10,924 to 16,201 rounds-plus-cycles can be
attributed to the differences in crack initiation and crack shape

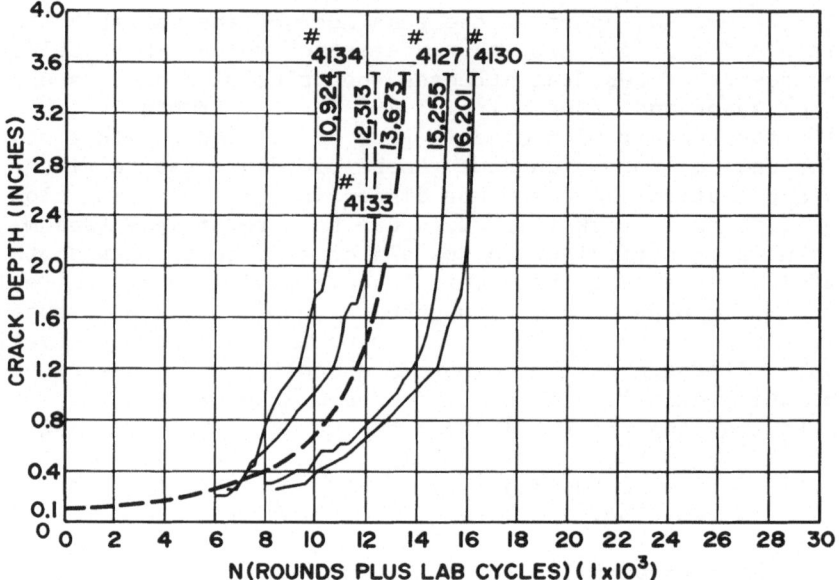

Figure 10.   Crack depth versus N, cycles, rifled 175mm tubes.   The
              graph shows crack depth versus fired rounds plus lab
              cycles from tests of four rifled 175mm tubes, showing
              data and a calculated representative curve to the aver-
              age fatigue life from an initial crack depth of 0.10
              inch.

among the sample of four tubes.   It appears that their average
fatigue crack growth is approximated well by the calculated curve.

A graph of the measured crack propagation rates and the calculated rate is shown on Figure 11 for the crack depths from 0.1
inch to failure.   As before, the calculated crack rates from 10^{-4}
to 10^{-3} inch per cycle provide a satisfactory approximation of the
measured fatigue crack growth shown on Figure 10.

A second example involves a pair of 105mm howitzer barrels [11]
which were fired to failure using rounds with a pressure of 42 ksi.
These barrels had a wall thickness of 1.32 inches and a diameter
ratio of 1.63.   They were autofrettaged to 100% overstrain, that is
yielding from the bore completely to the outer surface.   They had
adequate properties and gave "leak-before-fracture" type failures,
which were anticipated by ultrasonic monitoring of the crack depth.
Figure 12 shows the fracture of one of them which was intentionally
produced by continuing to fire it after the fatigue crack had
pierced the wall.   The continued firing caused the through-crack
to grow lengthwise until it became unstable and split the tube.

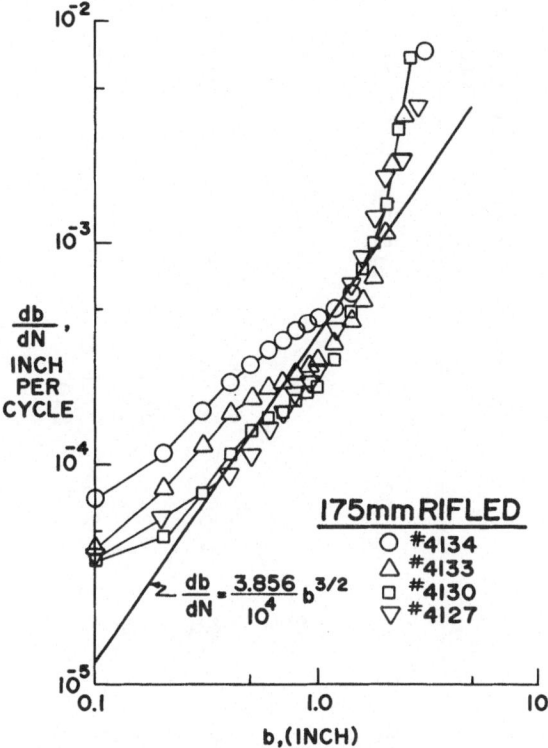

Figure 11.   Crack rate versus crack depth, rifled 175mm tubes.   The
             log-log graph shows fatigue crack propagation rate
             db/dN for four rifled partially overstrained autofret-
             taged cannon tubes, showing data and calculated line.

High speed movies, taken each round for several rounds before fracture, showed the progressive enlargement of the opening. A fracture
toughness of about 140 ksi $\sqrt{\text{in}}$ is required to sustain a crack of
this depth.

The curves on Figure 13 show the fatigue crack growth measured
by ultrasonics and a calculated curve shown as a dashed line.   This
was calculated using the values α = 0.832, b_i = 0.02 inch and b_f =
1.32 inches in Equation (10), giving a life of 14,336 cycles equal
to the average round life of the two tubes.   The spread from 12,091
to 16,582 rounds may be attributed to differences in crack initiation
under firing conditions.

The crack propagation rates shown in Figure 14 indicate that
approximation between 5 x 10^{-5} inches per cycle and 5 x 10^{-4} inches

TABLE IV

$$\mathcal{D} = \frac{2M}{\pi^{3/2} \alpha^3 \left(\frac{w^2+1}{w^2-1} p\right)^3 \sqrt{B}}$$

Tube Size & Crack Shape	P (ksi)	α	$\mathcal{D}$	$\left(\frac{\mathcal{D}}{\mathcal{D}_o}\right)^*$	N_f	N_i^{**}	Average $(N_f - N_i)$	Cycle*** Ratio
175mm Long Curved Crack	48	1.1	825	.22	1500	0	1500	.14
175mm Semi-elliptical	48	0.954	1264	.33	3430	0	3430	.32
105mm Rifled; Fired	42 42	0.832	2012	.53	12091 16582	8300 9550	5412	.51
175mm Rifl; Fired & Cycled	46 46 46 46	0.769	2746	.72	10924 12313 15255 16201	6700 6200 7500 8400	6473	.61
175mm Semicircular	48	0.660	3819	1.0	10617	0	10617	1.0

*$\mathcal{D}_o$ = 3819 for semicircular notch, 175mm
**N_i = N0.25" for 175mm and N_i = N0.092" for 105mm
***Cycle Ratio = $(N_f - N_i)/10617$

Figure 12. Fatigue failure in howitzer tube. The photo shows the fracture produced by fatigue crack growth during firing to failure of a 100% overstrained autofrettaged 105mm howitzer barrel.

per cycle, in this case, is sufficient to adequately approximate the fatigue crack growth curve shown on Figure 13.

The practical implication is that adequate fracture toughness and sufficient overstrain in a thinner tube produces satisfactory performance under severe service conditions.

The values of $\mathcal{D}$ from Equation (17) for fatigue crack propagation from 0.07 B to full wall thickness B are compared with the values for the three crack shapes in the 175mm non-autofrettaged cylinder specimens in Table IV. In the 175mm tubes this is from 0.25 inch depth to 3.575 inches, while in the 105mm tubes it is from 0.09 inch to 1.32 inches.

The ratio of $\mathcal{D}$ to that for the semicircular crack in the non-autofrettaged cylinder shows their relative fatigue crack propagation life. This ratio indicates that, on the average, both the

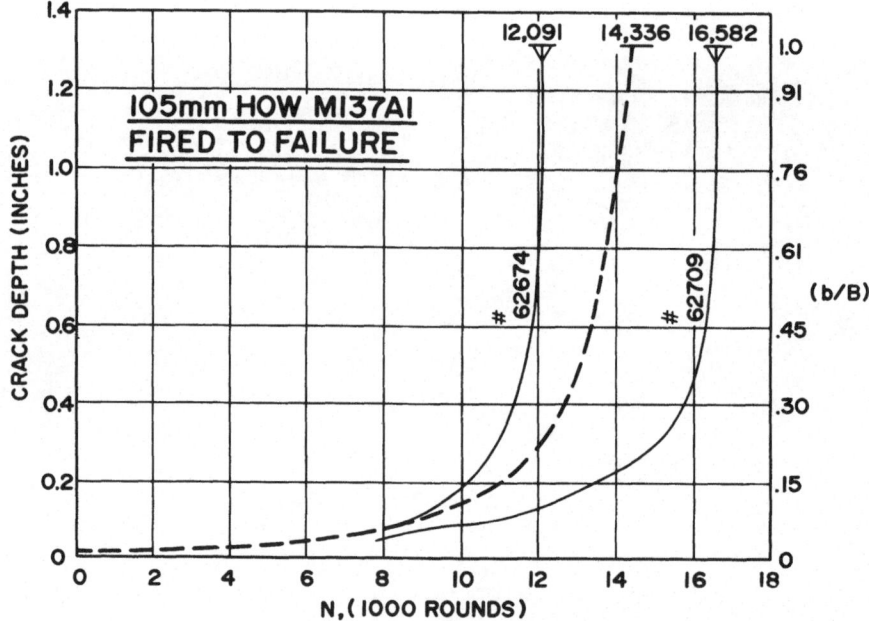

Figure 13.   Crack depth versus N, 105mm howitzer tube. The graph
             shows crack depth versus rounds fired at 43 ksi in two
             105mm barrels showing data and calculated representative
             curve to average fatigue life from initial crack depth
             of 0.02 inch.

autofrettaged 175mm tubes and the autofrettaged 105mm tubes are
equivalent to a non-autofrettaged 175mm cylinder with a crack shape
less severe than the 4-inch notch but more severe than the semicircle. This represents a considerable design improvement over
the fatigue test results from non-autofrettaged cannon tubes. For
any given configuration, the greater the percentage overstrain the
larger will be the compressive residual bore stresses, which provides larger enhancement of fatigue life.

    While the 105mm tubes were 100% overstrained and the 175mm
tubes were only 40% overstrained, their compressive residual bore
stresses were comparable because the residual stresses are a function of the percent overstrain, the diameter ratio and the yield
strength. Hence at comparable pressures their fatigue performance
should be about equal, as suggested by their values of $\mathcal{D}$ in Table IV.

    A plot of the values of $(N_f - N_i)$ versus $\mathcal{D}$ from Table IV is
shown on Figure 15 for all of the examples discussed, along with
the line $(N_f - N_i) = 2.78 \, \mathcal{D}$. This relationship comes from Equation

(16) for the ratio b_i/b_f equal to 0.07. Test results that plot below the line indicate tubes in which the α for crack shape was more severe than average or which failed early because of insufficient toughness for the crack to grow clear through the wall. Since fatigue life may be limited by fracture at less than the wall thickness in such cases, Equation (15) must also be considered in evaluating the performance.

The position of the points representing the average performance of the 105mm and the 175mm autofrettaged tubes on Figure 15 indicates that their behavior was nearer that of the semi-elliptical crack shape than that of the semicircular crack. This is to be expected since they were rifled tubes, which in the non-autofrettaged condition would have behaved more like the cylinder specimen with the long curved crack.

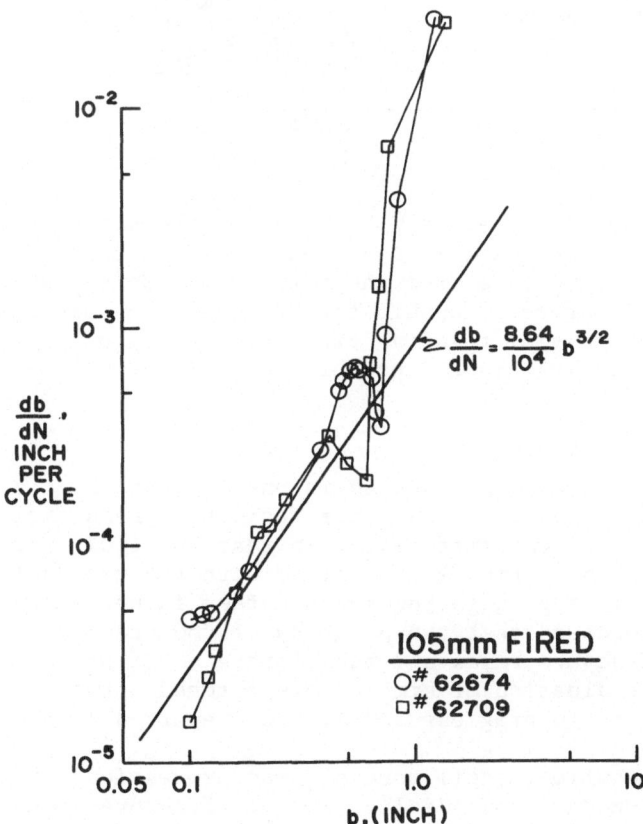

Figure 14.   Crack rate versus crack depth, 105mm Howitzer tubes. The log-log graph shows fatigue crack propagation rate db/dN in two rifled 105mm fully autofrettaged howitzer barrels, showing data and a calculated line.

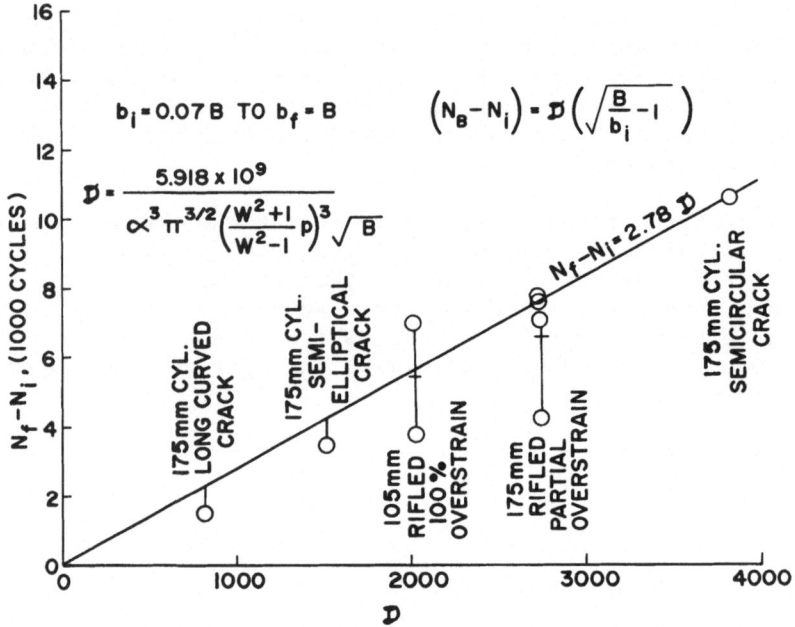

Figure 15. $(N_f - N_i)$ versus $\mathcal{D}$. The graph shows the remaining
fatigue crack life versus the similitude parameter, $\mathcal{D}$,
showing data from the cited examples and a calculated
linear relationship for fatigue crack propagation from
an initial relative crack depth of 0.07 to the final
relative crack depth of 1.0 in thick walled cylinders.

The relationships for stress intensity factors in autofrettaged
cylinders are now under study. The residual stress field for partial
overstrain is less and different than that for 100% overstrain. The
interaction of the active stress field with the residual stress field
while the crack is propagating through then is not yet well analyzed,
but it is recognized that the progress of the crack must permit the
residual stresses to relax and equilibrate ahead of the crack front.
Experimental evaluation of similar pre-notched autofrettaged cylinder specimens is in progress and will be reported later.

In the meantime, until more precise expressions become available, these concepts and idealizations of fracture mechanics and
fatigue crack propagation can provide the rationale for calculating
fatigue lives and estimating the effects of different design variables upon the fatigue and fracture of thick-walled cylinders. They
also point the way toward obtaining more valid equations in the
future.

CONCLUSIONS

1. The fatigue crack propagation rates for part-through wall cracks in thick-walled cylinders can be approximated by the product of a materials constant times the cube of the stress intensity factor range.

2. The stress intensity factor expression $K = Y\,S\sqrt{\pi b}$ can be approximated by $K = \alpha\,S\,\sqrt{\pi b}$, where the variable Y is replaced by a constant α which depends upon the crack shape and the stress state.

3. The upper bound of K is expressed by the Bowie and Freese equation which is for a straight fronted notch. For non-autofrettaged cylinders of 2.0 diameter ratio this is approximated by $K = 1.79\,S\,\sqrt{\pi b}$, where S is the tangential bore stress. The lower bound corresponds to a semicircular crack shape and is expressed by $K = 0.66\,S\,\sqrt{\pi b}$.

4. At any measurable crack depth, b, the remaining fatigue life for "leak-before-fracture" may be estimated by $(N_f - N_i) = \mathcal{D}[\sqrt{B/b} - 1]$ where $\mathcal{D}$ depends upon design variables. Adequate toughness must be provided to prevent fracture at crack depths less than the wall thickness.

5. Design variables such as materials properties, operating pressure, diameter ratio, wall thickness and residual stresses determine the stress intensity factor, which in turn controls the fatigue crack propagation rate and the failure crack depth. These design variables may be adjusted to optimize the fatigue life expectancy and to avoid brittle fracture.

ACKNOWLEDGMENTS

This work was performed under funding from Project Number 1T162105AH84 of the Army Materials and Mechanics Research Center, Watertown, Massachusetts.

REFERENCES

1. Paris, P. C., "The Fracture Mechanics Approach to Fatigue",
    in Fatigue, an Interdisciplinary Approach, Burke, J. J.,
    Reed, N. L., Weiss, V., Editors, Syracuse University Press
    (1964) pp 107-132.

2. Paris, P. C. and Sih, G. C., "Stress Analysis of Cracks", in
    Fracture Toughness Testing and Its Applications, ASTM-STP 381,
    Am. Soc. for Testing and Mats. (1965) pp. 38-40.

3. Timoshenko, S. and Goodier, J. N., "Theory of Elasticity",
    McGraw Hill Book Co., N.Y., 3rd Edition, p. 71.

4. Bowie, O. L. and Freese, C. E., "Elastic Analysis for a
    Radial Crack in a Circular Ring", Engineering Fracture Mechanics
    (1972) Vol. 4, pp. 315-321.

5. Underwood, J. H., Lasselle, R. R., Scanlon, R. D. and Hussain,
    M. A., "A Compliance K-Calibration for a Pressurized Thick-
    Wall Cylinder with a Radial Crack", Engineering Fracture
    Mechanics (1972) Vol. 4, pp. 231-244.

6. Smith, C. W., Jolles, M. and Hu, T., "Stress Intensities for
    Low Aspect Ratio Surface Flaws in Pressurized Thick Walled
    Tubes", Virginia Poly. Inst. Report VPI-E-75-19, Aug. 1975,
    Blacksburg, Va. 24061.

7. Hussain, M. A., Pu, S. L., Scanlon, R. D. and Throop, J. F.,
    "Stress Intensity Factor for a Pressurized Thick-wall Cylinder
    with Part through Circular Flaw, Collocation Method and Compli-
    ance Calibration", in preparation (1976).

8. Barsom, J. M., "The Dependence of Fatigue Crack Propagation on
    Strain Energy Release Rate and Crack Opening Displacement", in
    Damage Tolerance in Aircraft Structures, ASTM-STP 486, Am. Soc.
    for Testing and Mats. (1971) pp. 1-15.

9. Throop, J. F. and Miller, G. A., "Optimum Fatigue Crack Re-
    sistance", in Achievement of High Fatigue Resistance in Metals
    and Alloys , ASTM-STP 467, Am. Soc. for Testing and Mats.
    (1970) pp. 154-168.

10. Davidson, T. E., Reiner, A. N., Brown, B. B., Throop, J. F.,
    Miller, J. J. and Austin, B. A., "The Fatigue Life Character-
    istics of the 175mm M113E1 Gun Tube", Watervliet Arsenal
    Technical Report WVT-6912, March 1969.

11. Brown, B. B., Reiner, A. N. and Davidson, T. E., "The Fatigue
    Life Characterisitcs of the 105mm M137A1 Howitzer Barrel",
    Watervliet Arsenal Technical Report WVT-7202, Jan. 1972.

THE APPLICATION OF FRACTURE MECHANICS TO PIPELINE DESIGN

A. A. Wells

The Welding Institute
Abington Hall
Cambridge, England CB1 6AL

ABSTRACT

Gas and oil pipelines are characteristically designed to accept yielding under hydrotesting and are operated at high design pressure stresses in relation to yield. The service defects which must be tolerated include stress corrosion cracks and mechanical impact damage from construction machinery, such that elastic-plastic bulging may precede fracture. These conditions lead to use of fracture mechanics in fracture control, although mainly through the ubiquitous Charpy V test.

These aspects of fracture mechanics will be reviewed in this chapter, with separate references to the conditions for crack initiation under static conditions, and the control of running brittle and ductile shear cracks. Reference will also be made to recent work on relationships between scatter in notch toughness tests, and probabilities of failure.

INTRODUCTION

Fluid pipelines represent one of the most concentrated energy transport systems known to man, whether for gas, oil or suspended solids. For comparison with other systems, a 36 inch diameter natural gas pipeline can easily transmit 15,000 megawatt of thermal power of excellent quality and utility, and a corresponding oil pipeline could transmit much more at normal pumping rates. Such pipelines are laid in trenches with backfill in temperate zones, but they may be exposed under dry temperate conditions, or placed on columns with insulation in permafrost territory. They may be wrapped, coated and cathodically protected against external corrosion, or additionally coated with concrete ballast and mechanical protection for use undersea.

This chapter is concerned with the scientific aspects of fracture control in long straight lengths of welded steel pipeline, and thereby pays tribute to practical pipeline technology developed over many years in the USA, Europe and USSR, as outlined in the brief, but not authoritative, description which appears immediately below. Much of the underlying philosophy is already embodied in codes and standards, recognized by national governments with international agreement. The very existence of these codes, embodying a vast collective experience, may be credited with the degree of safe operation that has now been achieved, following a development period during which a few spectacular casualties occurred. It should be emphasized that pipelines in general, and the larger diameter gas transmission pipelines in particular, depend for their safety in the public domain on the continuous and watchful application of fracture control procedures.

Pipeline construction and operation also presents a formidable economic problem, since one pipeline can represent both the capacity and the investment characteristic of a fleet of supertankers. The solution of this problem has made considerable use of controlled rolled medium strength steels of the microalloyed type, employed at fairly high design stress levels. These steels embody a considerable technology of their own, but are eminently suited to the restricted thicknesses used in pipelines, and are readily fusion welded with certain processes and consumables. Fracture control is much concerned with the properties of these materials and their welded joints.

These steels exhibit yield at a suitably defined limit of proportionality and a small degree of strain hardening consistent with their elevated yield resistances. The plate material is usually rolled and longitudinally (or spirally) welded on the mill, and then subjected to a shop hydraulic expansion operation which both proof tests the welded joints and improves the circularity. Whereas the diametrical strain in hydraulic expansion can be controlled in individual pipes on the mill, variations of yield strength must be accommodated through application of tolerances. Strings of pipe assembled taking into account these tolerances are then field welded at the circumferential joints and again hydrostatically tested. It is usual to perform these tests very carefully with regard both to exclusion of air and of differences in elevation, and to carry each test to yield of the pipeline section as a whole, at a pressure equivalent to as much as 100% of the nominal yield of the steel, and defined by the pumped volume-pressure curve obtained during the proof test. The final pressure in a successful test must be sustained without leakage for some time to justify qualification of the pipeline section. When there is a break in the field hydrotest, it is usually confined to isolated pipes in the string in which yielding has been concentrated, so that repair and retest is quite feasible. The field hydrotest is a conditioning as well as a rigorous proof test and, in particular, it further improves the circularity, so that bending stress concentrations are removed before they can cause fatigue or stress corrosion failures in service, and it attenuates residual stresses from welding and other causes. Every effort is made to anticipate and to avoid field hydrotest failures, even though they are restricted with regard to damage, since they are expensive in time and effort (a length of ten miles of pipe contains, during the test, a large quantity of high quality water, possibly remote from a supply, and it must be swabbed and dried out before rewelding). Nevertheless, one or two such local failures will be expected to occur in a long pipeline construction. The overall importance attributed to the field hydrotest is thereby demonstrated. The importance with regard to fracture control is that the hydrotested pipeline is made susceptible to accurate stress prediction during service, and is cleared to the maximum possible extent of stress concentrations and possible critical defects. Of course, a pipe string would not sustain an indefinitely large number of such proof tests without growth of indigenous small defects, but this aspect has not proved to be limiting.

Substantial effort is made to survey for cracklike defects before and in service by application of non-destructive detection methods, such as magnetic particle, radiography and ultrasonic, both from outside the pipe, and inside from equipment mounted on travelling trolleys or pigs, whose positions along the pipeline can be controlled. With regard to the growth of defects by stress corrosion cracking most pipelines are also wrapped externally with a hessian type material, coated with bitumen, and then cathodically protected. Coating gaps or 'holidays' may then be detected electrically and reinstated.

Of the types of fracture casualty that have occurred, the fast running brittle crack is probably the most damaging, since the propagation may be sustained, at least in gas pipelines, along many pipe lengths. The fracture surface is characterized by a flat crystalline appearance which may be accompanied by shear lips. Additional features which have been successfully reproduced in full scale fracture tests, notably those conducted by the Battelle Memorial Institute and UK Gas Corporation, are crack branching and fragmentation under very brittle conditions, and sinusoidal undulation of cracks under less brittle conditions. These will be further discussed. In general, brittle cracking is a less likely event in pipelines less than 12 inches in diameter for several reasons. Firstly, the hoop stresses are less in small diameter pipes because of thicknesses limited at proportionally higher ratios to diameters by handling and general corrosion requirements. Secondly, the smaller thicknesses of material are intrinsically less brittle for metallurgical and other reasons.

Brittle cleavage fractures may be initiated at undetected and unrepaired weld faults, at stress corrosion cracks, or even at fatigue cracks. However, it has been the experience that dents and bruises incurred during manufacture or service are a more potent source. Untreated dents lead to substantial stress concentration under internal pressure and the associated surface bruises to intense local material damage of the strain aging type. While it is possible to exclude such damages by care in pipe manufacture and installation, it has proved to be quite difficult by surveillance to defend even buried pipelines from such damage inflicted unknowingly by third parties engaged in construction operations using heavy earth moving equipment. This can happen even when pipelines are patrolled daily, whether using helicopters or vehicles, and illustrates the importance to be attributed to fracture control aimed at preventing brittle crack propagation under the conditions of use of pipelines, and particularly in those for high pressure gas transmission.

Whereas the risk of brittle cracking is accentuated by low temperatures which cause the steel to approach its brittle transition temperature, it is also important to recognise that there can be relatively fast running ductile tear failures at higher atmospheric temperatures in gas filled pipes made of material exhibiting a low energy relative to this mode. Such failures are usually characterized by an oblique fracture surface, exhibiting little or no crystallinity and considerable thinning and frilling of the edges of material local to the fracture. Otherwise, such tears usually run straight (although they may be abruptly terminated by running round the pipe) unless affected by material anisotropy.

For practical purposes, it may be argued that the crack propagation resistance of a gas filled pipe is not significantly increased by containment of the pipeline. Although it is claimed that there is some improvement from optimally consolidated earth backfill, it is known that a poured concrete backfill can diminish the crack propagation resistance under some circumstances, notably where conditions for straight running fracture are restored. Outer continuously welded steel sleeves placed with an annular gap, while providing containment for short sections where there is more than usual proximity to public places, or likelihood of mechanical damage to the pipeline itself, introduce difficulties in maintaining the effectiveness of cathodic protection. There is probably no equally dependable short cut or alternative to ensuring the highest possible integrity for the pipeline by the methods under discussion.

Liquid filled pipelines, while not so hazardous with regard to crack propagation as compared with gas transmission lines, do have to be considered with regard to a further effect, which is high cycle fatigue from the pump piston oscillations of small magnitude, superimposed on static pressures and relatively slow fluctuations. The particular problem is related to cumulative damage, and leads to considerations of national surveillance and licensing for limited periods, which may typically vary from one to ten years, depending upon loading.

FRACTURE CONTROL STRATEGY

The Charpy V notch test still dominates quality control of the notch ductility of steels. It is also an essential feature of contemporary fracture control for pipelines and in this case the standard specimen thickness of 10mm is close to the plate thicknesses used, which mainly lie between 3/8 in. and 3/4 in. It is, nevertheless, customary also to use a sub-standard thickness of 6.7mm in this application. Despite its many questionable features, such as the blunt notch with regard to brittle crack initiation, and the measurement only of an energy absorption associated with the whole of the fracture over a very short path, this test is preferred because it is thoroughly standardized and suited to repetitive manufacture and test of small specimens. The Charpy V test does not measure a notch toughness that can be immediately interpreted in terms of the fracture criteria to be considered below, but it is susceptible to calibration over narrow ranges, making use of the energy absorption per unit area of fracture, w/a, which has the same units as crack extension force, G_c. Many attempts have been made to correlate G_c and w/a, leading to results varying from $G_c = 2(w/a)$, usually under very brittle conditions, to $G_c = 1/2(w/a)$ for very ductile conditions. This topic has been expertly reviewed by Sailors and Corten [1].

The Charpy V test is preferred in the pipeline fracture control strategy because it is suited to steel quality control. It is well known that an alternative that would better represent crack propagation conditions would employ a specimen sufficiently deep to contain the characteristic of the fully developed propagating fracture, but this would eliminate its use for quality control. The alternative is to use the Charpy V test on the one hand, with sufficiently broad sampling, and burst tests on full scale and model pipeline sections on the other hand, leading to a direct calibration between the two. There is ample scope to use intermediate scale notch ductility tests, together with fracture mechanics modelling between these two extremes, with the object of exploring the effects of the many secondary variables.

With regard to intermediate scale notch toughness tests, brief mention may be made of the family of drop weight tear tests explored by several investigators with reference to (1) fracture appearance,

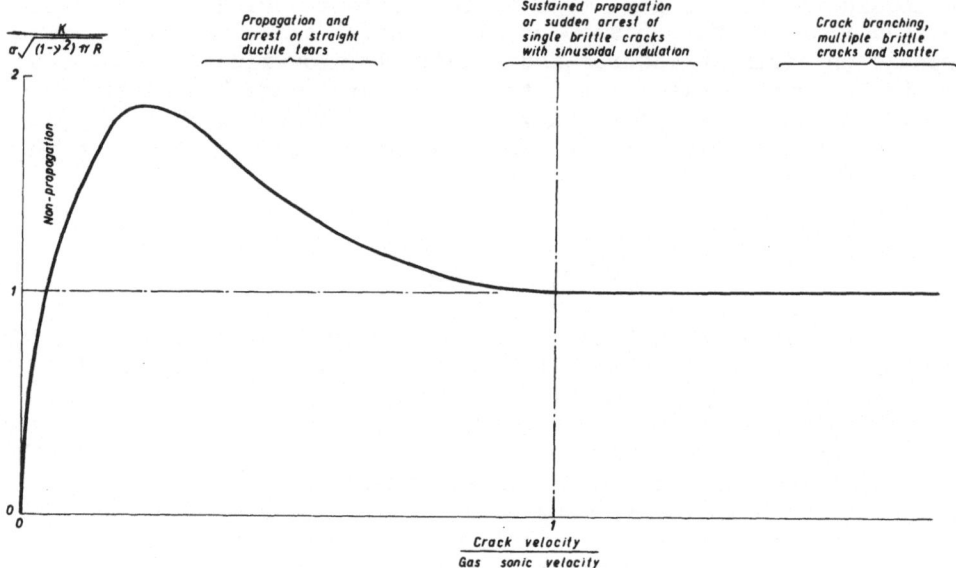

Figure 1. The role of fracture toughness in brittle cracking and
ductile tearing of pipes.

which has been shown to be an accurate if somewhat conservative
indicator of propensity to cleavage fracture, and (2) energy absorption, which may be shown to be incrementally proportional to incremental specimen depth, hence crack length as in N.R.L. tests, if the
effect of a crack starting length is disregarded. Further mention
is made below of static loading techniques which also evaluate this
(R curve) phenomenon [2]. Such effects lend perspective to the
limitation of a direct calibration between Charpy and full scale
test results, however directly useful and essential the latter may
be for practical purposes.

        The introduction of fracture mechanics modelling to the analysis
of full scale pipeline burst tests leads to a diagrammatic representation of crack propagation results as in Figure 1, in which the stress
intensity is plotted as ordinate, representing the fracture toughness
which must be exhibited to achieve various sustained crack speeds,
plotted as abscissae. Bearing in mind that such results should
represent at least limited ranges of internal fluid pressures, pipe
diameters and thicknesses, together with the sonic velocity in the
fluid, ordinate and abscissa are each non-dimensionalized, as dis-

cussed in detail below. Brittle cleavage fracture is represented to
the right hand side of the diagram, ductile tearing near the center
and non-propagation on the left hand side. The phenomena exhibited
for decreasing crack speeds are approximately as follows:

1. crack branching and fragementation, or multiple sinusoidal
    cracking
2. steady state brittle propagation of one crack with sinu-
    soidal undulations (transition temperature criterion)
3. steady state straight line ductile tearing (upper shelf
    energy criterion)
4. non-propagation or positive arrest due to gas discharge
    unloading.

Charpy V notch quality control criteria may be extracted from conditions 2 and 3, either directly from the full scale tests, or by correlation with the stress intensity data. The latter indicate that
there should be small but significant increases of toughness for
increased pipe sizes as well as pressures, but the former effect does
not yet seem to have been fully confirmed by full scale tests.
Although many such have been performed, the number is still too
limited for all such variables to be encompassed. Figure 1 is as yet
qualitative, but it may be taken as an indication that the suppression
of sustained propagation of brittle and ductile cracking in gas pipelines up to 36 inch diameter, with 1000 lb/in^2 design pressure, of
steel with 65 ksi yield point, depends upon an energy absorption with
the sub-standard Charpy V specimen of approximately 15-40 ft.lb.

   Although pipelines which are designed on a basis of proof
against crack propagation may also be expected to possess considerable
resistance to crack initiation, much attention has been given in the
published literature to the latter, especially with regard to the
effect of internal pressures on bulging, whereby the notch effect of
a longitudinal crack is supplemented. Recent work has embraced both
elastic-plastic conditions, and part-through thickness precracks,
which may be marginally more severe in some circumstances than through
cracks. In spite of the bulging supplementation to crack extension
force it is apparent from all of these investigations that the
critical crack sizes at design pressure remain quite large compared
with the detection capabilities of non-destructive testing methods,
and it is thus confirmed that local damage to the material must
usually be an important factor in producing crack initiations. Nevertheless, this work makes an important contribution to the assessment
of defects exposed by N.D.T. during surveillance, with respect to
decisions on repair. It is part of the fracture control strategy in
this respect, and is further discussed.

FRACTURE CRITERIA

Mention has already been made of a Charpy V measure of notch toughness, w/a, critical crack extension force or fracture toughness, G_c, and critical stress intensity, K (Figure 1). The summary relationship between fracture criteria that has been exhaustively studied with the aid both of classical mechanics and elastic-plastic finite element treatments, further introducing the path independent integral J_c, and the crack opening displacement, δ_c (COD), hence-forth disregarding the suffix denoting critical conditions, is

$$G = J = \frac{K^2}{E} = m\sigma_Y \delta \qquad (1)$$

where E is the elastic modulus for plane stress or plane strain, respectively, σ_Y is the uniaxial yield stress, and m is a constant which is approximately 2 for plane strain and 1 for plane stress, the latter being representative of most of the following. The limitations which may be recognized are as follows:

1. G is usually confined to linear-elastic situations.
2. J, being applicable to non-linear (including linear) elastic behavior, may be extended with reasonable approximation to elastic-plastic situations involving crack initiation with monotonic loading, with or without strain hardening.
3. K has usually been reserved for linear-elastic situations, but has recently been demonstrated also to have significance in circumstances to which J may be applied.
4. δ has limitations corresponding with those of J. It is more readily measured under plane stress than plane strain conditions. In spite of the latter the COD has been widely and successfuly applied to assessment of the longitudinal crack in the pressurized cylinder in the presence of bulging effects. It has also been extensively compared with conditions of plane elastic-plastic loading, and the degree of mutual proportionality that has been obtained has permitted consistent determinations of the constant m [Equation (1)] to be made [3-5] and has thus enhanced the fundamental significance of the COD. Recent work by Hutchinson [6] has also extended understanding of the local stress and strain conditions at a crack tip in terms of an equivalent stress intensity K which may be applied under non-linear conditions.

It is demonstrated with these reservations that the four criteria may be used exchangeably through Equation (1), dependent upon convenience in calculation, measurement and presentation.

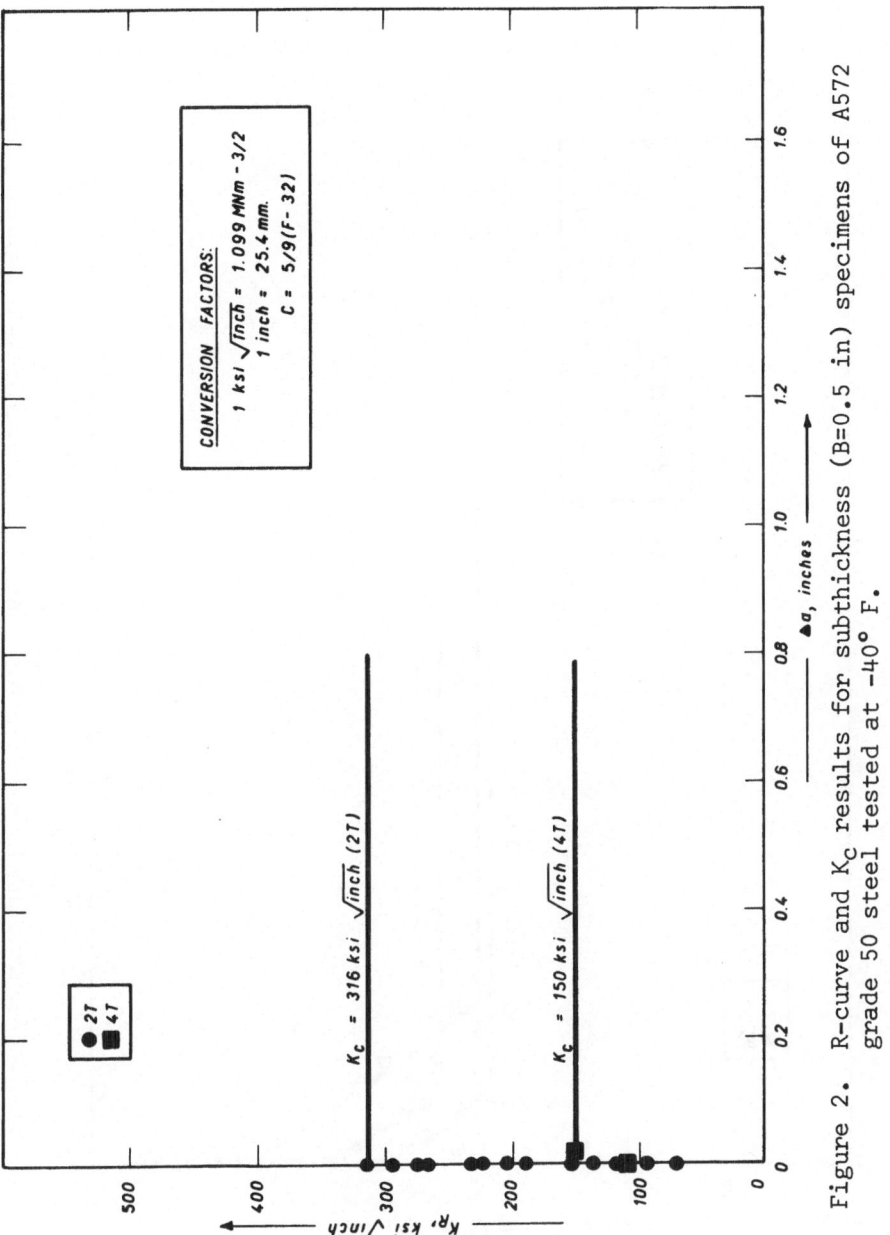

Figure 2. R-curve and K_C results for subthickness (B=0.5 in) specimens of A572 grade 50 steel tested at -40° F.

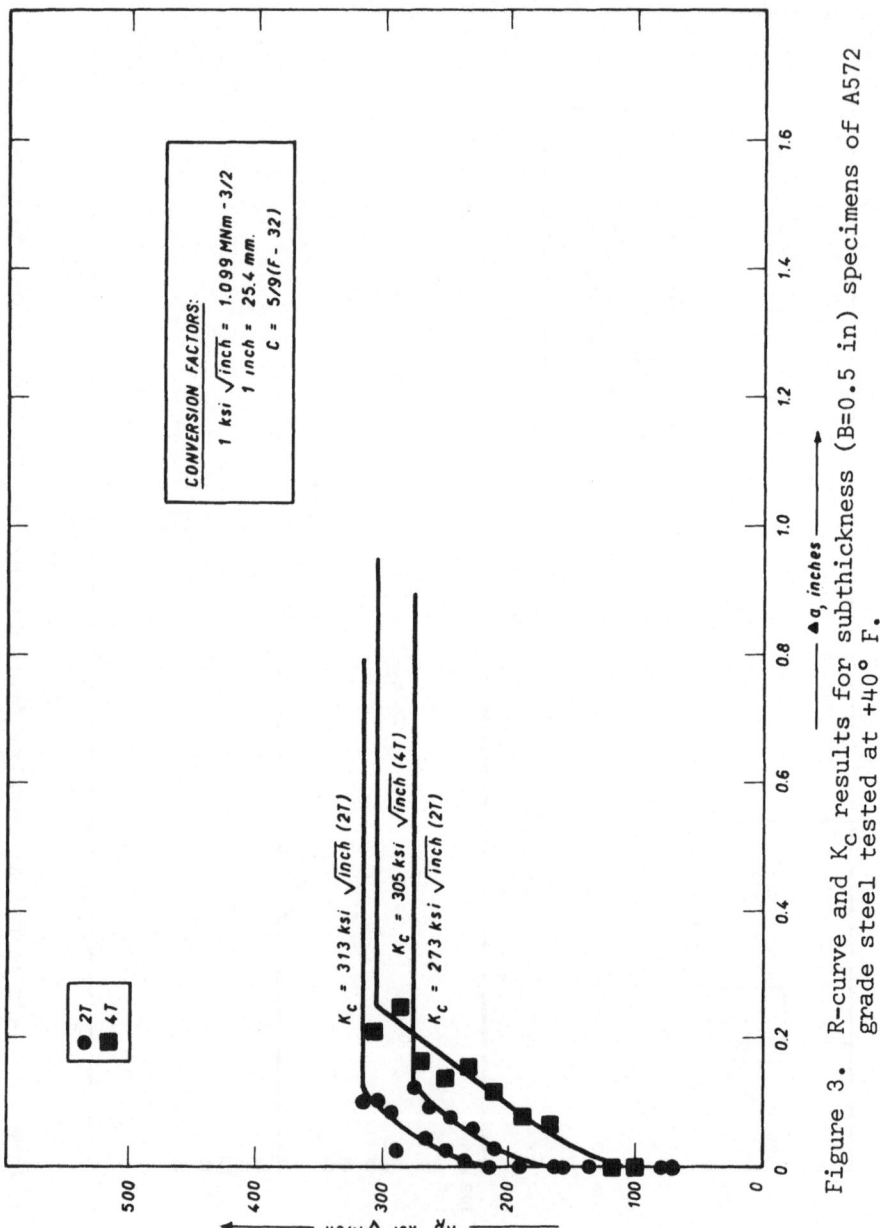

Figure 3. R-curve and K_c results for subthickness (B=0.5 in) specimens of A572
grade steel tested at +40° F.

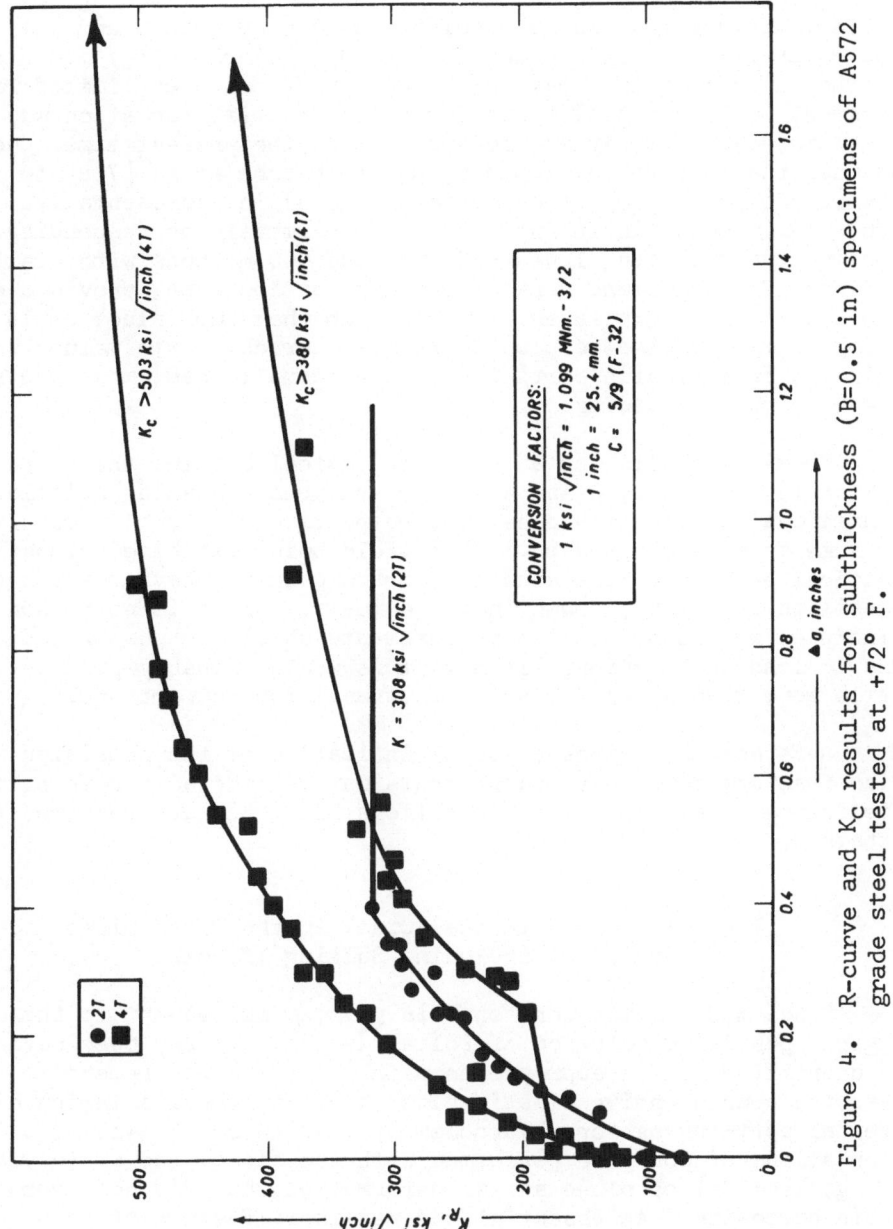

Figure 4. R-curve and K_C results for subthickness (B=0.5 in) specimens of A572 grade steel tested at +72° F.

Definitive Measurements of Fracture Toughness

Mention has been made above of dynamic drop weight tear methods
of fracture toughness measurement, relative to the running fracture
problem. This method may use sufficiently long cracks and full thickness materials in order to make standard fracture energy measurements
for the purpose of calibrating the Charpy V test over limited ranges.
The growth of crack resistance during its initial formation may only
be estimated indirectly by such methods at the present time. Novak [2]
has used the methods developed by Heyer, McCabe et al [7,8] to
measure the whole resistance curve (R curve) in representative
structural steels (including steels approximately corresponding to
pipeline types), using linear-elastic and COD methods with static
loading. The specimens were of compact tension type, provided each
with two pairs of displacement gauges, so that COD values could be
estimated in the elastic-plastic ranges. Crack growth values were
estimated from elastic partial unloading compliances, measured at
intervals during loading.

Novak's results for an A572 grade steel 0.5 in. thick are sum-
marised in Figures 2, 3 and 4 at temperatures embracing brittle,
transitional and ductile behavior, respectively. Further results on
the same steel processed to 62 ksi yield point exhibited almost
identical behavior. The change in crack growth behavior through the
transition is clearly shown in these tests. There is no reason, in
principle, why such detailed measurements should not be extended to
dynamic loading behavior, although this would probably produce
little more than an upward shift in transition temperature.

Reference [2] gives a useful indication of the precision with
which fracture toughness measurements may be made with respect to
pipeline materials under the conditions that they are required to
withstand.

The Static Longitudinal Crack in the Pressurized
Cylinder, including Bulging Effects

Since much of the work on this problem subsequent to the
original published solution of Folias [9] has already been reviewed
and quantified, it is appropriate mainly to note the recently
presented comprehensive elastic-plastic plane stress solution of the
external part-through crack problem by Erdogan and Ratwani [10]. The
calculations of COD were performed with the aid of the strip yielding
or Dugdale model of plane stress deformation, but did not embrace
strain hardening. An essential and realistic feature of this calculation is that the crack ligament is assumed to have yielded under
plane stress conditions, and a feature of the results in much of the
range explored is that the COD at the center of length of the crack
under those conditions is somewhat in excess of that at the ends of
the crack (Figures 5 and 6). This confirms the tendency noted

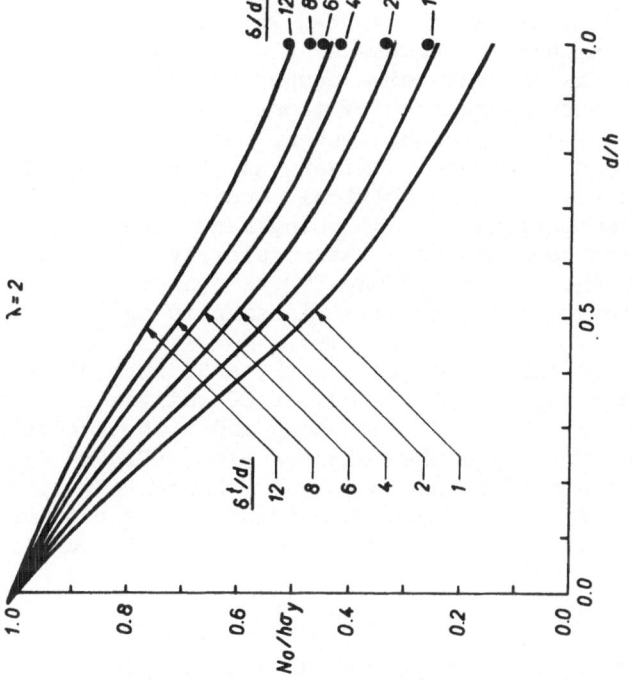

Figure 6. N_0 vs. d for constant total crack opening stretch, δ_t at the leading edge of the crack in the mid-section (x=0, z=h/2-d) in a cylindrical shell with an external surface crack, λ=2.

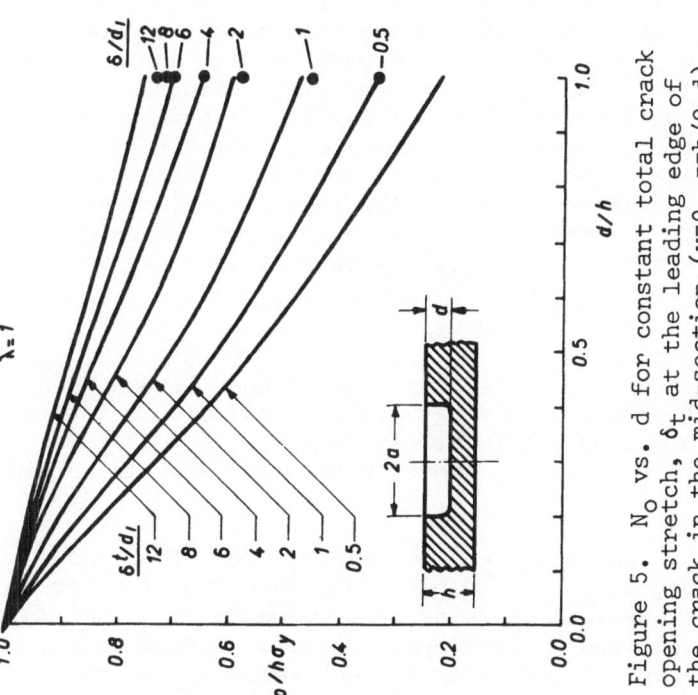

Figure 5. N_0 vs. d for constant total crack opening stretch, δ_t at the leading edge of the crack in the mid-section (x=0, z=h/2-d) in a cylindrical shell with an external surface crack, λ=1.

experimentally for there to be a leak-before-break condition. It is clear when these COD values are converted into stress intensities that they are substantially in excess of elastic values. A unique feature of the paper is the extensive comparison of calculated COD values for through cracks with most of those of the values measured under elastic-plastic loading conditions by Fearnehough and Watkins, Bowen and Langford, Anderson and Sullivan, and Kiefner et al, with a wide variety of materials and geometrical conditions. The agreement is remarkably precise throughout, including many cases extending to plastic collapse, and confirms both the completely definitive nature of the solution and the success of the COD as a fracture criterion for practical problems where extensive yielding is present.

As an example of the use of this work, a 36 inch diameter pipe of 0.5 inch thickness may be considered, of 65 ksi yield strength steel, with an external crack 3.3 inch long and 0.25 inch deep. At 1000 lb/in^2 internal pressure, the value of $N/h\sigma_y$ (Figure 5) is 0.55, $\lambda = 1.0$, and the value of δ/d_1 becomes 0.50. Hence, $\delta = 7.2 \times 10^{-3}$ in., and K = 119 ksi$\sqrt{in}$. This case is evidently closer to critical than the Folias elastic treatment for an equivalent through crack would indicate.

The Running Crack in a Pressurized Cylinder

The earliest studies of brittle crack propagation in gas pipelines confirmed what had been learned from casualty pipelines; that these fractures travel at speeds comparable with sonic speed in the compressed gas, and that there is sometimes a sinusoidal crack path. Model tests on unplasticized PVC pipes [11,12] have shown that a critical condition, below which there is arrest, is closely associated with a crack velocity corresponding with sonic speed, and that fracture energy represents the whole elastic strain energy in the pipe wall without supplementation from work done after fracture by the expanding gas on the unfolding pipe walls, such that

$$K = \sigma\sqrt{(1-\nu^2)\pi R} \qquad (2)$$

where σ is the static tensile component of hoop stress, and R is pipe radius. The term involving Poisson's ratio ν is involved if the axial component of static stress in the pipe wall is assumed to be $\nu\sigma$. A criterion for high speed brittle cracking is obtained thereby.

A mechanism has also been postulated from close observations of high speed photographs taken during this investigation for the occurrence of sinusoidal cracking by a resonant torsional mode of oscillation, energized by local gas pressures acting on the flaps of the unfolding pipe. A lower limit crack velocity for the occurrence is confirmed.

The phenomenon of fast ductile fracture, extensively studied at the Battelle Memorial Institute and elsewhere, is more complex. Here, the crack velocity is less than the sonic speed of the gas and there is some distributed axial decompression preceding fracture, when the event is viewed as a quasi-stationary state. There is also local radial decompression near the crack tip, but a considerable amount of work can still be done by the gas on the unfolding pipe, so that the elastic strain energy release at the crack is above that of Equation (2). The author devised an elementary theory based upon elastic bulging in the unfolding pipe walls, so as to have a characteristic wavelength, which would correspond with an axial flexural wave velocity equal to the crack velocity. By determining the propagation wavelength in this way, the supplemented K could be calculated using the static bulging analysis of Folias [9] and of Erdogan and Kibler[13]. Using a simple approximation for the results of the latter investigators, the values of the stress intensity at these lower crack velocities, V, compared with sonic gas speed, U, and longitudinal elastic wave velocity in the pipe wall, C, becomes:

$$K = p\sqrt{\pi R}\ \left[\ \frac{\pi C}{2\sqrt{3}V}\ \right]^{3/2} \left[\ \frac{2 + (\gamma-1)\frac{V}{U}}{\gamma + 1}\ \right]^{\frac{2\gamma}{\gamma-1}} \tag{3}$$

compared with Equation (2). p is steady state internal pressure, γ is the ratio of specific heats of the gas, and the bracketed term containing γ accounts for the axial decompression recognized by most investigators. Equation (3) does not encompass the effect of radial decompression, and therefore fails to predict a limitation on K at low fracture velocities. Nevertheless, Equation (3) correlated fairly well with small scale tests on steel pipes (Figure 7), and has since been shown to correlate with those of the Battelle full scale tests for which there were estimates of the dynamic K_c values (Figure 8).[11]

Notwithstanding these earlier attempts to describe the mechanics of fast ductile fracture, the most recent analysis of Kanninen, Sampath and Popelar [14] is preferred compared with those of several earlier investigators, since they have additionally accounted for plasticity effects, which are probably substantial, and the radial decompression. Their method of calculation is based upon a combination of gas dynamic and dynamic shell theory, and satisfies an energy balance for the assumed quasi-stationary state. Their result, in Figure 9, exhibits similar characteristics to Figure 7, with respect to the existence of a maximum K at low crack speeds. Although the magnitude of the maximum is not explicit with respect to the prediction of an arrest condition, the authors quote an interesting result for zero crack speed, namely,

$$K = \sigma_Y \sqrt{\pi R} \tag{4}$$

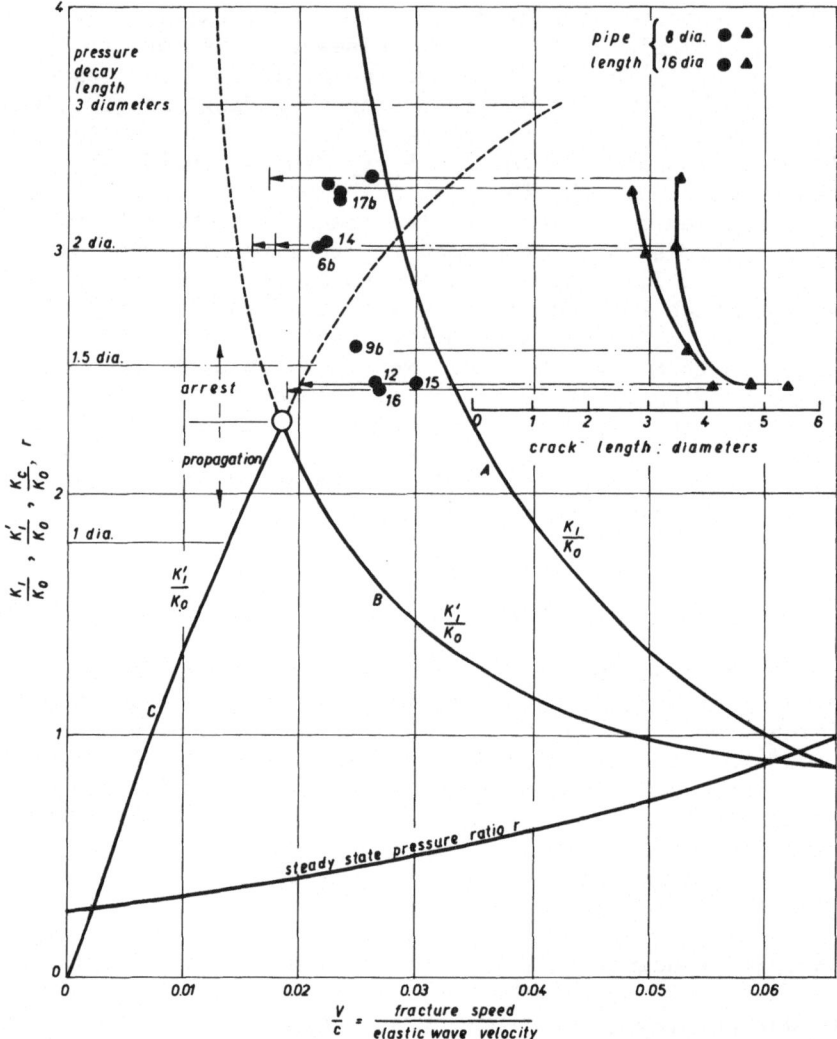

Figure 7. Arrest of cracks in air pressured steel pipes: R/t=61.
            Curve A - no pressure reduction. Curve B - steady-state
            reduced pressure. Curve C - reduced pressure decay length
            at low velocities.

which may be compared with Equation (2), which it exceeds. Again, it
may be claimed that the mechanics of the problem are approaching a
definite understanding, even if the effect, if any, of the R/t ratio
on the conditions for arrest is still in doubt. It may be noted, in
contrast with earlier work of Kanninen et al, that Equation (4) does
not contain this ratio, while if Equation (3) is examined in relation
to the occurrence of a maximum, by introducing an elementary treatment of radial decompression, the dependence upon R/t does not exceed
$(t/R)^{1/4}$. A definite effect of this variable does not yet appear to
have emerged from experimental studies.

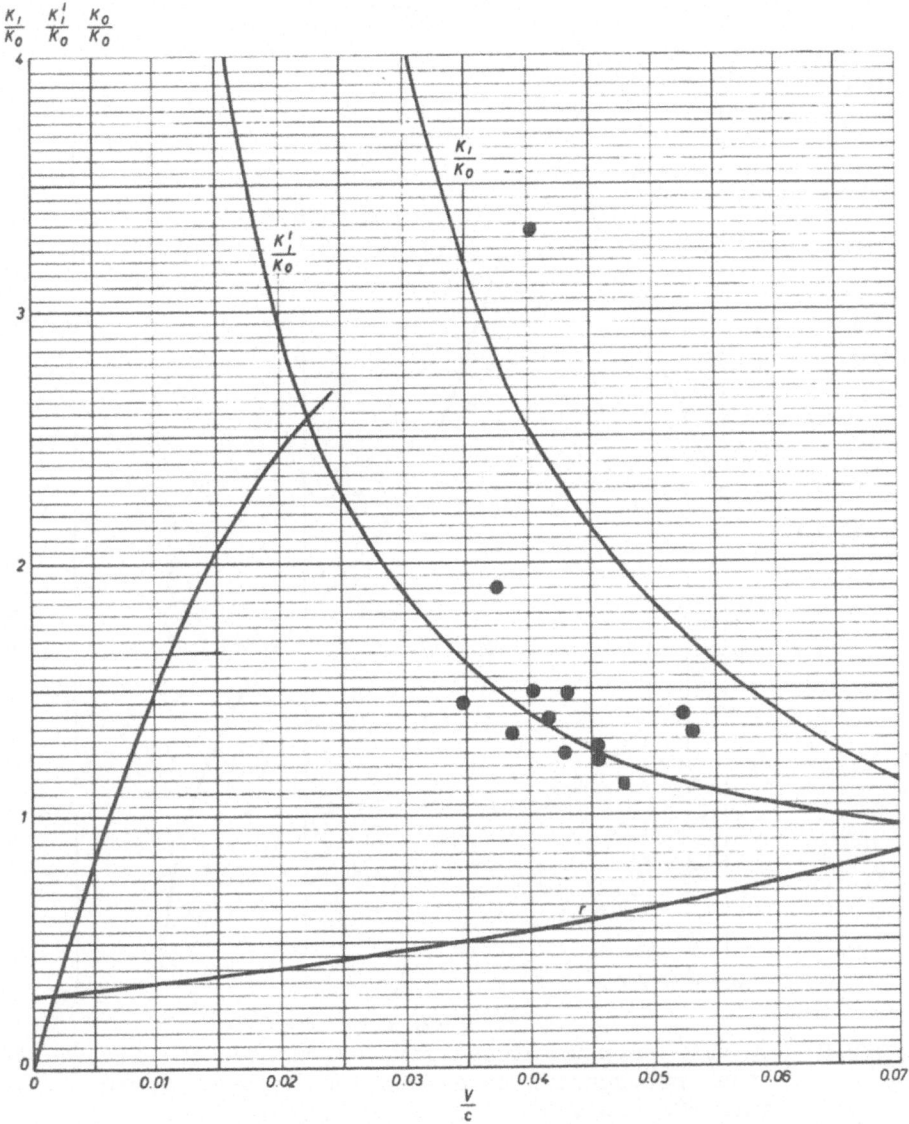

Figure 8. Arrest of cracks in methane pressurised steel pipes: R/t=46

Failure Probability Analysis

It should be evident from this review that dependence upon the
Charpy V test still represents a somewhat weak link in the chain with
respect to the applications of fracture mechanics to fracture control
in pipelines, since each of the foregoing analyses determines a
limiting stress intensity condition to be compared with the measured
toughness of the steel. Yet it has been maintained that the Charpy
V test is valuable for retention as a quality control test, and since

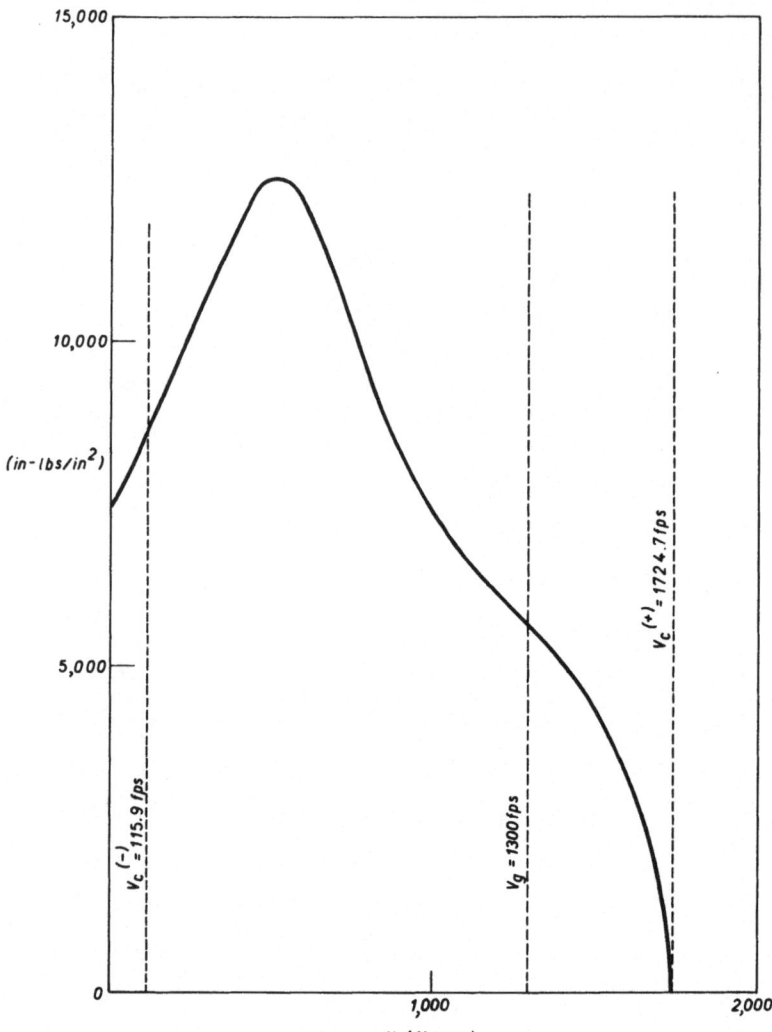

Figure 9.   Dynamic energy release rate calculated for steady-state
            crack propagation as a function of crack speed for test
            no. A32 (CA5).

all fracture toughness tests are subject to scatter, representative of
non-uniformity of the material (which is particularly evident near the
transition temperature), the test is invaluable for statistical assessments of this non-uniformity. Figure 10 illustrates the cumulative
frequency diagram from a world-wide collection of 291 tests of ship
plate at 0°C [18], and similar curves for much larger samples of
quenched and tempered plates at three temperatures in the transition
range. The behavior is regular on this scale, and it is of interest
to see that the scatter tends to be bimodal, as is the fracture appearance, but that otherwise the slopes of the curves are similar for all

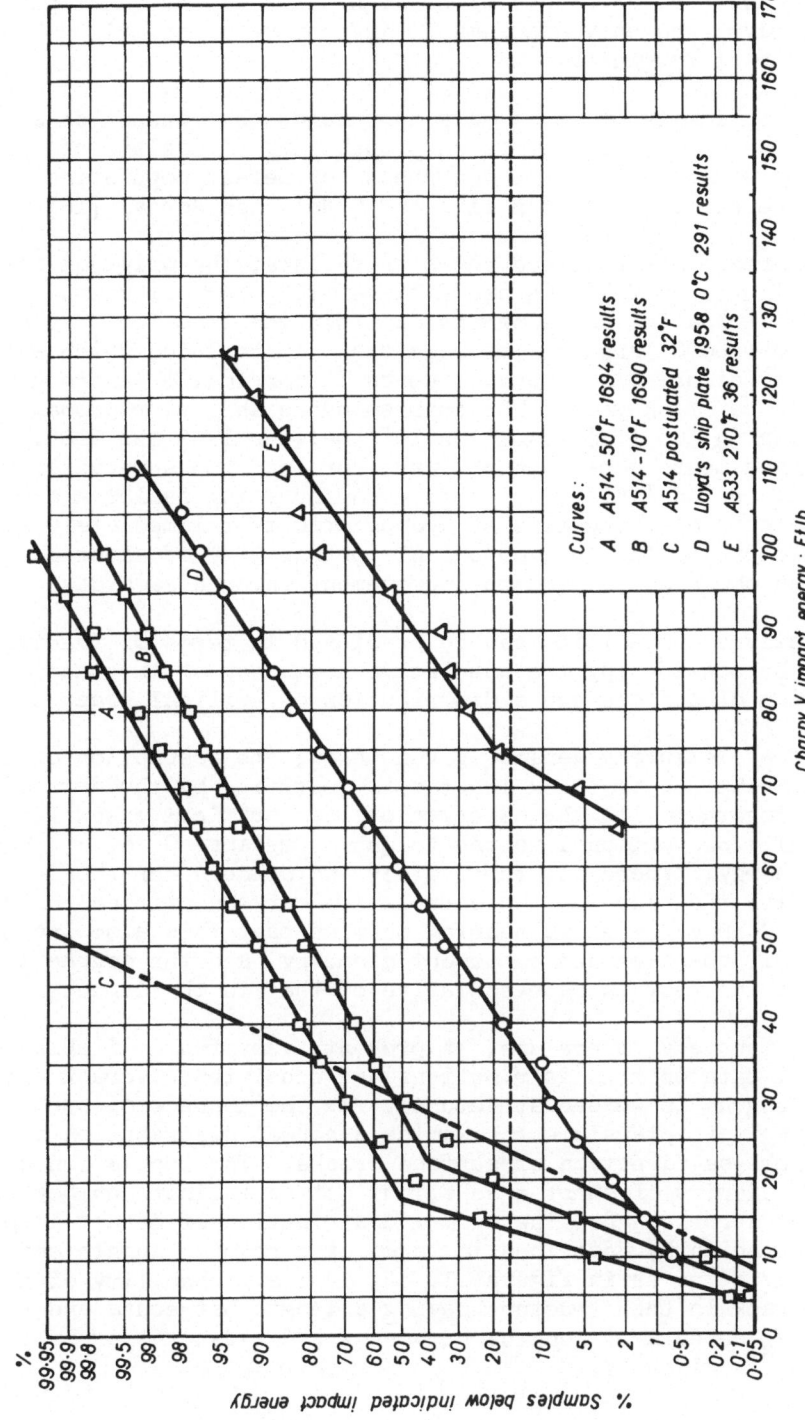

Figure 10. Cumulative frequency representation of Charpy V impact energies.

of the steels.   The existence of straight line experimental plots on
the cumulative frequency diagrams is indicative of normal or Gaussian
distribution of toughness.

The regularity of these data encourages the conduct of syntheses
in order to observe the effects on probability of failure of altering
acceptance levels of fracture toughness, or defect population, in the
manner of Becher and Pedersen [15] and O'Neil and Jordan [16].

With regard to recorded rates of failure, the original analysis
of ship failures by Williams and Ellinger [17] is well known.   Boyd[18]
also quotes actuarial and economic data derived from the U.S. population of about 6000 large oil storage tanks up to 1954, which exhibited
one brittle failure per 3400 tank years in service, or a probability
of 3×10^{-4} per tank year.   The corresponding ship year probability
for breaking in two [18] was 1.2×10^{-4} between 1949 and 1963.
Similar assessments more recently made for land pressure vessels
within that period [16,17,20] considering bursting failures almost all
of which occurred at the time of hydrostatic test immediately following manufacture, point to a comparable probability of about 10^{-5}.
There is an observable trend to improvement through each decade.

A failure probability can be predicted in terms of fracture
mechanics by considering the cumulative frequency of interaction of a
distribution of defects and a distribution of critical crack lengths.

In the elementary version given below, the assumption of a
normal distribution of crack lengths is justified by the data of
Figure 4, together with the observation that critical crack length is
dimensionally proportional to the toughness measure G_c of J_c, hence to
Charpy V energy.   There is, fortunately, no concern for the present
purposes with the slope of the correlation between critical crack
length and Charpy V energy, bearing in mind that this also involves
stress level, specimen and component geometry, and the degree to which
fracture in the test is representative of that in the structure.

Unfortunately, there are, at present, very few available
statistical data on the distributions and densities of crack defect
sizes or lengths in welded structures.   In the absence of such
information it is plausible to examine a normal distribution of lengths
in the thickness direction of surface cracks.   The population will
contain a majority of short cracks, but not an infinite number of those
approaching zero length.   The normal distribution of defect lengths may
then be conveniently described in terms of zero mean length and a
standard deviation as in Figure 11. P_a is the probability of a
critical crack in the structure having a length between a and a+da, and
there are N_x defects of lengths between x and x+dx; the probability of
interaction is given by,

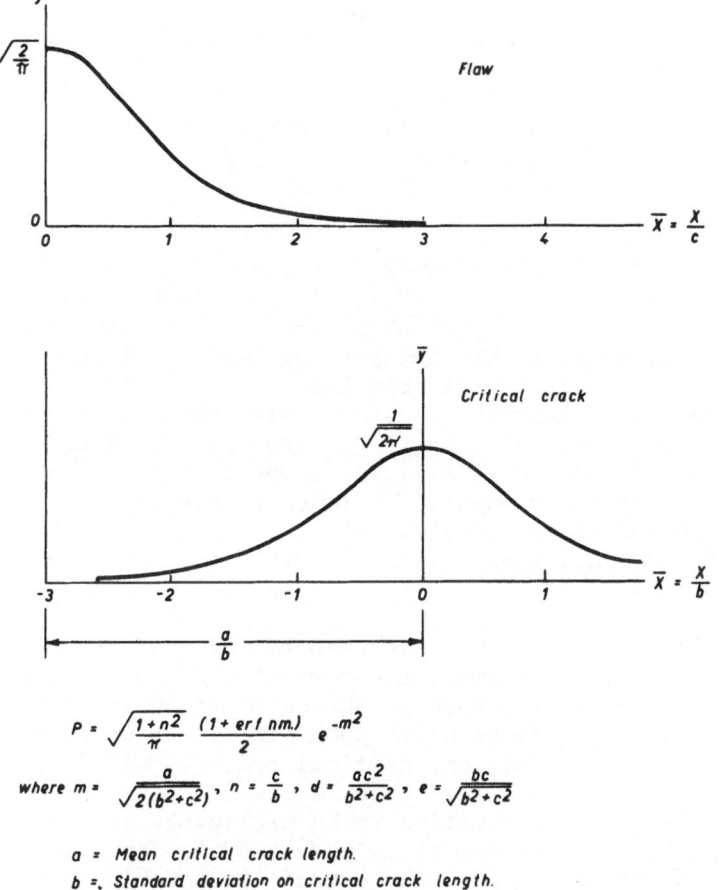

Figure 11.   Interaction of actual and critical crack lengths.

$$P = \int_{0}^{\infty} \int (P_a \, da) \, N_x dx \qquad (5)$$

Hence, for mean critical crack length a and standard deviations b and c on the critical crack lengths and defect lengths, respectively, with the two normal distributions, it may be shown that

$$P = \sqrt{\frac{1+n^2}{\pi}} \; \frac{(1+\text{erf } nm)}{2} \; e^{-m^2} \tag{6}$$

where

$$m = \frac{a}{\sqrt{2(b^2+c^2)}}, \quad n = \frac{c}{b}.$$

Maximum interaction between defect length and critical crack length occurs at depth $d = (ac^2)/(b^c+c^2)$, with the corresponding standard deviation $e = bc/\sqrt{(b^2+c^2)}$, which proves to be remarkably small. It is immediately evident that the primary variable is m, which is proportional to a reciprocal root mean square scatter of actual crack and critical crack lengths. The bracketed term can vary between 1 and 2 and represents the cut-off at zero crack length. P is plotted against the RMS scatter $\sqrt{(b^2+c^2)}/2a^2$, for various values of c/b in Figure 12. From the viewpoint of material development, Figure 12 usefully summarizes the relative effect on failure probability that increasing the consistency of the fracture toughness property can have.

For instance, in the cases reported by Boyd for oil storage tanks and ships [18], where the materials could be considered to conform to the relative fracture toughness distribution of Figure 10, the ratio b/a would be 0.35 at $0^{\circ}C$. Such structures are habitually operated at low stresses, and critical crack (semi) lengths are large, of the order a = 12 in. The standard deviation actual crack length c would be known from observation to be negligible in comparison, so that RMS scatter can be put at $0.35/\sqrt{2} = 0.25$, for which the calculated total probability of failure throughout life, for unit crack population from Figure 12, is 4.5×10^{-3}. But the observed failure probability per structure year is 3×10^{-4} for oil storage tanks, so that for an average life of 20 years this result is consistent with an average defect population of $(3 \times 10^{-4} \times 20)/(4.5 \times 10^{-5}) = 4/3$ defect per structure. This is quite plausible, and would also be compatible with an observed relative insensitivity to defect sizes, because most of the RMS scatter is due to material variability in these cases. In contrast, the variables b and c would be of comparable magnitude for pipelines.

The above is essentially an analysis based upon crack initiation. As such, it would probably be of greater interest with respect to oil than to gas pipelines, since the former are susceptible to crack growth by fatigue, and initiation is to be avoided, whereas gas pipelines are probably more suited to the crack propagation and arrest criteria. Data on defect populations are not at present available to the author for pipelines, so that an explicit numerical evaluation cannot be presented. However, the lessons of the analysis are clear, namely that (1) a sufficiently wide margin should always be maintained between

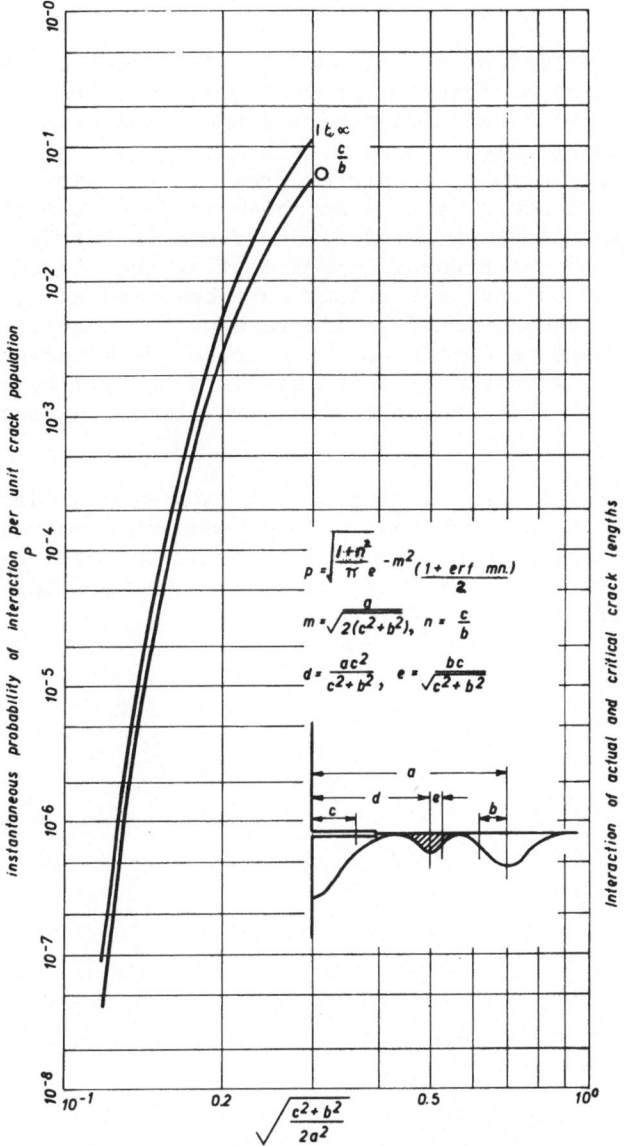

Figure 12. RMS scatter of actual and critical crack lengths.

actual defect lengths to be tolerated and mean critical crack lengths at the operating conditions, and (2) due attention should always be given to the lower end of the cumulative frequency curve in Charpy V testing for quality control.

CONCLUSIONS

It may be claimed on the basis of this review that the scientific application of fracture control procedures to both gas and liquid pipelines has developed firmly and in an orderly manner during the past decade to meet the responsibility placed upon these procedures. Pipeline structures differ from others, particularly with regard to the economic demands on the materials, and the nature of the surveillance task in service, and there is little doubt concerning the value of the rigorous proof testing that is conducted in maintaining the safety levels that have been achieved, to improve these safety standards, and to increase performances, as steel and welding developments continue. The further development of notch toughness testing methods and the statistical quality and other uses that are made of them are of obvious importance in reaching this target.

It is of interest to record that this area of development has provided some of the most challenging research problems in fracture mechanics; it has also contributed at least one notable success, which is the firm establishment of crack opening displacement as an elastic-plastic fracture criterion.

REFERENCES

1. Sailors, R.H. and Corten, H.T. "Relationship Between Material
    Fracture Toughness using Fracture Mechanics and Transition
    Temperature Tests", in Fracture Toughness, Special Technical
    Publication 514. Philadelphia: Am. Soc. for Testing and
    Materials (1972), 164-91.

2. Novak, S.R., "Resistance to Plane-Stress Fracture (R-Curve
    Behavior) of A572 Structural Steel", in Mechanics of Crack
    Growth, Special Technical Publication 590. Philadelphia: Am.
    Soc. for Testing and Materials (1976), 235-42.

3. Wells, A.A., "Crack Opening Displacements from Elastic-Plastic
    Analyses of Externally Notched Tension Bars", Eng. Fract. Mech.,
    1 (1969), 399-410.

4. Boyle, E.F., "Calculation of Elastic Plastic Crack Extension
    Forces", unpublished Ph.D. dissertation, Queen's University of
    Belfast, 1972.

5. Sumpter, J.D.G., "Elastic-Plastic Fracture Analysis and Design
    using the Finite Element Method", unpublished Ph.D. dissertation,
    University of London, 1973/74.

6. Hutchinson, J.W., "Singular Behaviour at the End of a Tensile
    Crack in a Hardening Material", J. Mech. Phys. Solids, 16 (1968),
    13.

7. Heyer, R.H., "Crack Growth Resistance Curves (R-Curves) -
    Literature Review", in Fracture Toughness Evaluation by R-Curve
    Methods, Special Technical Publication 527. Philadelphia: Am.
    Soc. for Testing and Materials (1973), 3-16.

8. Heyer, R.H. and McCabe, D.E., "Crack Growth Resistance in Plane-
    Stress Fracture Testing", Eng. Frac. Mech., 4 (1972), 413-30.

9. Folias, E.S., "An Axial Crack in a Pressurized Cylindrical Shell",
    Int. J. Fract. Mech., 1 (1965), 104-11.

10. Erdogan, F. and Ratwani, M., "The Use of COD and Plastic Instabi-
    lity in Crack Propagation and Arrest in Shells, Paper 6, Crack
    Propagation in Pipelines", presented at the British Gas Corporation
    and Institution of Gas Engineers Symposium, Newcastle, 1974.

11. Wells, A.A., "Fracture Control: Past, Present and Future", Exp.
    Mech., 13 (1973), 401-10.

12. Shannon, R.W.E. and Wells, A.A., "Brittle Crack Propagation in Gas Filled Pipelines - A Model Study using Thin Walled Unplasticised PVC Pipe", Int. J. Fract., 10 (1974), 471-86.

13. Erdogan, F. and Kibler, J.J., "Cylindrical and Spherical Shells with Cracks", Int. J. Fract. Mech., 5 (1969), 229-37.

14. Kanninen, M.F., Sampath, S.G. and Popelar, C., "Steady-State Crack Propagation in Pressurized Pipelines without Backfill", ASME Paper 75-PVP-39, 1975.

15. Becher, P.E. and Pedersen, A., Paper presented at 2nd Intl. Conf. on Structural Mechanics in Reactor Technology, Berlin, 1973, Paper No. M6/4.

16. O'Neil, R. and Jordan, G.M., "Safety and Reliability Requirements for Periodic Inspection of Pressure Vessels in the Nuclear Industry", in Proceedings of Conference on Periodic Inspection of Pressure Vessels, London, England, May 9-11, 1972. London: Institute of Mechanical Engineers (1972), 140-146.

17. Williams, M.L., "Analysis of Brittle Behavior in Ship Plates", in Symposium on Effect of Temperature on the Brittle Behavior of Metals with Particular Reference to Low Temperatures, Special Technical Publication 158. Philadelphia: Am. Soc. for Testing and Materials (1954), 11-44.

18. Brittle Fracture in Steel Structures, Ed. by G.M. Boyd. London: Butterworth & Co. (Publishers), Ltd. (1970), 5-11, 89 (Fig 6.4).

19. Phillips, C.A.G. and Warwick, R.G., "A Survey of Defects in Pressure Vessels", AN5B (s) R162, 1968.

20. Kellerman, O. and Tietzle, A., IAEA Technical Report 99, 1969.

FRACTURE DESIGN FOR STRUCTURAL STEELS

R. Roberts

Lehigh University
Bethlehem, Pennsylvania

ABSTRACT

Fracture mechanics had found wide application in many engineering designs prior to 1968. The catastrophic failure on December 15, 1967, of the Point Pleasant Bridge, an eyebar chain suspension bridge over the Ohio River connecting Ohio and West Virginia, and the subsequent questions raised by the Federal and State Governments helped thrust mechanics upon the designer of steel structures. Structural grades of steel, typified by yield strengths between 36 ksi (248 MN/m^2) and 120 ksi (827 MN/m^2), heretofore considered immune to fracture in terms of normal usage, were suddenly suspect. This chapter sets forth some of the fracture mechanics related problems associated with structural grade steels within the specific framework of bridge design. To this end basic crack initiation, subcritical crack propagation, and fracture behavior of structural grade steels are presented. Furthermore, to illustrate how fracture mechanics currently affects bridge design with structural steels, the current fatigue design rules and toughness requirements for bridge steels of the American Association of State Highway and Transportation Officials are given and discussed. Some of the results of research on fatigue and fracture of structural steels at Lehigh University over the past five years are also presented. Finally, comments are made as to some of the directions fracture mechanics and structural steel design might take in the near future.

INTRODUCTION

Fracture Mechanics and Structural Steels

The body of knowledge known as fracture mechanics has evolved over the last twenty years as a direct result of the technical community's attempt to measure and quantify the ability of a material to withstand fracture. During this twenty year period great progress has been made in terms of both theory and experiment. A major portion of this work has been guided by ASTM Committee E24 on Fracture Testing of Metals which has developed a number of standard methods for measuring fracture toughness [1].

The application of fracture mechanics to the design process has also received considerable attention during the last twenty years. In fact, it was the ability of fracture mechanics to be applied to design problems and to effectively solve these problems which spurred and justified a great deal of the more esoteric fracture mechanics work in applied mechanics and material behavior. Fracture mechanics methodology has been successfully applied to the design of space vehicles, missles, aircraft structures, gas pipelines, pressure vessels and ship structures, to name just a few.

For the most part, fracture mechanics found useful application in situtations where the ratio of the fracture toughness, as measured by the critical plane strain stress intensity factor K_{IC} [1], to yield strength was one or less. The performance of structural grade steels does not generally fall in this category and besides the structural grade steels, characterized in this chapter as those having yield strengths between 36 ksi (248 MN/m^2) and 120 ksi (827 MN/m^2), did not appear to have major fracture mechanics related problems. This was due to a number of factors. The particular application and environments in which these structural steels were used was certainly a major factor. Bridge design had avoided using welded construction in areas where the nominal stress field was positive or tensile. The live loads in bridges were also relatively small, thus producing very small cyclic stresses. In building construction the basic static nature of the load and the fabrication procedures tended to preclude fracture related problems. In the truck industry which uses structural type steel for chassis rails and connecting members, bolted construction avoids some of the fatigue and fracture problems recently encountered in welded truck frames.

Another factor which certainly contributed to the apparent lack of fracture mechanics methodology in design with structural steels was pointed out in a paper by Madison and Irwin [2]. They call attention to the observation of Shank [3] that engineers learn slowly from fractures in service in the case of structures because the information necessary for careful fracture analysis is rarely recorded and made available for open study. The truth of this observation is quite evident to anyone who has been involved with failure analysis of structural steel. Finally, the body of know-ledge known as fracture mechanics within its linear elastic frame-work prior to about 1965 just did not seem appropriate for the analysis of fracture of structural steels.

All of the above has changed within the last five or so years. The subject of fracture mechanics and fracture safe design has become a topic of great interest to designers who employ structural steels and to the steel industry. In 1971 the American Society of Civil Engineers formed a Task Committee on Fracture as part of their Committee on Metals. At the initial meeting of April 20, 1971, the purpose for the commitee was described by Dr. S. T. Rolfe as:

"The function of the task committee will be to review, analyze, and synthesize available information from research studies on the fracture behavior of structural steels to establish specific fracture-safe design of civil engineering structures, primarily bridges. Liaison with existing organizations that are active in the field of brittle fracture and fracture mechanics will be established. Recommendations for needed research on criteria for fracture-safe design of structures will be made and state-of-the-art reviews will be presented at technical session.

"Although rare, civil engineering structures and particularly bridges, sometimes do experience brittle fracture. At present, the designer does not have any guidelines regarding the relative im-portance of various material, design, and fabrication factors that affect the brittle fracture of civil engineering structures. The purpose of this committee would be to provide such information through technical sessions, preparation of state-of-the-art papers or design guides, and bibliographies."

With these introductory comments on fracture mechanics and design with structural steels in mind, the remaining portions of this chapter will deal with fracture mechanics and design as related to structural steel. The specific example of bridge design will be employed as the means through which actual fracture problems can be cited and also how fracture mechanics has and is being employed to help achieve fracture safe design.

It is noteworthy to point out that although the structural engineer is arriving late in the fracture mechanics game, he is starting on an equal footing with his fellow designers who have previously employed fracture mechanics. The truth of this is borne out unfortunately by the report in the January 24, 1972, issue of Newsweek of the January 1972 failure in the Port Jefferson Harbor on New York's Long Island of the Martha Ingram, a 620 foot ship owned by the Ingram Corporation or more recently the report in the August 11, 1975, issue of Aviation Week and Space Technology on the May 23, 1974, crash of a Saturn Airways' Lockheed L-382. The

preliminary finding on probable cause of the failure of the Lockheed
L-382 as found by the National Transportation Safety Board indicated
that the failure was due to pre-existing fatigue cracks, which
reduced the strength of the left wing to the degree that it failed
as a result of positive aerodynamic loads created by moderate tur-
bulence. Thus time teaches that as long as there are engineering
structures, the potential of failure is real and must be constantly
guarded against.

Bridge Service History

The service record of steel bridges has been excellent over
the years. In fact, when one associates with designers and erectors
of monumental bridge structures the general picture presented is
that if the bridge successfully withstands the rigors of erection
then nothing will affect it. This general belief is based in some
part on the past successful operating records of steel bridges and
the fact that the erection stresses are generally the most severe
stresses imposed on the structure on a one time basis.

In spite of the past excellent operating history of steel
bridges, many real concerns exist with respect to safety of older
bridges and new construction. There are many different reasons
for these concerns. With time both the frequency and weight of
bridge traffic has increased, thus subjecting the structures to
loads which were not considered in the initial design. Fatigue was
not a major concern in older steel bridges. Welded construction
and new higher yield strength steels have also entered the picture
in terms of newer bridge construction. Fatigue and fracture problems not previously encountered are suddenly showing up. These
problems are in part due to the new materials and fabrication
techniques. Furthermore, new understandings of basic material
behavior raise questions as to the possibility of subcritical flaw
growth in older bridges which might make them unsafe as of this time
or some time in the future. Certainly bridges designed forty years
ago were not designed for inspection for fatigue or stress corrosion
cracks.

In order to more fully appreciate some of the existing and
potential fracture problems associated with bridges, the failures
of five major structures will be briefly examined. Also the tem-
porary removal from service of one major bridge will be reviewed.

King's Bridge Failure. One span of the King's Bridge [4] in
Melborne, Australia, after 15 months of service as part of a four-
lane freeway system failed on the morning of July 11, 1962. The
actual failure, although expensive and inconvenient was not a major
catastrophe as the failed span only dropped 18" at which point it

was caught and successfully supported by concrete retaining walls.
Figures 1 and 2 copied from Madison and Irwin [2] show some of the
more important details of the failed four girder span.

The Royal Commission Report [4] indicated that the failure was
due to the use of a steel with poor reliability and poor welding
procedures and controls. When the girders left the fabrication
shop, all four contained weld shrinkage cracks at the coverplate
termination welds. It was determined that one center girder had
failed prior to the July 11 failure of the span and at the time of
the failure the cracks in the remaining three girders were growing
by fatigue. At the time of complete failure the remaining center
girder had reached a critical crack size at which point it failed
and caused the outer two girders to fail.

CROSS SECTION OF THE
WEST PORTION OF SPAN #14

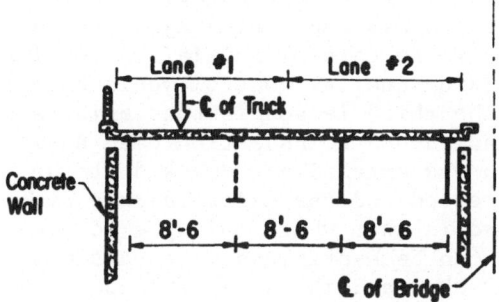

LONGITUDINAL SECTION OF SPAN W14

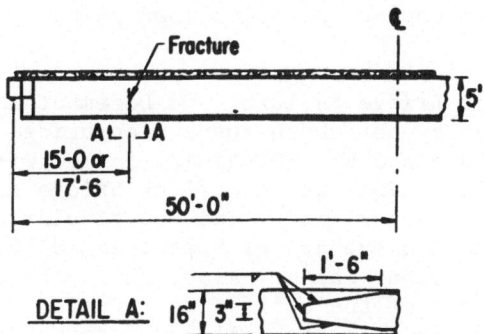

Figure 1.   Failed span of King's Bridge coverplate fracture detail.
            (1" = 25.4mm)

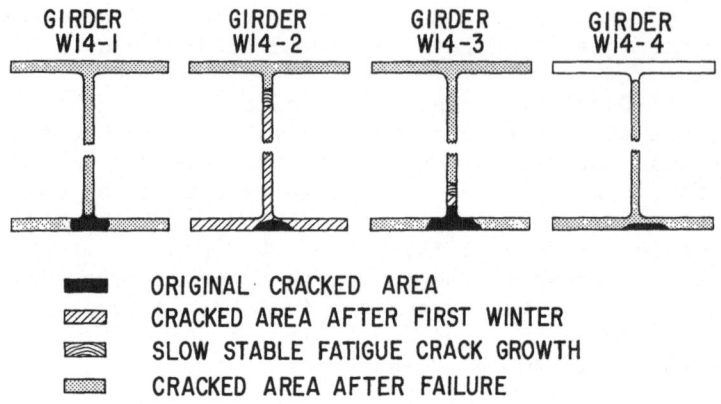

Figure 2.   Condition of failed girders of King's bridge.

     There are a number of lessons to be learned from this failure.
The first and most obvious is that materials with poor weldability
and poor welding procedures can cause major problems.   The second
lesson is that proper and careful initial inspection can eliminate
problems.   Good welding quality control would have rejected those
beams initially.   The third lesson is that bridges should be built
for inspection.   Had adequate inspection been available, the one
center girder which was extensively cracked at the time of failure
might have been detected and the major failure avoided.   The fourth
lesson to be learned is that where multiple load paths exist some
fracture toughness can be sacrificed.   In this case, one out of the
four girders was cracked and the span still performed for some
period of time.   Also the concrete retaining walls acted to catch
and support the span and avoid a major tragedy.   Although the concrete wall was not designed to catch a failed bridge span, it does
emphasize the importance of parallel load paths so that things can
fail-safe.

     Point Pleasant Bridge Failure.   On December 15, 1967, the Point
Pleasant Bridge, an eyebar chain suspension bridge over the Ohio
River connecting Ohio and West Virginia, failed with the subsequent
loss of 46 lives [5].   This was the first bridge failure in the U.S.
to result in major loss of life.   This particular design was unique
in that it employed two eyebars of heat treated 1060 steel to make
up one link of the eyebar chain.   Also, the eyebars formed a portion of the top chord of the stiffening trusses.   A failure of one
eyebar would cause the complete failure of the bridge structure.

     The findings of the commission which investigated the failure
[5] point toward a conclusion that the failure of one eyebar caused

complete failure of the bridge. The particular failure was postulated as being caused by subcritical crack growth due to stress
corrosion. The stress corrosion caused the crack to grow to a
critical length where it became unstable. The 1060 steel had
very low toughness at the failure temperature of 32°F (0°C).

In this particular failure the question of stress corrosion in
steel bridges is raised. Are older structures sitting and just
waiting to fall down? How will new ones perform in aggressive
environments? The problems of inspection of bridges also is
raised as it would be impossible to detect by normal techniques a
crack of critical length in the Point Pleasant eyebar before
failure occurred. The lesson of King's Bridge is also relearned
in that parallel load paths can save lives. Other eyebar chains
employ four eyebars to form a link, thus giving some degree of
stability if one link fails.

Bryte Bend Bridge Failure. In June 1970 a 2-1/4 inch (57mm)
thick plate of A517 steel cracked while concrete was being placed
for the deck of the Bryte Bend Bridge north of Sacramento, California. The crack extended across the 30 inch (762mm) flange plate
and 4 inches (102mm) into the web [6,7]. This particular failure
emanated from a region where a lateral attachment was made to the
flange.

While the Bryte Bend failure did not cause any loss of life,
it did represent a tremendous loss of time and money. The details
of the failure are not readily available at this time due to liti-
gation concerning the failure. One did, however, get a general
feeling from rubbing elbows with experts warming up for the trial
that the failure was due in part to material which had low toughness and welding procedures which left a lot to be desired. The
specific details might never be known and reinforce the statement
offered earlier in the section on fracture mechanics and structural
steels about why it is difficult to learn from structural failures.

The failure of the Bryte Bend Bridge teaches that better
material controls (better than 1970) and good welding procedures
and controls are needed for satisfactory structures.

Fremont Bridge Failure. In the latter part of 1971 a fracture
of the main arch of the Fremont Bridge in Portland, Oregon, occurred
while the bridge was under construction [8]. The 3 inch (76mm)
bottom flange of A588 steel fractured across the full 66 inch
(1676mm) bottom flange width. A recent report [9] indicated that
the failure was due to low toughness as measured by Charpy V-notch
tests and welding details which used thick plates and massive welds.
Very good reports on the details of the Fremont Bridge failure as
with the Bryte Bend Bridge failure are not available.

This failure and delay of a 21.8 million [8] construction project also points toward the need for better material and welding
controls. It appears that the Fremont Bridge and the Bryte Bend
Bridge failures could have been avoided with proper material and
welding controls.

Sidney Lanier Bridge Failure. On Tuesday, November 7, 1972,
the U.S. freighter African Neptune was heading out to sea on the
Brunswick River in Georgia when it hit the Sidney Lanier Bridge
[10]. The impact of the crash caused three 150 foot (45.7m) spans
to collapse into the 40 foot (12.2m) water. Approximately ten
vehicles were sent into the water, with a loss of life estimated
at five people.

This particular failure is mentioned here not because better
fabrication or better material would have saved the bridge, but
because it emphasizes that loading rates can be high in bridge
structures due to unusual events.

Carquinez Bridge Closing. The October 24, 1974, issue of
Engineering News Record [11] reported the impending closing of the
older of two parallel Carquinez Strait bridges for inspection and
repairs. The reason cited for the closing was apparent corrosion
in some of the steel members. The Carquinez Bridge was constructed
47 years ago about the same time as the Point Pleasant bridge. It
also is an eyebar suspension bridge and uses material similar to
that found in the Point Pleasant bridge. The actual bridge design
is not identical to Point Pleasant but it does have some areas with
low redundancy.

This incident is cited here because it emphasizes that answers
are needed quickly for many questions which are now arising relative
to structural steels. In the case of the Carquinez bridge, the
State of California is taking the bull by the horns by closing the
bridge. California, the Federal Highway Administration, and the
American Bridge Division of U.S. Steel Corporation are presently
studying the stress corrosion, fatigue, and fracture behavior of
materials from a number of eyebars removed from the bridge. Based
on these tests, the future of the bridge will be determined.

Design and Fracture Mechanics

It is certainly clear from the brief review of bridge fracture
experience given above that fracture mechanics has a real and major
role in bridge safety and most structural steel applications. This
role is to provide a fracture control plan whereby the designer can
produce new fracture safe structures. Its role is also to provide
plans whereby existing structures can be assessed, modified, and

maintained in a fracture safe condition. Barsom [12] describes a fracture control plan as one which is used

1. "To identify all the factors that may contribute to the fracture of a structural detail or to the failure of the entire structure.

2. "To assess the contribution of each factor and any synergistic contribution of these factors to the fracture process.

3. "To determine the relative efficiency and trade-off of various methods used to minimize the probability of fracture.

4. "To assign responsibility for each task that must be undertaken to ensure the safety and reliability of the structure."

The four objectives described by Barsom are statements of a general nature which should be followed for any structure. The specific details of a fracture control plan will vary from structure to structure and should be tailored to the individual situation. A discussion of some of the finer details of fracture control plans for structural steels will be put off at this point until after the basic material behavior of structural steels has been discussed.

FATIGUE OF BRIDGE STEELS

Material Behavior

A number of different steels have been used in bridge construction. These include ASTM designations A7, A8, A36, A242, A373, A440, A441, A514, A517, A572, A588, and ASA 1035 with many of these steels being further divided into sub grades. These steels can be conveniently grouped as follows:

(a) Structural Steels (Carbon Steels)
     Low carbon, pearlitic steels with yield strengths in the range of 30 to 40 ksi (207 to 276 MN/m^2) which are commonly used in bridge construction. These include A7, A8, A36, A373.

(b) High-Strength Structural Steels and High-Strength Low-Alloy Structural Steels
     Pearlitic steels with yield strengths in the range of 40 to 65 ksi (276 to 448 MN/m^2) which are commonly used in bridge construction. These include A242, A440, A441, A572, and A588.

(c) High Yield Strength Quenched and Tempered Alloy Steels
     Tempered martensitic and bainitic steels with yield strengths in the range 80 to 130 ksi (552 to 896 MN/m^2). Used in more recent bridge construction. These include A514 and A517.

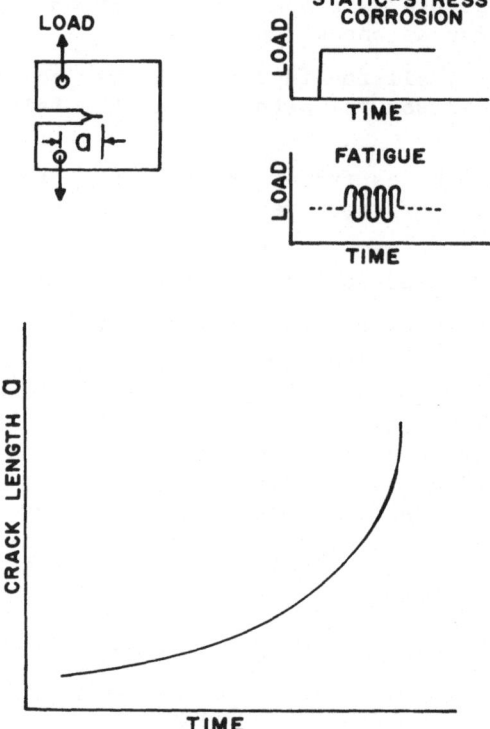

Figure 3.   Crack length-time behavior for fatigue or stress
corrosion.

     The useful life of a structure fabricated with any of the
above steels can be thought of as being composed of a period of
crack initiation and a subsequent period of subcritical crack propagation.   In some instances the structure will be fabricated, as
was the King's Bridge, such that flaws or cracks exist initially,
thus precluding the crack initiation period.

     Subcritical crack growth in metals is generally understood to
be a slow stable increase in crack length with time.   This growth
can occur at constant load or due to cyclic loads.   The former is
termed stress corrosion cracking while the latter is termed fatigue
crack propagation.   It has been shown that both forms of subcritical crack extension can be readily dealt with by fracture mechanics
methods through the use of the prevailing stress intensity factor
[13,14].

     Consider first subcritical crack growth due to cyclic load.

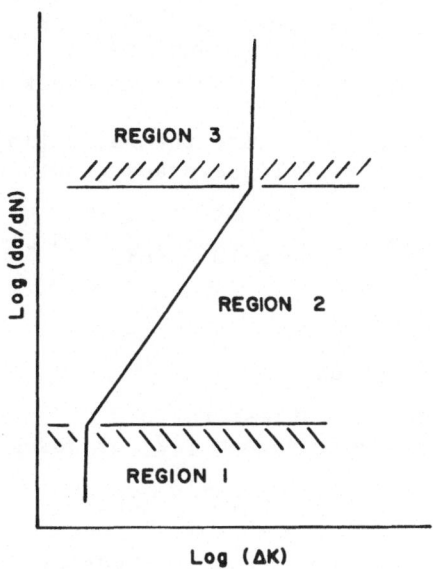

Figure 4. Typical fatigue crack growth rate behavior.

Figure 3 schematically represents the behavior of a compact tension sample, as found in ASTM Standard E399-74, when subjected to a constant amplitude sinusoidal load. Paris [13,15] proposed that such behavior could be best handled in terms of the crack tip stress intensity range ΔK where ΔK is the difference between the maximum stress intensity level and the minimum stress intensity level per cycle. Figure 4 shows a schematic log-log plot of the specimen behavior in Figure 3 in terms of the growth rate per cycle versus the ΔK level.

Three distinct regions are visible in Figure 4 and for convenience these are labelled as Regions 1, 2, and 3. In Region 1 there is a threshold level of ΔK below which the crack will not propagate. For martensitic steels and ferrite-pearlite steels, Barsom [12] suggests that a conservative estimate of the fatigue crack propagation threshold ΔK_{th} can be made for R values greater than +0.1 from the relationship

$$\Delta K_{th} = 6.4(1 - 0.85\ R) \tag{1}$$

where ΔK_{th} = fatigue crack propagation stress intensity threshold
           in ksi$\sqrt{in}$
      R = ratio of maximum stress to minimum stress in a cycle

For R values less than +0.1, R < +0.1, ΔK_{th} can be taken as constant and equal to 5.5 ksi$\sqrt{\text{in}}$ (6MN m$^{-3/2}$). In a cracked structure reducing ΔK below ΔK_{th} guarantees no crack growth.

In Region 2 Barson [12] using the Paris [15] model finds that for martensitic steels the fatigue crack growth rate can be expressed as

$$\frac{da}{dN} = 0.66 \times 10^{-8} \ (\Delta K)^{2.25} \qquad (2)$$

where   a = inches
ΔK = ksi$\sqrt{\text{in}}$
N = number of cycles

It was also found that the stress ratio R did not affect the results. For ferrite-pearlite steels Barsom [12] proposes

$$\frac{da}{dN} = 3.6 \times 10^{-10} \ (\Delta K)^{3.0} \qquad (3)$$

As with martensitic steels the ferrite-pearlite steels do not show a significant mean stress or R effect. Figure 5 is a plot of Equations (2) and (3). Barsom [12] also notes that Equations (2) and (3) give reasonable results for weld metal and the heat affected zone of weldments.

Region 3 in Figure 4 shows that there is a point at which an increase in ΔK produces very large increases in the crack growth per cycle. Generally this region is not of interest for design purposes as a structure does not have much life left when da/dN is greater than 10^{-4} inches per cycle (4 x 10^{-6} mm/cyc.). The type of fatigue response described here has been for cyclic loads which can be represented as constant amplitude sinusoidal patterns. A large number of studies have been carried out on other load histories from single overloads [16,17], block loading [18,19] to fully random loads [20,21]. Figure 6 shows some of these loading patterns. The details of these works are beyond the scope of this chapter but they generally show that the stress intensity factor is still the best correlation parameter for the observed growth rates.

Stress corrosion cracking which is also schematically represented in Figure 3 will not receive much coverage in this chapter in spite of it being a contributing factor in the Point Pleasant bridge collapse. As with fatigue crack propagation, stress corrosion cracking is effectively correlated by the prevailing stress intensity level as described elsewhere [22-25]. Similar to the threshold ΔK_{th} for fatigue crack propagation there is a threshold stress intensity level K_{ISCC} below which crack propagation will not occur for the particular material and environment. Currently it is felt that stress corrosion cracking is not a serious factor in the design of new bridges or the safety of older structures. Sinclair

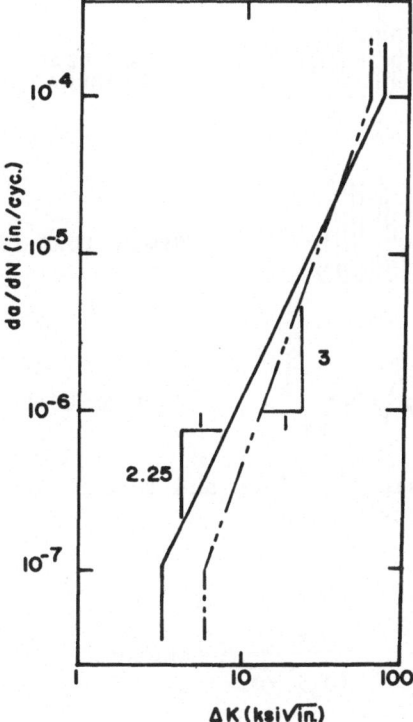

Figure 5. Fatigue crack growth rates for ferrite-pearlite steels and martensitic steels (1" = 25.4mm, 1 ksi√in = 1.099MN/m$^{3/2}$).

[26] reports that steels of strength levels similar to those found in bridges are relatively insensitive to stress corrosion cracking. A major program [27] related to stress corrosion cracking is being carried out by the Boeing Company for the Federal Highway Administration at this time. Initial results of this program also point to a lack of sensitivity in typical service environments to stress corrosion cracking on the part of most bridge steels.

The remaining topic to be dealt with in terms of basic material response is that of crack initiation. This is one subject which has defied rational analysis previously due to the difficulty of defining the point at which crack initiation occurs. To date, in this writer's views, the only consistent piece of work relative to crack initiation is that of Barsom and McNicol [28] on notched specimens. Their results suggest that there exists a fatigue crack initiation threshold for martensite steel subjected to zero to tension load of the form:

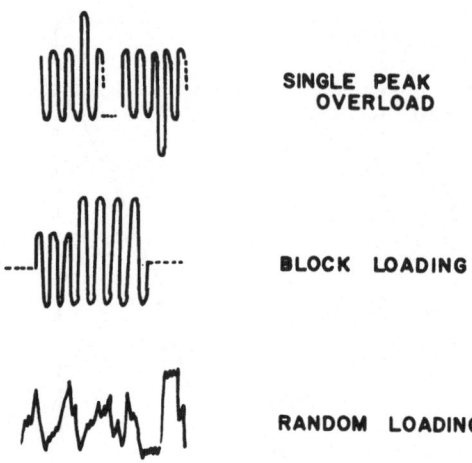

Figure 6.   Typical single, block and random load patterns used in
fatigue tests.

$$\frac{\kappa}{(\sigma_y)^\alpha \sqrt{\rho}} = \text{constant} \qquad (4)$$

where κ = stress intensity fluctuation calculated as though the
total notch depth was the crack length
ρ = notch tip radius
σ_y = 0.2% yield strength
α = a constant less than 1.0

Barsom and McNicols' [28] work is relatively new and as yet
has not found wide application. With time their type of approach
to fatigue crack initiation should find wide usage.

1974 AASHTO Fatigue Rules

The current American Association of State Highway and Transportation Official Rules for fatigue design in bridges are described in detail by Fisher [29]. These rules are the direct result
of the work of Fisher and his co-workers at Lehigh University [30,
31]. In essence they found after fatigue testing many different
types of welded beam details such as are found in bridge design
that mean stress did not affect fatigue life and that solely stress
range, the difference between the maximum stress and the minimum
stress in a cycle, controlled total fatigue life. Typical results
of the tests are shown in Figures 7 through 9. Figure 10 shows some
of the welded details tested and Table 1 shows how these are

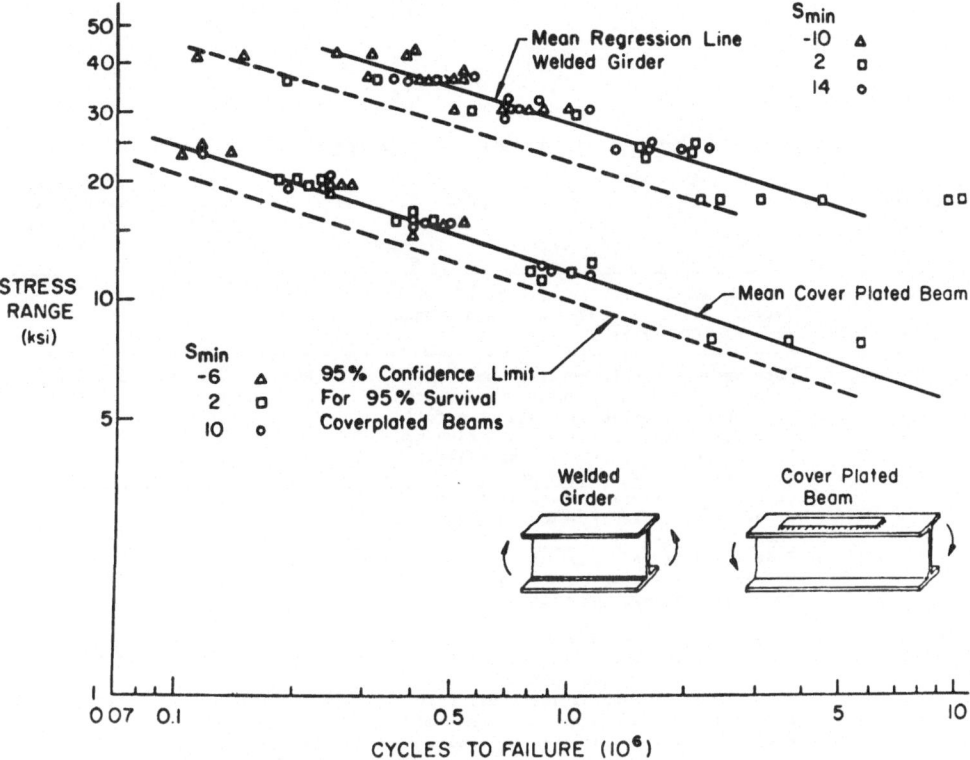

Figure 7. Effect of minimum stress and stress range on the cyclic life for the welded end of coverplated beam and plain welded beam (1 ksi = 6.895 MN/m^2).

classified while Table 2 gives the design stresses for different details. Figure 11 shows the S-N curve for the different details. Figures 12 through 16 show some of the failures.

The actual fatigue rules are set up so as to minimize the effort on the part of a designer. In real life the actual load distribution is quite complex, as shown in Figure 17. This fact is accounted for in the design rules by the use of Miner's rule [32]. To do this the distribution of weight shown in Figure 17 is reduced to an equivalent load at a specific number of cycles. Thus, for a desired traffic pattern, cyclic life and welded detail the designer has only to pick the proper stress range out of the tables.

The approach used by Fisher [29-31] is basically an S-N type approach to fatigue life. The primary reason for this is that it

TABLE 1

WELD DETAIL CLASSIFICATION

General Condition	Situation	Kind of Stress*	Stress Category (see Table 2B)	Illustrative Example No. (see Fig. 10)
Plain material	Base metal with rolled or cleaned surfaces. Flame cut edges with ASA smoothness of 1000 or less	T or Rev.	A	1,2
Built-up members	Base metal and weld metal in members without attachments, built-up of plates or shapes connected by continuous full or partial penetration groove welds or by continuous fillet welds parallel to the direction of applied stress	T or Rev.	B	3,4,5,7
	Calculated flexural stress at toe of transverse stiffener welds on girder webs or flanges	T or Rev.	C	6
	Base metal at end of partial length welded cover plates having square or tapered ends, with or without welds across the ends	T or Rev.	E	7
Groove welds	Base metal and weld metal at full penetration groove welded splices of rolled and welded sections having similar profiles when welds are ground flush and weld soundness established by nondestructive inspection	T or Rev.	B	8,9,13
	Base metal and weld metal in or adjacent to full penetration groove welded splices at transitions in width or thickness, with welds ground to provide slopes no steeper than 1 to 2-1/2, with grinding in the direction of applied stress, and weld soundness established by nondestructuve inspection	T or Rev.	B	10,11
	Base metal and weld metal in or adjacent to full penetration groove welded splices with or without transitions having slopes no greater than 1 to 2-1/2 when reinforcement is not removed and weld soundness is established by nondestructive inspection	T or Rev.	C	8,9,10,11,13
	Base metal at details attached by groove welds subject to transverse and/or longitudinal loading when the detail length L, parallel to the line of stress, is between 2 in. and 12 times the plate thickness, but less than 4 in.	T or Rev.	D	12,13
	Base metal at details attached by groove welds subject to transverse and/or longitudinal loading when the detail length L is greater than 12 times the plate thickness or greater than 4 in. long	T or Rev.	E	12,13

TABLE 1 (continued)

Fillet welded connections	Base metal at intermittent fillet welds	T or Rev.	E	
	Base metal adjacent to fillet welded attachments with length L in direction of stress less than 2 in. and stud-type shear connectors	T or Rev.	C	13,14,15,16
	Base metal at details attached by fillet welds with detail length L in direction of stress between 2 in. and 12 times the plate thickness but less than 4 in.	T or Rev.	D	13,14,15
	Base metal at attachment details with detail length L in direction of stress (length of fillet weld) greater than 12 times the plate thickness or greater than 4 in.	T or Rev.	E	13,15
Mechanically fastened connections	Base metal at gross section of high-strength bolted slip resistant connections, except axially loaded joints which induce out-of-plane bending in connected material	T or Rev.	B	17
	Base metal at net section of high-strength bolted bearing-type connections and other mechanically fastened joints	T or Rev.	B	17
Fillet welds	Shear stress on throat of fillet welds	Shear	F	8a

* In Table 2, "T" signifies range in tensile stress only; "Rev." signifies a range of stress involving both tension and compression during a stress cycle.

TABLE 2

FATIGUE STRESSES

Stress Cycles

Main (Longitudinal) Load Carrying Members

Type of Road	Case	(ADTT)*	Truck Loading	Lane Loading†
Freeways, expressways, major highways and streets	I**	2500 or more	Over 2,000,000	500,000
	II	less than 2500	500,000	100,000
Other highways and streets not included in Case I or II	III	--	100,000	100,000

Transverse Members and Details Subjected to Wheel Loads

Type of Road	Case	(ADTT)*	Truck Loading
Freeways, expressways, major highways and streets	I**	2500 or more	Over 2,000,000
	II	Less than 2500	2,000,000
Other highways and streets	III	--	500,000

 * Average daily truck traffic
 † Longitudinal members should also be checked for truck loading
** This condition corresponds to an extremely heavily traveled artery

Stress Range

Allowable Range of Stress, F_{ST} (ksi)

Category (see Table 1.7.3C)	For 100,000 Cycles	For 500,000 Cycles	For 2,000,000 Cycles	Over 2,000,000 Cycles
A	60	36	24	24
B	45	27.5	18	16
C	32	19	13	10,12*
D	27	16	10	7
E	21	12.5	8	5
F	15	12	9	8

* For transverse stiffener welds on girder webs or flanges

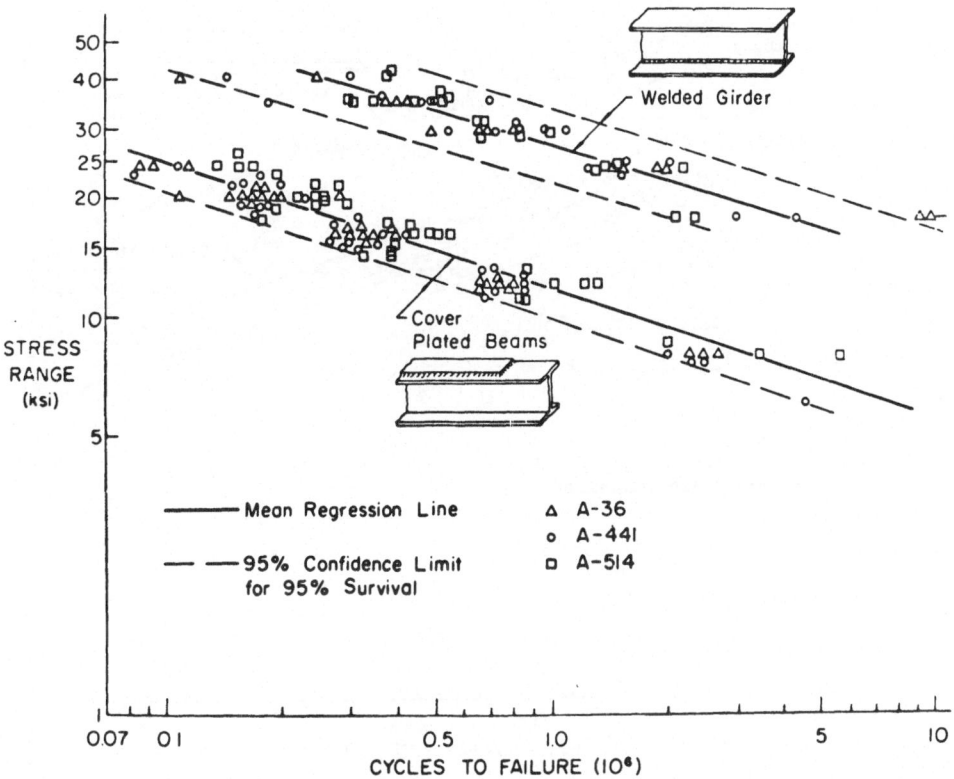

Figure 8. Effect of stress range and type of steel on cycle life of coverplated and plain welded beams (1 ksi = 6.895 MN/m^2).

simplifies design rules. The general results basically showed that above a detail's endurance limit the fatigue life could be represented by

$$N = A\, S_r^{-3} \tag{5}$$

where   N = number of cycles of life
      A = a constant
      S_r = stress range

This result was found to be in agreement with fracture mechanics results [29-31,33,34] for crack growth rates such as proposed by Barsom [12].

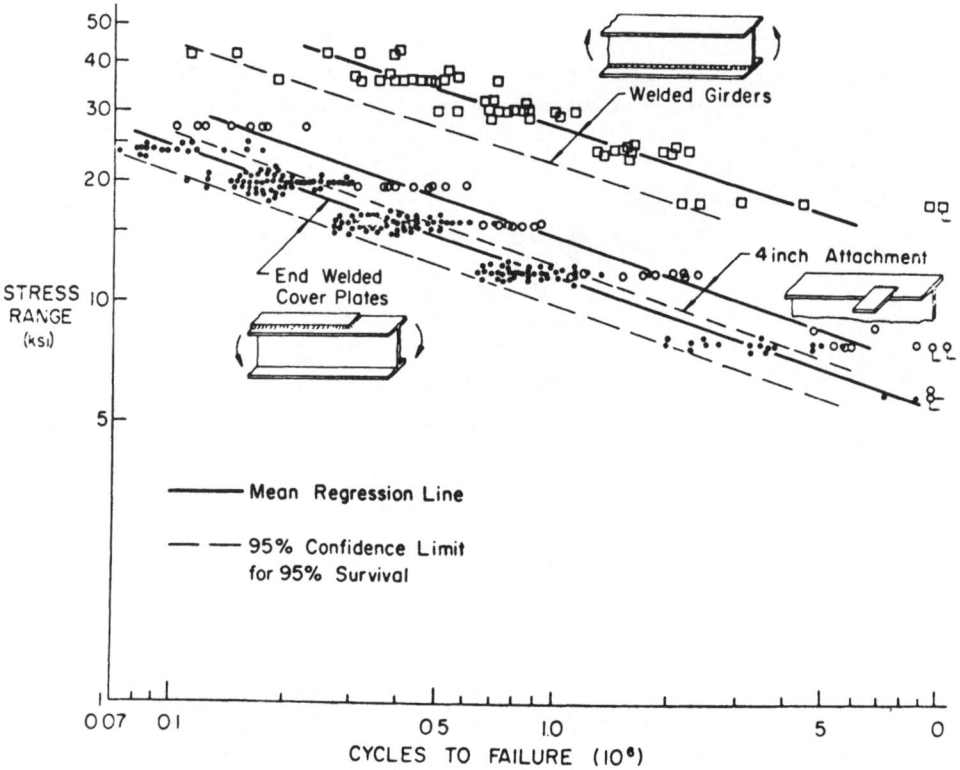

Figure 9.   Comparison of short welded attachments with coverplated
            and plain welded beams (ksi = 6.895 MN/m^2).

FRACTURE OF BRIDGE STEELS

Material Behavior

The fracture behavior of bridge steels is affected by temperature, strain rate, and plate thickness [35,36]. Figures 18 through
20 show the basic effect of these three variables on fracture toughness. Figure 18 shows that as temperature increases the measured
fracture toughness herein designated as K_c slowly rises from a
plateau value until a transition temperature is reached at which
point K_c rises very rapidly. Figure 19 shows that the variation of
K_c with temperature is similar for static loads and very fast rates
of loading. Generally the dynamic curve is shifted to the right so
that the transition region where K_c is increasing rapidly occurs at
a higher temperature. Barsom and Rolfe [37] have shown that this
shift in transition temperature with loading rate can be predicted

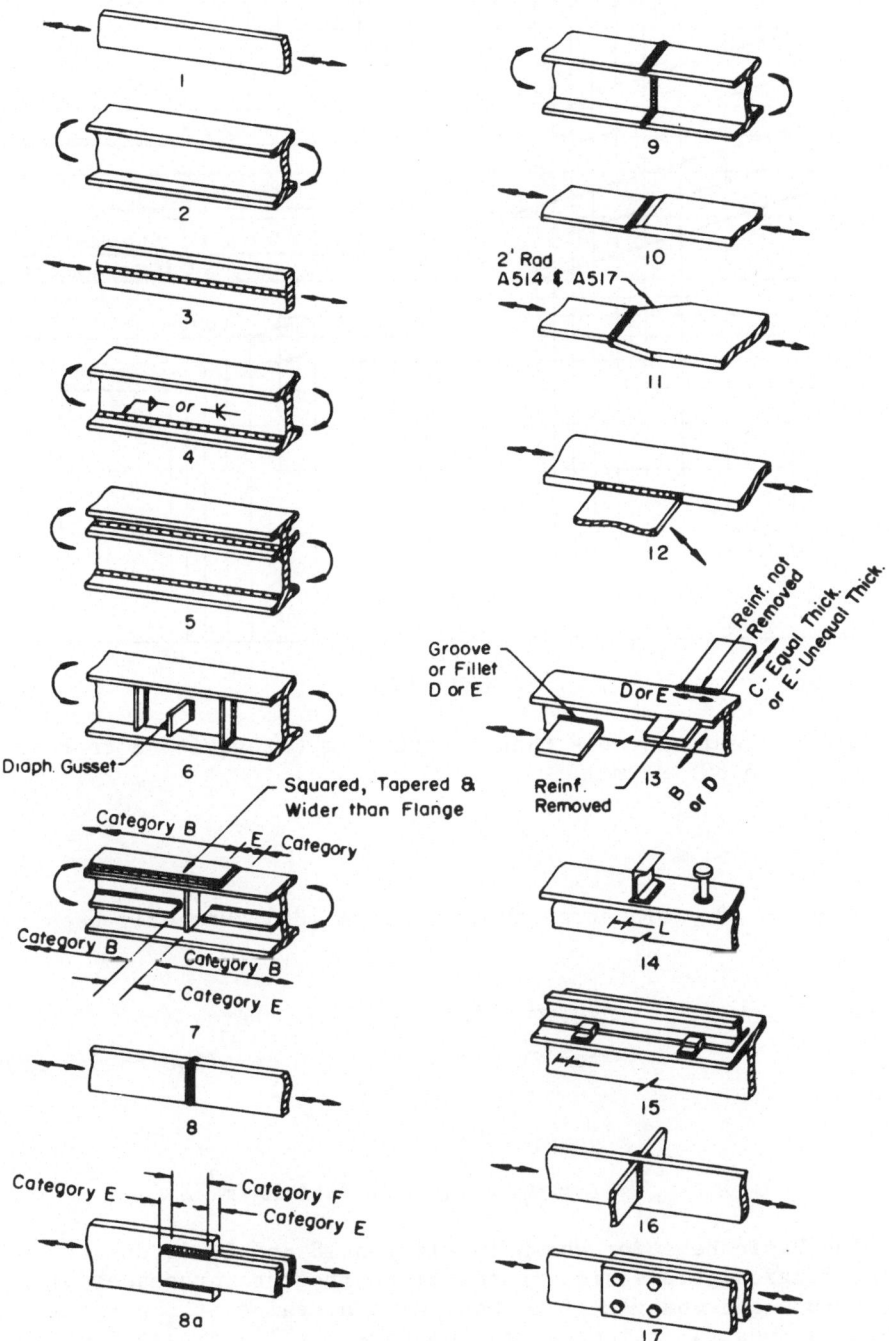

Figure 10.   Typical details in AASHTO Fatigue Rules.

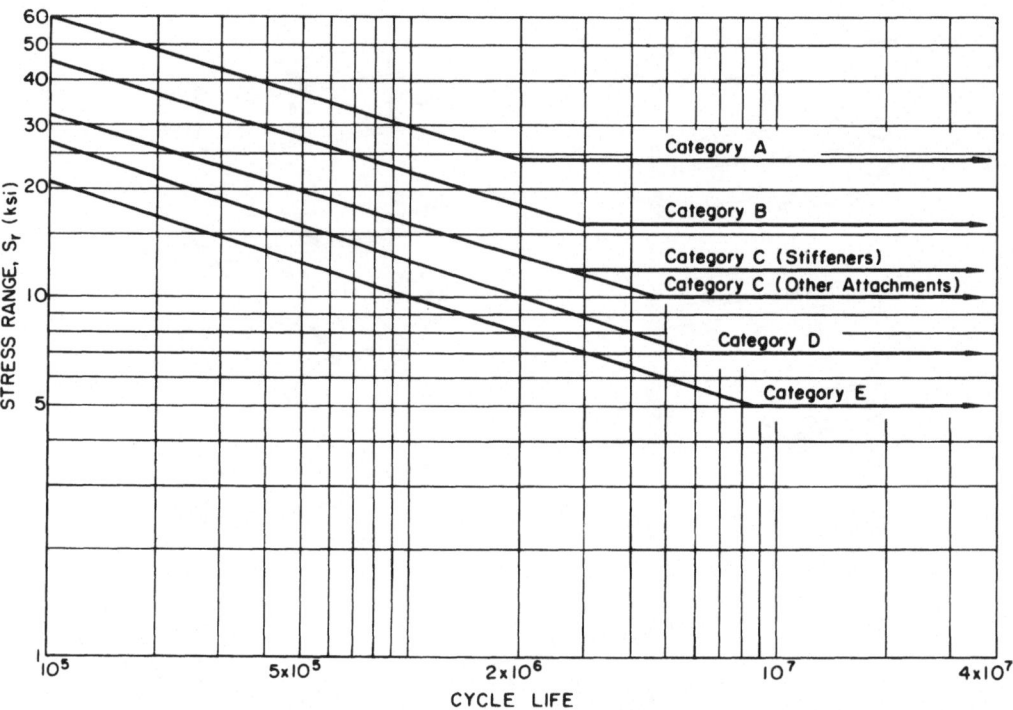

Figure 11.   Design stress range curves for Categories A to E (1 ksi =
             6.895 MN/m^2).

on the basis of yield strength and Barsom [38] currently uses the
formula

$$T_{SHIFT} = 215 - 1.5\ \sigma_y \qquad\qquad (6)$$

$$\text{for } 36 \text{ ksi} < \sigma_y < 140 \text{ ksi} \ (248 \text{ to } 965 \text{ MN/m}^2)$$

$$T_{SHIFT} = 0.0$$

$$\text{for } \sigma_y > 140 \text{ ksi} \ (965 \text{ MN/m}^2)$$

The basic behavior shown in Figures 18 and 19 is due to both
a metallurgical transition in the micro fracture processes with
temperature and changes of mechanical constraint at the crack tip
[38]. As constraint increases, toughness will decrease. Constraint
is a function of yield strength which is a function of both temperature and strain rate. For bridge steels [35,36], yield strength
increases as temperature decreases and strain rate increases. Thus,

Figure 12. Fatigue crack at end of coverplate (not welded across).

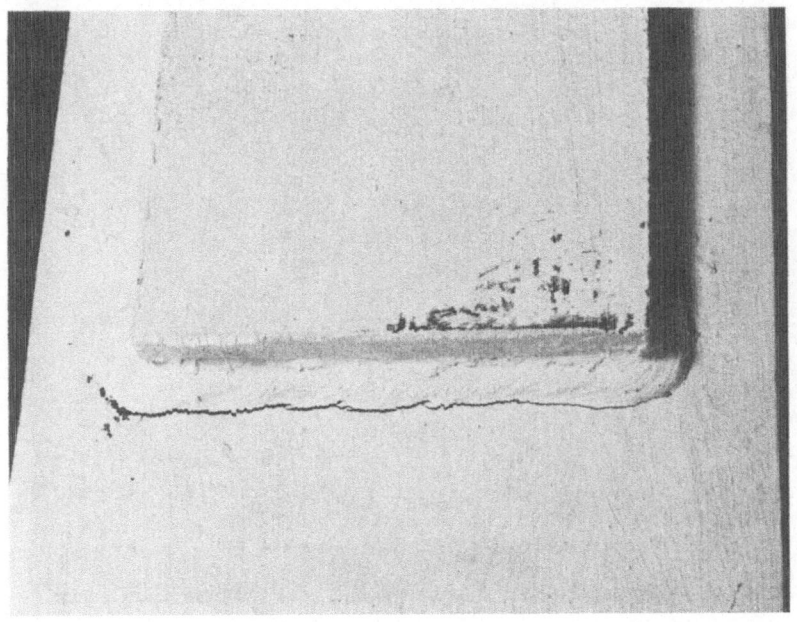

Figure 13. Fatigue crack at end of coverplate (welded across).

Figure 14.   Fatigue crack in welded beam at the weld toe of a
             transverse stiffener.

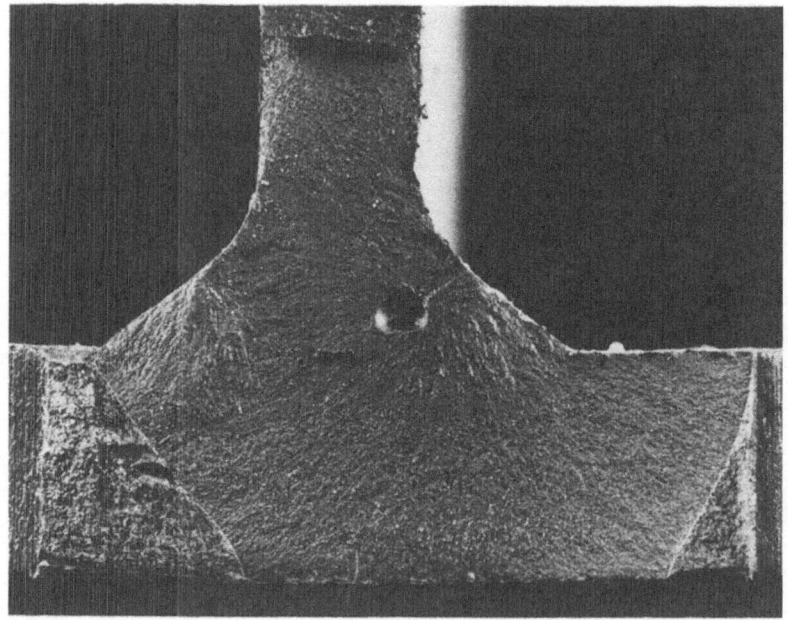

Figure 15.   Crack initiation from gas porosity in web-flange fillet
             weld.

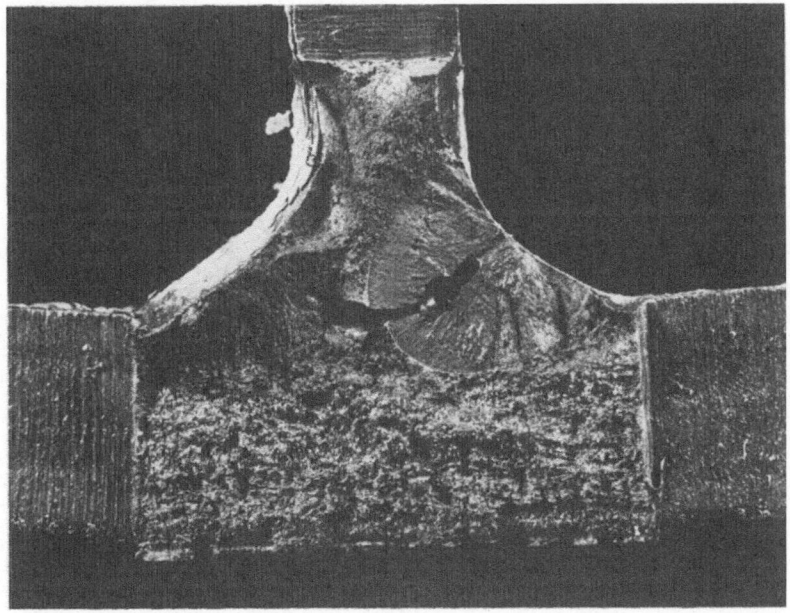

Figure 16. Two small fatigue cracks that initiated from pores in the flange-web fillet welds.

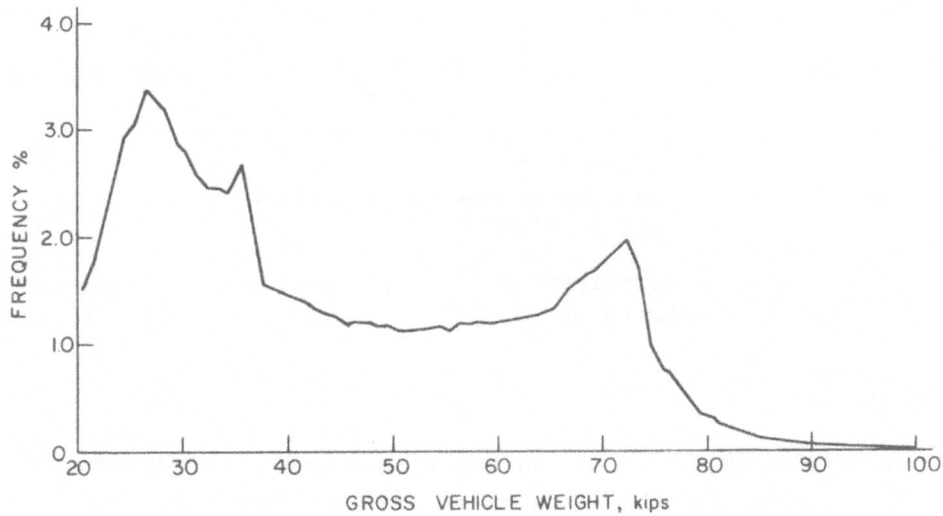

Figure 17. Gross vehicle weight distribution from 1970 FHWA National Loadometer Survey (1 Kip = 4450 N).

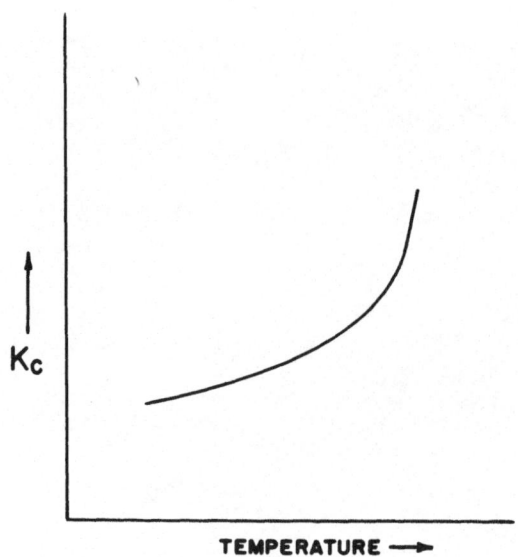

Figure 18.   Schematic representation of typical K_c response of a structural steel.

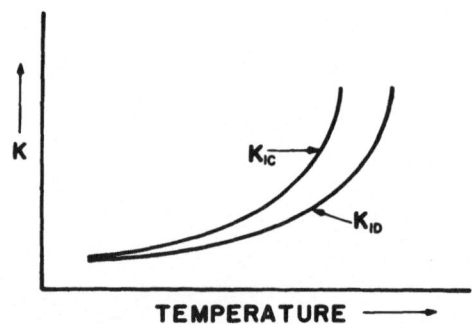

Figure 19.   Schematic representation of K_c response as a function of loading rate.

as yield strength increases constraint increases, so toughness will decrease.

The behavior shown in Figure 20 is due primarily to changes in constraint [35,36] due to the differences in thickness.   The thicker plates exhibit more constraint at a given temperature and therefore lower toughness.

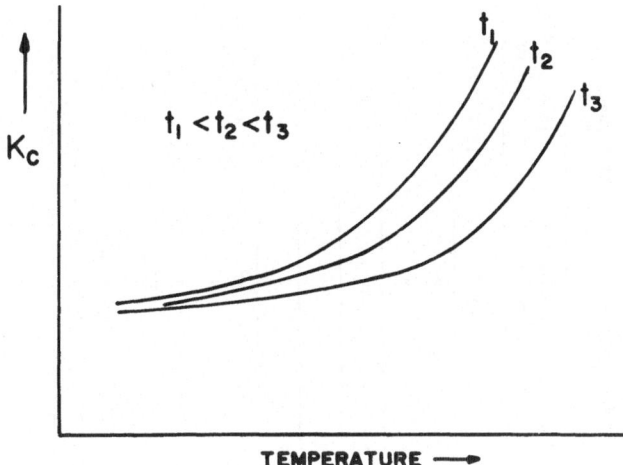

Figure 20. Schematic representation of the effect of thickness on
            K_c behavior.

The basic behavior shown in Figures 18 through 20 was determined using specimens similar to those found in ASTM E399-74 [1].
Barsom and Rolfe [37] noted for dynamic loading or very high strain
rates that the transition temperature of the dynamic K_c curves
corresponded to the transition temperature in a standard Charpy
V-notch (CVN) test. They also noted that the transition in CVN
energy levels for standard CVN specimens tested at very slow loading
rates corresponded to the transition temperatures for static K_c
measurements. This is schematically shown in Figure 21. This led
Barsom and Rolfe to attempt a correlation between K_c and CVN values.
This correlation is now given by Barsom [38] as

$$K_{IC}^2 = 5E \ (CVN) \tag{7}$$

where K_{IC} = the plane strain fracture toughness in $ksi\sqrt{in}$
         E = Young's Modulus in ksi
       CVN = Charpy V-notch energy in foot pounds

TABLE 3

AASHTO SPECIFICATIONS FOR BRIDGE STEELS

ASTM Designation	Thickness (in.)	Originally Proposed by FHWA Based on Minimum Service Temperature of -30° CVN ft.-lbs.	CVN Recommendations of AISI Bridge Steel Specifications Group (6/28/73)		
			Group 1 See Footnote	Group 2 See Footnote	Group 3 See Footnote
A36		15 @ 40°F	15 @ 70°F	15 @ 40°F	15 @ 10°F
A572*	Up to 4" mechanically fastened	"	"	"	"
	Up to 2" welded	"	"	"	"
A440		"	"	"	"
A441		"	"	"	"
A242		"	"	"	"
A588*	Up to 4" mechanically fastened	"	"	"	"
	Up to 2" welded	"	"	"	"
	Over 2" welded	20 @ 40°F	20 @ 70°F	20 @ 40°F	20 @ 10°F
A514	Up to 4" mechanically fastened	25 @ 0°F	25 @ 30°F	25 @ 0°F	25 @-30°F
	Up to 2-1/2" welded	25 @ 0°F	25 @ 30°F	25 @ 0°F	25 @-30°F
	Over 2-1/2" to 4" welded	35 @ 0°F	35 @ 30°F	35 @ 0°F	35 @-30°F

Footnotes: Group 1: Minimum Service Temperature 0°F and above
             Group 2: Minimum Service Temperature from -1° to -30°F
             Group 3: Minimum Service Temperature from -31° to -60°F

* If the yield point of the material exceeds 65 ksi, the temperature for the CVN value for
  acceptability shall be reduced by 15°F for each increment of 10 ksi above 65 ksi.

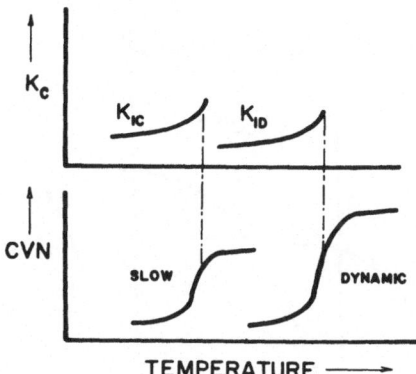

Figure 21. Schematic representation of regions for CVN-K_c correlations.

To estimate a dynamic K_c value from Equation (7) at a specific temperature, one uses the dynamic CVN value at that temperature. Similarly, to estimate a static K_c value, one would use a CVN energy level obtained from slow loading tests of CVN specimens. Barsom's [38] equation has been generally substantiated by the work of Roberts [36].

AASHTO Fracture Toughness Requirements

The basic American Association of State Highway and Transportation Officials fracture toughness requirements are an outgrowth of a proposal by Frank and Galambos [39]. These requirements are not imposed in terms of a fracture control plan but as a quality assurance check on the material. Table 3 gives the current requirements for primary tension members. As can be seen in Table 3, the requirements take the form of CVN requirement on the steels based on the expected operating temperature of the bridge. This type of requirement is similar to the Nil Ductility Temperature, NDT, requirements for ship steel [40] and Fracture Appearance Transition Temperature, FATT, requirement for line pipe [41].

The basis for the current AASHTO requirement can be found in the work of Barsom [35] and Roberts [36]. The basic concept as developed by Barsom [38] was to make sure that the material had a temperature transition behavior so that at the particular operating temperature and loading rate the K_c level of the steel was rapidly rising. This will assure that the steel will not fracture in a plane strain mode. To more fully understand this consider Figure 22 after Barsom [38] which shows the CVN behavior of a hypothetical

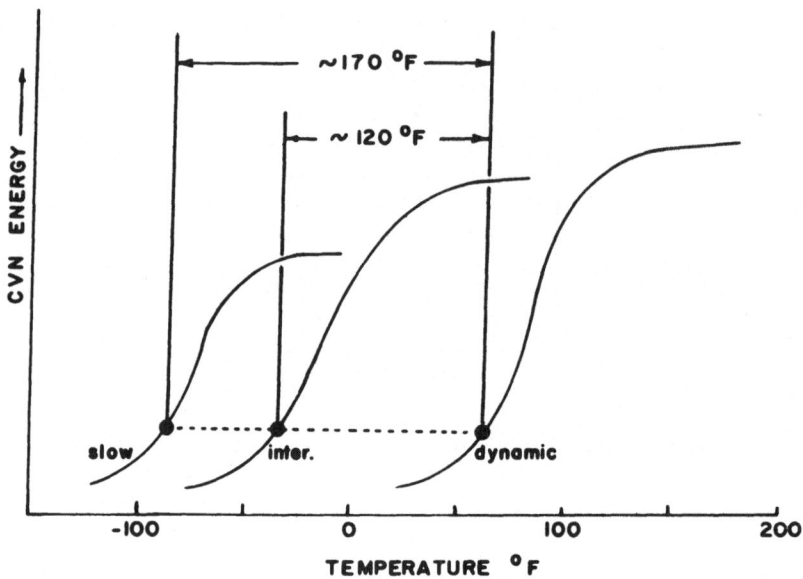

Figure 22.   Typical CVN response for a 36 ksi (248 MN/m^2) steel.

36 ksi (248 MN/m^2) steel.   Three CVN curves are shown for slow,
intermediate and dynamic rates of loading.   In the particular case
of bridges the normal rate of loading is best represented by intermediate loading rates [42].   Based on Barsom's work [35], the
temperature shift for slow to dynamic will be ∿170°F (81°C) as
shown and the shift between intermediate and dynamic 120°F (57°C).
Now Barsom [38] argues that if non plane strain behavior is desired,
this occurs at about a temperature level equal to the transition
temperature plus 50°F (24°C).   Roughly this is the 15 ft-lb. (20J)
CVN temperature plus 50°F (24°C).   Thus non plane strain behavior
is guaranteed at the minimum service temperature of the bridge if
the 15 ft-lb. (20J) CVN levels falls 50°F (24°C) below the service
temperature.

        Since it is difficult to run slow or intermediate strain rate
CVN tests, it is more appropriate to make any specification in
terms of standard CVN tests.   This can be done by utilizing the
temperature shift between the intermediate and dynamic CVN curves.
Barsom states that this is about 0.75 of T_{SHIFT} given by Equation
(6).   In the case of the 36 ksi (248 MN/m^2) yield strength steel
this is 120°F (57°C).   Thus, to specify a CVN and testing temperature to guarantee non plane strain behavior one takes the service
temperature and subtracts 50°F (24°C) and then adds 120°F (57°C) to
this.   This then gives the standard dynamic CVN 15 ft-lb. (20J)
temperature requirement for the steel in question.   Slight modifica-

tions of the above arguments were made for steels with $\sigma_y > 50$ ksi
(345 MN/m^2). Barsom gives complete detail of this in reference 38.

It should be evident at this point, as already mentioned, that
the current AASHTO toughness requirements are essentially a quality
assurance program for material rather than a fracture control plan
or system. All that the requirement does is guarantee some minimum level of toughness. It is important to note that the actual
level of toughness is not known since in the region of the specification the intermediate loading rate produces a toughness level which
was beyond the measurement capabilities of both the work of Roberts
[36] and Barsom [35]. It is also important to note that this
specification is designed for situations where intermediate loading
rates exist. If dynamic loading rates occur then this specification
can prove to be inadequate.

LEHIGH RESEARCH ON BRIDGE STEELS

Over the years a great deal of research has been carried out
at Lehigh University relative to fracture mechanics. In terms of
bridge steels, the fatigue work of Fisher et al. [30,31] has led
to the AASHTO fatigue rules. The fracture related work of Roberts
et al. [36] formed part of the basis for the AASHTO fracture requirements. This work was sponsored by the Federal Highway Administration in response to the Point Pleasant Bridge failure [5].

Recognizing the limitations of fracture mechanics as applied
to steel common to bridge construction, the U.S. Department of
Transportation Federal Highway Administration initiated at Lehigh
University a research project to study the fracture toughness of
bridge steels. The primary objectives of this project were to
establish meaningful measures of fracture toughness for bridge
steels, practical methods of obtaining these measurements and to
collect sufficient data on past and present bridge steels to
establish the fracture resistance of these steels. In particular
the research was separated into three phases.

Phase 1 - Phase 1 was concerned with the establishment of meaning-
         ful measures and tests for fracture toughness. The
         materials considered in this portion were ASTM grades
         A36, A441, and A514 in thicnesses of 1/2" (12.7 mm), 1"
         (25.4 mm), and 2" (51 mm).

Phase 2 - Phase 2 was concerned with the collection of fracture
         resistance data for ASTM grade steels A7, A242, A440,
         A588 and SAE 1035 in thicknesses of 1/2" (12.7 mm), 1"
         (25.4 mm), and 2" (51 mm) where possible. Phase 2 also
         included the fracture testing of four (4) precracked
         A36 beams.

TABLE 4

DATA SUMMARY

Material	Thickness	T_{15} (°F)	T_{15}^{P} (°F)	ΔT (°F)	ΔT^{P} (°F)	ΔT_{B} (°F)	$T_{15}-T_{5}$ (°F)	CVN_{SH} (ft-lb)	K_{ID}^{15} (ksi√in)	K_{ID}^{*} (ksi√in)	σ_{y} (ksi)	σ_{u} (ksi)	Elongation (%)
A7	1/2	100	--	125	--	162	60	73	> 80	35	35.5	63.0	27.5
A36	1/2	50	70	160	100	158	60	> 85	70	50	37.5	62.0	30.0
A36	1	60	100	150	80	152	60	62	60	50	42.5	65.0	31.0
A36	2	60	50	175	60	148	60	>100	60	50	45.0	76.0	--
SAE 1035	1/2	70	--	175	--	147	70	55	70	50	45.3	80.7	23.0
SAE 1035	1	55	--	135	--	155	70	52	70	30	39.7	76.2	25.0
SAE 1035	2	50	--	0	--	148	125	> 60	60	50	44.3	89.7	--
A242	1/2	- 65	--	200	--	134	50	>120	50	40	53.9	73.5	26.0
A242	1	25	--	200	--	138	50	> 90	50	30	50.9	74.8	23.0
A242	2	35	--	150	--	148	60	88	50	40	45.0	72.0	--
A440	1/2	- 25	0	125	100	121	50	65	90	50	62.6	83.2	23.0
A440	1	0	50	135	60	137	60	> 80	50	40	51.8	78.8	28.0
A440	2	70	30	210	130	121	120	> 60	62	40	62.5	82.0	--
A441	1/2	- 15	60	115	45	130	40	100	60	40	56.7	82.3	27.2
A441	1	- 10	65	60	40	130	70	>100	40	40	55.9	87.0	--
A441	2	40	80	65	40	130	90	> 80	40	40	55.0	94.0	29.0
A558-B	1/2	- 10	50	160	70	112	90	80	80	40	68.5	94.0	20.0
A558-B	1	- 75	100	--	75	111	75	95	45	30	69.1	80.5	20.0
A558-B	2	- 15	115	85	135	121	60	90	43	30	62.5	87.0	--
A514-M	1/2	-150	--	110	0	24	50	35	--	--	127.0	131.5	--
A514-P	1	-100	-145	200	0	52	125	33	--	--	108.0	121.8	--
A514-M	2	-125	- 50	95	105	55	75	58	--	--	106.0	117.0	--

TABLE 4 (continued)

T_{15} — Temperature corresponding to the 15 ft.-lb. energy level from a test of standard CVN specimens

T_{15}^{P} — Temperature correspondong to the 15 ft.-lb. energy level from a test of pre-cracked CVN specimens

ΔT — Temperature shift at 15 ft.-lb. level between dynamic and slow bend (.02 in./min. crosshead speed) standard CVN specimens

ΔT^{P} — Temperature shift at 15 ft.-lb. level between dynamic and slow bend (.02 in./min. crosshead speed) pre-cracked CVN specimens

ΔT_{B} — $(215 - 1.5 \, \sigma_y)$ where σ_y is the room temperature yield strength

$T_{15}-T_5$ — Difference in temperature going from 5 to the 15 ft.-lb. energy level in a standard CVN test

CVN_{SH} — Upper shelf Charpy energy

K_{ID}^{15} — K_{ID} at the temperature corresponding to the 15 ft.-lb. level in a standard CVN test

K_{ID}^{*} — Apparent K_{ID} lower shelf

σ_y — Yield strength

σ_u — Ultimate strength

Phase 3 - Phase 3 was a study of the applicability to bridge steel
    weldments of the techniques developed in Phase 1.

In this section of the chapter a summary of the Phase 1 and 2 CVN
and K_c test results will be given.

The following CVN and K_c test program was carried out as a
result of Phase 1 of the project:

1. Standard and precracked V-notch Charpy tests were conducted
    over a broad range of temperatures for most materials and
    thicknesses. The tests were generally conducted at three
    loading rates, impact, very slow and intermediate. For
    the slow and intermediate tests, the tests were performed
    on an Instron Test Machine at crosshead speeds of 0.02
    in/min. (8×10^{-4} mm/min.) and 2.0 in/min. (8×10^{-2}
    mm/min.), respectively.

2. Dynamic and Slow Bend (1 sec to maximum load) K_c values
    were measured for most materials and thicknesses over a
    wide range of temperature.

Details of the test procedures can be found in reference [36].

In general it was found for a given material and thickness
that the standard CVN or precracked Charpy data took the form
shown in Figure 23. Figure 23 schematically shows that as the
testing speed is increased the transition temperature and the upper
shelf energy tend to increase. At the lower temperature ranges the
curves merge to a low energy level usually below 5 ft.-lbs. (7 J).
Some of the more important values relative to the Charpy data are
summarized in Table 4. It can also be stated in general that the
precracked Charpy results for a given material, thickness and testing speed fall to the right of the standard Charpy results. This

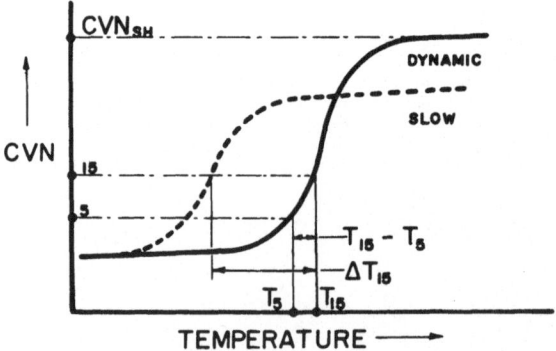

Figure 23. Typical static and dynamic CVN behavior.

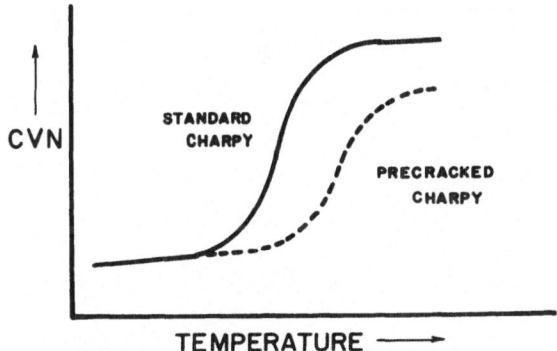

Figure 24. Typical standard and precracked CVN behavior.

is schematically shown in Figure 24. The upper shelf energy levels
were lower for the precracked specimens. The effect of material
thickness for a given test type, precracked or standard and testing
speed are shown in Figure 25.

In general it was found for the K_c tests that the results for
a given material and thickness took the form shown in Figure 19,
where it is shown that the dynamic K_c curve falls to the right of
the static curve. It should be noted, in the overall test program,
that greater attention was paid to dynamic K_c measurements than the
static K_c measurements. Figures 26 through 29 present the K_c data
for the 1/2" (12.7 mm) and 1" (25.4 mm) tests of the A440 and A588
materials. These results are typical of the behavior of the other
material in the program. Figures 30 through 36 show the trends for
the materials as found in Reference [36]. Various levels of K are

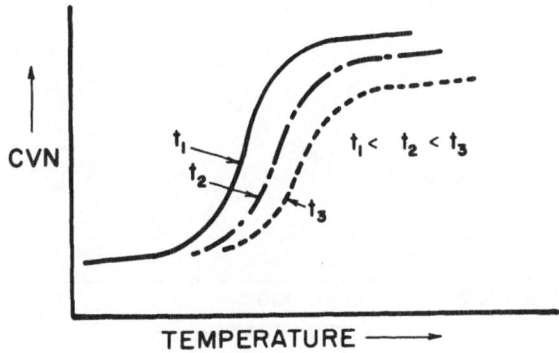

Figure 25. Typical thickness effect on CVN behavior.

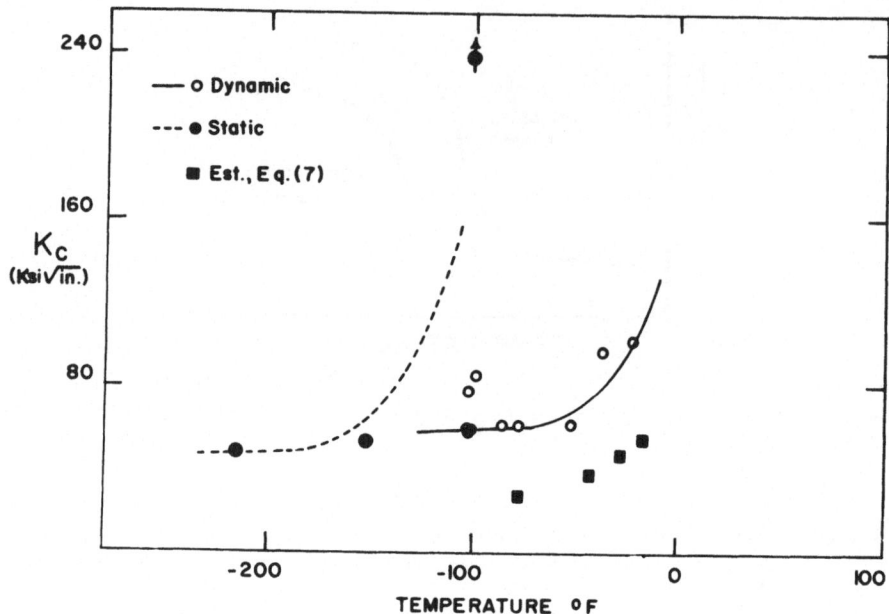

Figure 26.   1/2" (12.7 mm) K_c behavior for A440 steel (1 ksi$\sqrt{in}$ = 1.099 MN/m^2).

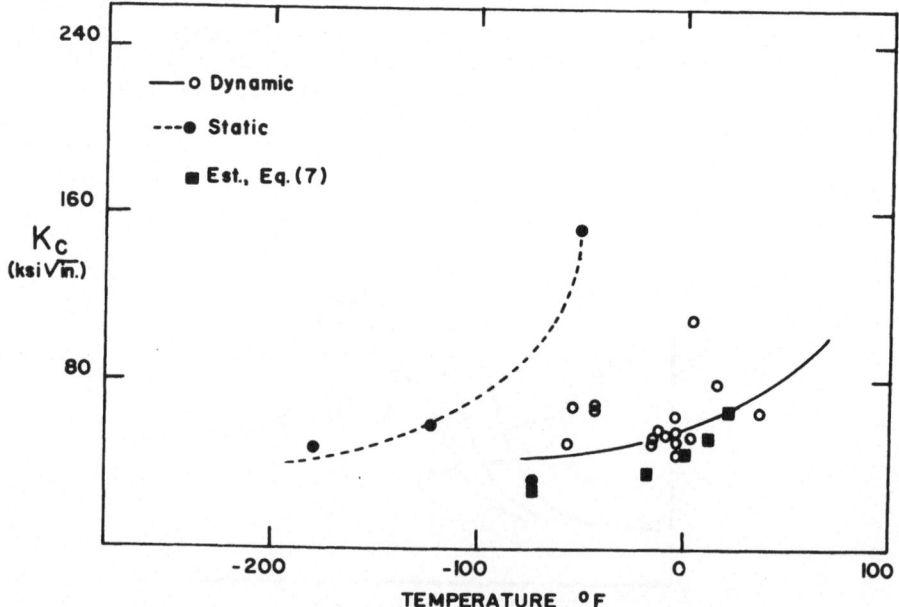

Figure 27.   1" (25.4 mm) K_c behavior for A440 steel (1 ksi$\sqrt{in}$ = 1.099 MN/m^2).

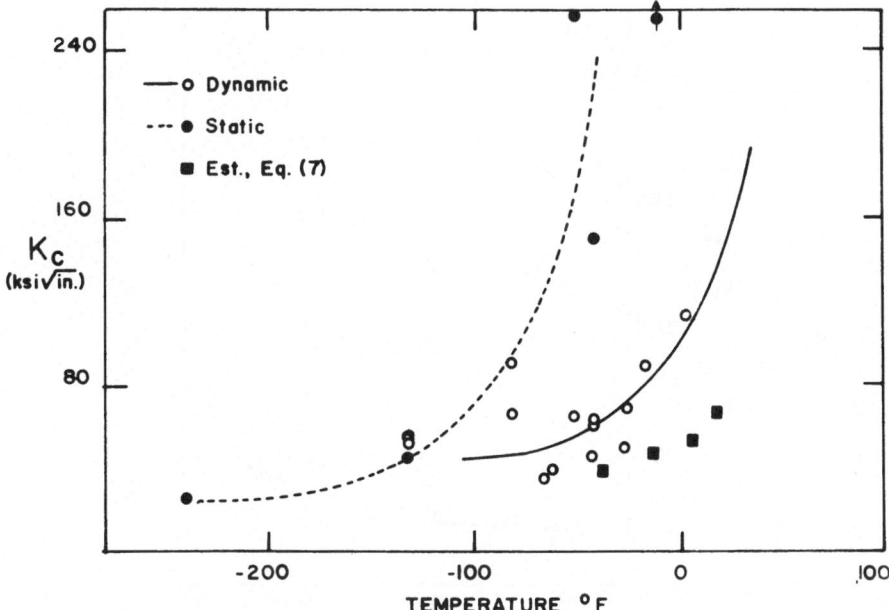

Figure 28.   1/2" (12.7 mm) K_c behavior for A588 steel (1 ksi$\sqrt{in}$ = 1.099 MN/m^2).

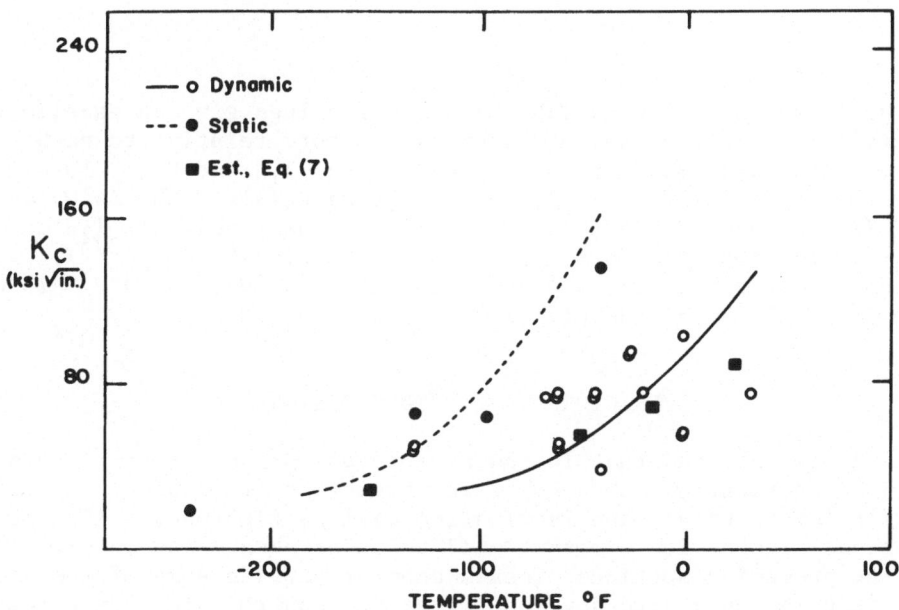

Figure 29.   1" (25.4 mm) K_c behavior for A588 steel (1 ksi$\sqrt{in}$ = 1.099 MN/m^2).

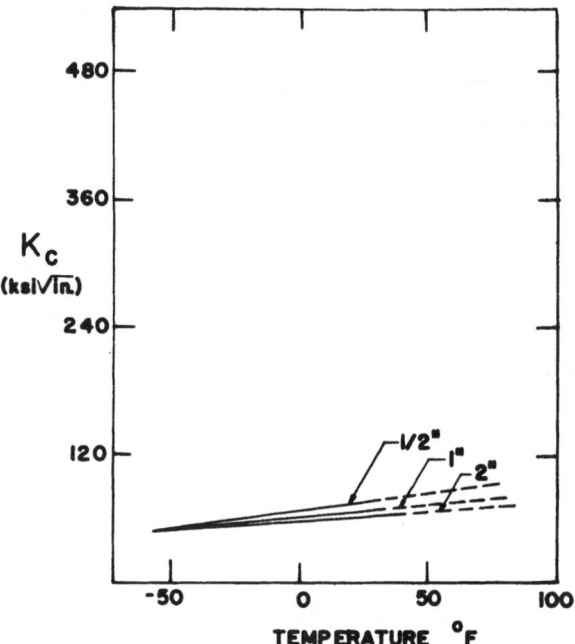

Figure 30.   Composite K_c behavior for A36 steel (1 ksi$\sqrt{in}$ = 1.099 MN/m^2).

given in summary form in Table 1. These values give an excellent overview of how the various materials perform relative to each other. As an example of this, consider the T_{15} and K_{Id}^{15} levels for the 1/2 in. (12.7 mm) A440 and A588 materials. The A440 has 0°F (-15°C) and 90 ksi$\sqrt{in}$ (99 MN/m$^{-3/2}$) respectively for T_{15} and K_{Id}^{15}. The A588 has 50°F (+8°C) and 80 ksi$\sqrt{in}$ (88 MN/m$^{-3/2}$). This indicates that for a given temperature above, say 0°F (-15°C), the 1/2" (12.7 mm) A440 will prove to be tougher.

FRACTURE SAFE BRIDGE DESIGN

It is expected that in the near future all major structures will be fabricated with a specific fracture control plan. Specification for materials and fabrication will be tightened. Also more rigorous inspection programs will be undertaken. The information in the preceding sections of this chapter provide some of the specifics necessary to develop a fracture control plan for bridges. Obviously any such plan must take into account many diverse factors. In spite of this one can take the simplistic view that a fracture control plan is one which provides a structure which will not fail

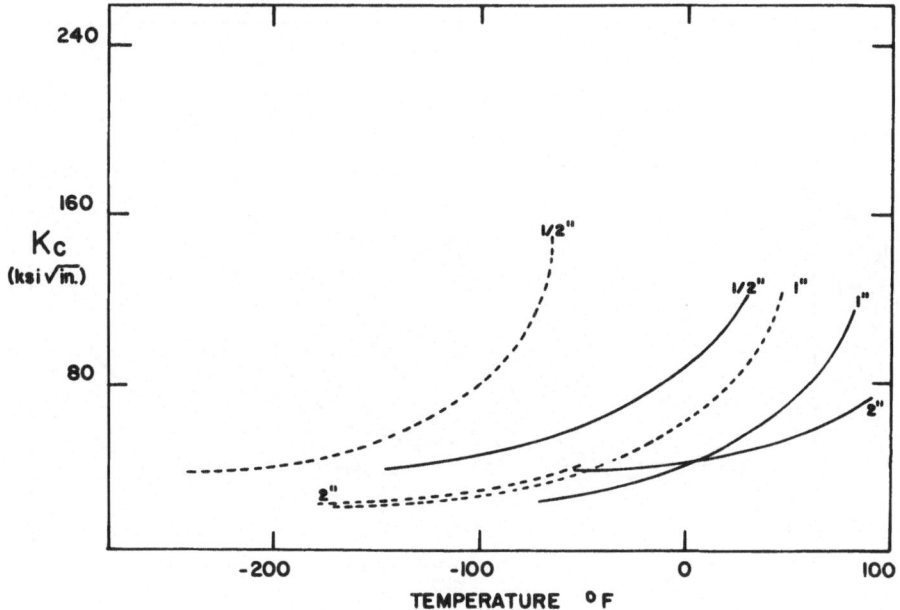

Figure 31. Composite K_c behavior for A242 steel (1 ksi$\sqrt{in}$ = 1.099 MN/m^2).

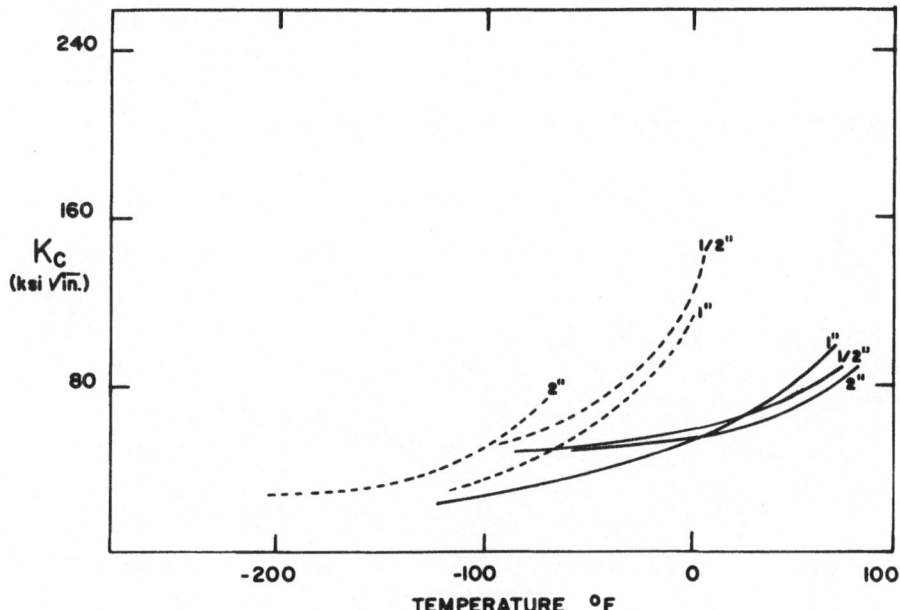

Figure 32. Composite K_c behavior for ASA 1035 steel (1 ksi$\sqrt{in}$ = 1.099 MN/m^2).

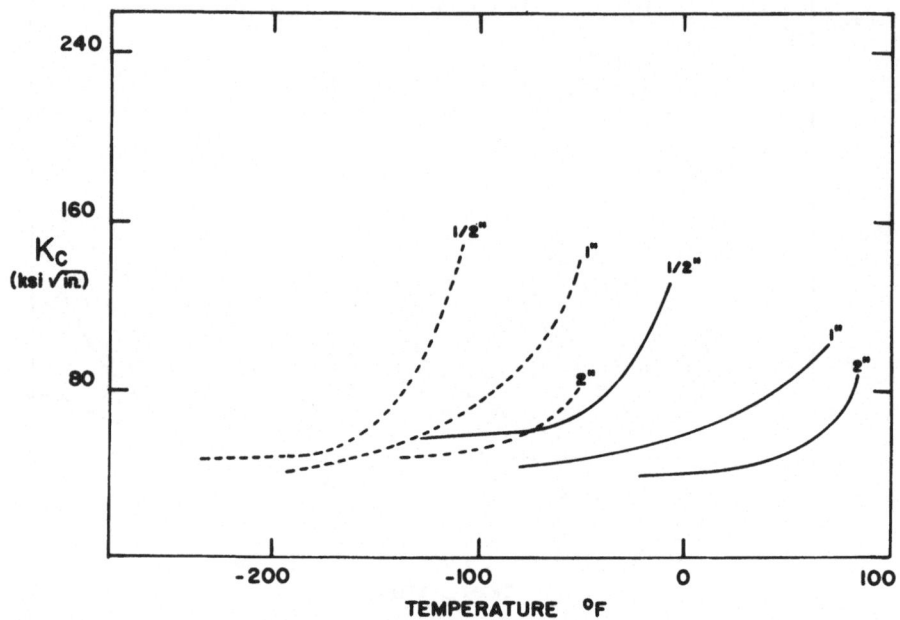

Figure 33. Composite K_c behavior for A440 steel (1 ksi$\sqrt{in}$ = 1.099 MN/m^2).

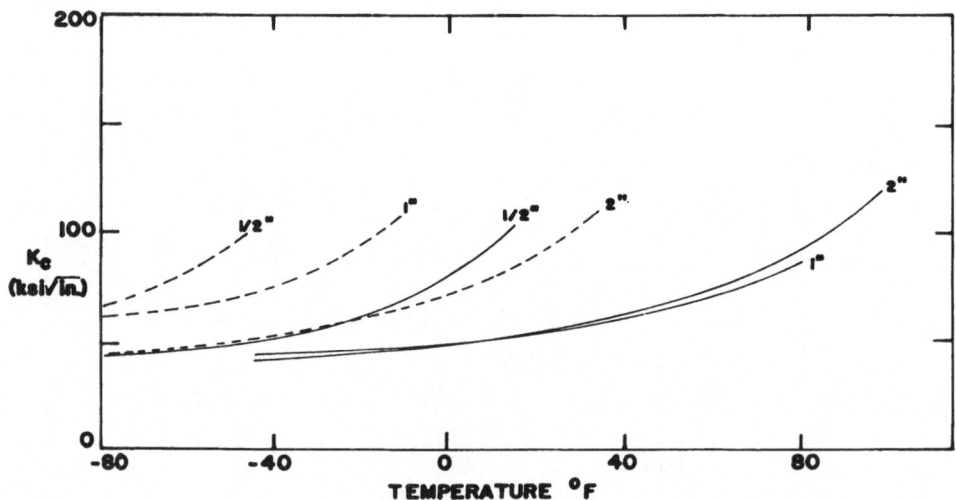

Figure 34. Composite K_c behavior for A441 steel (1 ksi$\sqrt{in}$ = 1.099 MN/m^2).

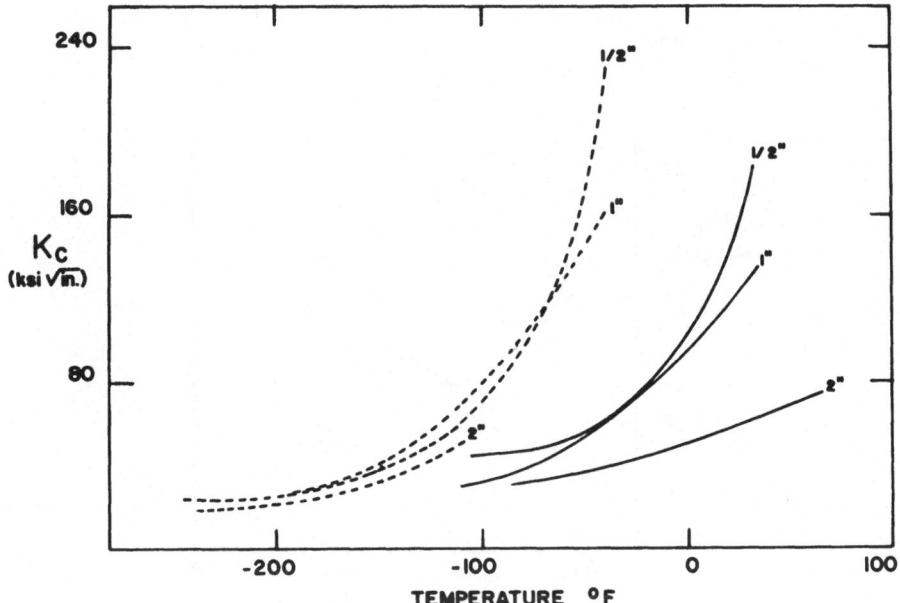

Figure 35.   Composite K_c behavior for A588 steel (1 ksi$\sqrt{\text{in}}$ = 1.099 MN/m^2).

during the design life. To obtain such a goal for a fixed design life, the designer can change only five fundamental items. These are:

1.   Basic structural configuration
2.   Specific detail design
3.   Operating stress levels
4.   Material of construction
5.   Inspection and maintenance program

These five items interact continuously in the design plan. A change of one will affect the others. To more fully appreciate the roles played by the various components they will be examined separately.

Basic Structural Configuration

The basic structural configuration will to a great degree determine the ability of a structure to withstand isolated failures. A structure with many parallel load paths will usually provide fixed displacement loading in the individual members. As one fails it will transfer its burden to the others.

In such systems fatigue cracks will arrest with time. Thus the

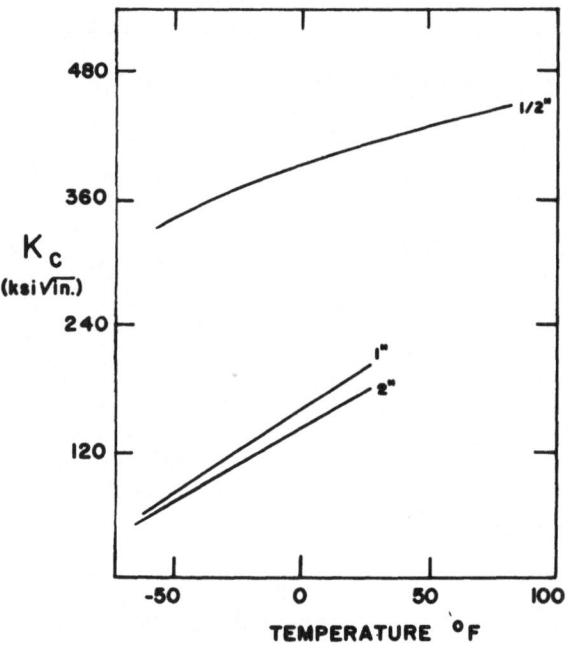

Figure 36.   Composite K_c behavior for A514 steel (1 ksi$\sqrt{in}$ = 1.099 MN/m²).

basic type of structure can be chosen so that many of the components can fail safe.

Specific Detail Design

The design of a specific detail plays a major part in a fracture safe design. Designing with thick sections connected by massive welds can produce situations where constraint and residual stresses are high. This can make a tough steel behave in a brittle fashion. The detail design also directly affects fatigue life as discussed relative to the ASSHTO fatigue rules [29].

Operating Stress Level

The operating stress levels will affect total fatigue life and critical crack size for a specific material. Probably reducing the cyclic stress has more to do with safe design than any other factor. Life is proportional to the inverse cube of stress range [29].

Materials of Construction

The choice of materials of construction will greatly affect toughness and only slightly affect cyclic life. However, the actual choice of toughness is a difficult decision. The choice will depend on how well the maximum stresses are known and what type of inspection and maintenance is planned for the structure. As pointed out, strain rate, temperature, and plate thickness affect toughness and their effect must be accounted for.

Inspection and Maintenance Program

The initial inspection and maintenance program for a structure will affect fatigue life and final crack size at failure. Improper inspection can eliminate the crack initiation portion of fatigue life which is felt to be significant by some researchers [38]. Better maintenance and periodic inspection will uncover potential failures and allow for timely repairs to be made.

Recommendations for Future Research

Although not necessarily justified by the material presented in this chapter, it is this author's opinion that the following research areas would be useful to fracture safe bridge design:

1. Crack initiation behavior of bridge steels
2. Crack initiation behavior of welded bridge details
3. Effect of warm prestressing or proof loading on subsequent cold fracture behavior of bridge steels and weldments
4. Inspection methods
5. Non plane strain fracture mechanics
6. Development of a full thickness quality control test to replace the CVN specimen for material quality control purposes.

SUMMARY

In summary, a great deal of information has been presented in this chapter. First, the failures of a number of bridges were briefly examined. These failures pointed to the potential of bridge failure due to poor detail design, poor welding procedures, poor inspection, misunderstanding design loads and environment, and material with inadequate toughness.

The fatigue properties of bridge steels were reviewed and the AASHTO fatigue design rules were seen to depend on stress range. Above the endurance limit fatigue life of various welded details was shown to be proportional to the inverse cube of stress range.

The fracture behavior of bridge steels was shown to be dependent on strain rate, temperature and plate thickness. The AASHTO requirements for toughness were examined and seen to be a means of assuring a minimum level of fracture toughness. The intent of the AASHTO toughness requirement was to produce non plane strain fracture behavior at the minimum design temperature for intermediate rates of loading. A summary of the toughness behavior of eight bridge steels tested at Lehigh University was given.

Finally, it was indicated that a fracture safe design will depend in an interactive way on the basic structural configuration, specific detail designs, operating stress level, materials of construction, and inspection and maintenance. Areas for future research were indicated as crack initiation behavior in bridge steels and bridge steel weldments, warm prestressing and fracture behavior of bridge steels, inspection methods, non plane strain fracture mechanics and new full thickness measurements techniques of fracture toughness to replace CVN testing for material quality control purposes.

ACKNOWLEDGEMENTS

The author would like to acknowledge the support of the Office of Research, Federal Highway Administration, U.S. Department of Transportation through contract DOT-FH-11-7664 for parts of the work reported here. The contents of this chapter reflect the views of the author at Lehigh University who is responsible for the facts and the accuracy of the data presented herein. The contents do not necessarily reflect the official views or policy of the Department of Transportation. This chapter does not constitute a standard specification or regulation.

The author would also like to acknowledge the many fruitful conversations with his colleagues at Lehigh, Dr. G. Irwin and Dr. J. Fisher, dealing with bridge design and fracture. A dialogue on bridge steel over the past few years with Dr. J. Barsom of the U.S. Steel Corporation Research Laboratory has proved invaluable to the author.

REFERENCES

1. "Standard Test Method for Plane-Strain Fracture Toughness of
    Metallic Materials", Designation: E399-74 in 1975 Annual Book
    of ASTM Standards, Pt. 10. Philadelphia: Am. Soc. for Testing
    and Materials (1975), 561-80.

2. Madison, R.B. and Irwin, G.I., "Fracture Analysis of King's
    Bridge, Melbourne", Proc. ASCE, J. Struct. Div., 97, no. ST9
    (1971), 2229-44.

3. Shank, M.E., "A Critical Survey of Brittle Failure in Carbon
    Plate Steel Structures Other than Ships", Weld Res. Counc.
    Bull., No. 17, 1954.

4. Report of the Royal Commission into Failure of King's Bridge,
    Victoria, Australia, 1963.

5. "Collapse of U.S. 35 Highway Bridge, Point Pleasant, West
    Virginia, December 15, 1967", National Transportation Report
    No. NTSB-HAR-71-1.

6. Engineering News Record, August 20, 1970.

7. Engineering News Record, January 7, 1971.

8. Engineering News Record, March 30, 1972.

9. Czyzewski, H., "Brittle Failure: The Story of a Bridge", Metal
    Progr. (West), 1, no. 1 (1975), W6-W12.

10. Philadelphia Bulletin, November 9, 1972.

11. Engineering News Record, October 24, 1974.

12. Barsom, J.M., "Fatigue Behavior of Pressure-Vessels Steels",
    Weld. Res. Counc. Bull., No. 194, 1974.

13. Paris, P.C., Gomez, M.P. and Anderson, W.E., "A Rational
    Analytic Theory of Fatigue", Trend. Eng., Wash. Univ., 13,
    no. 1 (1961), 9-14.

14. Smith, H.R., Piper, D.E. and Downey, F.K., "A Study of Stress-
    Corrosion Cracking by Wedge-Force Loading", Eng. Fract. Mech.,
    1 (1968), 123-28.

15. Paris, P.C., "The Fracture Mechanics Approach to Fatigue", in
    Fatigue - An Interdisciplinary Approach, ed. by J.J. Burke,
    N.L. Reed and V. Weiss. Syracuse, N.Y.: Syracuse University
    Press (1964), 107-32.

16. Schijve, J., "Significance of Fatigue Cracks in Micro-Range and Macro-Range", in Fatigue Crack Propagation, Special Tech. Publ. 415. Philadelphia: Am. Soc. for Testing and Materials (1967), 415-57.

17. von Euw, E.F.J., Hertzberg, R.W. and Roberts, R., "Delay Effects in Fatigue Crack Propagation", in Stress Analysis and Growth of Cracks, Special Tech. Publ. 513. Philadelphia: Am. Soc. for Testing and Materials (1972), 230-59.

18. Trebules, V.W., Jr., Roberts, R. and Hertzberg, R.W., "Effect of Multiple Overloads on Fatigue Crack Propagation in 2024-T3 Aluminum Alloy", in Progress in Flaw Growth and Fracture Toughness Testing, Special Tech. Publ. 536. Philadelphia: Am. Soc. for Testing and Materials (1973), 115-46.

19. Mills, W.J., "Load Interaction Effects on Fatigue Crack Growth in 2024-T3 Aluminum Alloy and A514F Steel Alloys", unpublished Ph.D. dissertation, Lehigh University, 1975.

20. Schijve, J. and De Rijk, P., "The Effect of 'Ground-to-Air-Cycles' on the Fatigue Crack Propagation of 2024-T3 Alclad Sheet Material", National Aero- and Astronautical Research Institute, Amsterdam, Netherlands, Report No. NLR-TR-M-2148, July 1966. (N66-39867)

21. Effects of Environment and Complex Load History on Fatigue Life, Special Tech. Publ. 462. Philadelphia: Am. Soc. for Testing and Materials, 1970.

22. Brown, B.F., "A New Stress-Corrosion Cracking Test for High-Strength Alloys", Mater. Res. Stand., 6 (1966), 129-33.

23. Novak, S.R. and Rolfe, S.T., "Modified WOL Specimen for K_{Iscc} Environmental Testing", J. Mater., 4 (1969), 701-28.

24. Wei, R.P. and Landes, J.D., "Correlation Between Sustained-Load and Fatigue Crack Growth in High-Strength Steels", Mater. Res. Stand., 9, no. 7 (1969), 25-27.

25. Barsom, J.M., "Corrosion-Fatigue Crack Propagation Below K_{Iscc}", Eng. Fract. Mech., 3 (1971), 15-25.

26. Sinclair, G.M., "Relation of Sub-Critical Crack Growth to Inspection Requirements", paper presented at ASM Conference on Fracture Control, Philadelphia, Pa., January 26-28, 1970.

27. Carter, C.S., Hyatt, M.V. and Cotton, J.E., "Stress-Corrosion
     Susceptibility of Highway Bridge Construction Steels", Boeing
     Company, Renton, Washington, Department of Transportation
     Contract Report No. FHWA-RD-73-46, April 1972. (PB 222 453)

28. Barsom, J.M. and McNicol, R.C., "Effect of Stress Concentration
     on Fatigue-Crack Initiation in HY-130 Steel", in Fracture Tough-
     ness and Slow-Stable Cracking, Special Tech. Publ. 559.
     Philadelphia: Am. Soc. for Testing and Materials (1974), 183-
     204.

29. Fisher, J.W., "Guide to 1974 AASHTO Fatigue Specifications",
     Am. Inst. of Steel Construction, 1974.

30. Fisher, J.W., Frank K.H., Hirt, M.A. and McNamee, B.M., "Effect
     of Weldments on the Fatigue Strength of Steel Beams", NCHRP
     Report 102, Highway Research Board, 1970.

31. Fisher, J.W., Albrecht, P.A., Yen, B.T., Klingerman, D.J. and
     McNamee, B.M., "Fatigue Strength of Steel Beams with Welded
     Stiffeners and Attachments", NCHRP Report 147, Highway Research
     Board, 1974.

32. Miner, M.A., "Estimation of Fatigue Life with Particular Em-
     phasis on Cumulative Damage", in Metal Fatigue, ed. by G. Sines
     and J.L. Waisman. New York: McGraw-Hill Book Company (1959),
     278-89.

33. Albrecht, P. and Fisher, J.W., "An Engineering Analysis of
     Crack Growth at Transverse Stiffeners", Int. Assoc. Bridge
     Struct. Eng. Publ., 35, Pt. I (1975), 1-22.

34. Hirt, M.A. and Fisher, J.W., "Fatigue Crack Growth in Welded
     Beams", Eng. Fract. Mech., 5 (1973), 415-29.

35. Barsom, J.M., "Toughness Criteria for Bridged Steels", Tech.
     Report No. 5 for AISI Project 168, February 1973.

36. Roberts, R., Irwin, G.R., Krishna, G.V. and Yen, B.T., "Fracture
     Toughness of Bridge Steels - Phase II Report", Lehigh University,
     Bethlehem, Pa., Dept. of Transportation Contract Report No.
     FHWA-RD-74-59, September 1974. (PB 239 188)

37. Barsom, J.M. and Rolfe, S.T., "Correlation Between K_{Ic} and
     Charpy V-Notch Test Results in the Transition-Temperature
     Range", in Impact Testing of Metals, Special Tech. Publ. 466.
     Philadelphia: Am. Soc. for Testing and Materials (1970),
     281-302.

38.   Barsom, J.M., "The Development of AASHTO Fracture-Toughness
      Requirements for Bridge Steels", paper presented at the U.S.-
      Japan Cooperative Science Seminar, Tohoku University, Sendia,
      Japan, August 1974. (Available from AISI)

39.   Frank, K.H. and Galambos, C.F., "Application of Fracture
      Mechanics to Analysis of Bridge Failures", in Safety and
      Reliability of Metal Structures. New York: Am. Soc. of Civil
      Engineers (1972), 279-306.

40.   Rolfe, S.T., "Fracture-Control Guidelines for Welded Steel Ship
      Hulls", in Significance of Defects in Welded Structures, ed.
      by F. Kanazawa and A.S. Kobayashi. Tokyo: University of
      Tokyo Press (1974)., 318-39.

41.   Eiber, R.J., Duffy, A.R. and McClure, G.M., "Fracture Control
      on Gas Transmission Pipelines", paper presented at ASM Conf.
      on Fracture Control, Philadelphia, Pa., January 26-28, 1970.

42.   Highway Research Board, the AASHTO Road Test, Report 4, Bridge
      Research, Special Report CID, National Academy of Science -
      National Research Council, Publication No. 953.

APPLICATION OF FRACTURE MECHANICS TO PREVENTION AND CONTROL OF
SUBCRITICAL CRACK GROWTH AND FRACTURE IN ADVANCED HIGH-PERFORMANCE
SHIP STRUCTURES

R. J. Goode and R. W. Judy, Jr.

Naval Research Laboratory, Washington, D. C.

ABSTRACT

Advanced surface craft and ships such as hydrofoils and
surface-effect ships represent one area in the ship structures
design field where fracture mechanics must be applied for prevention
of fracture and control of crack growth. These vehicles feature the
unique problem of weight-critical, monolithic structures fabricated
of high-strength metals operating in an aggressive environment at
exceptionally high levels of performance.

The integrity of high-performance surface craft and ships depends on both the resistance to propagation of cracks and resistance
to fracture of the structural material. Design procedures based on
engineering application of fracture mechanics principles to assure
structural integrity have been established. These procedures enable
designers to systematically take into account applicable metal crack
tolerance parameters and their relation to structural performance.
The three-part Ratio Analysis Diagram (RAD) system, developed to
provide an analysis technique for determining the significance of
stress-corrosion cracking, sustained load cracking, and fracture in
terms of critical flaw size and stress level, is presented. Considerations on the effects of electrochemical coupling on corrosion
fatigue and stress-corrosion cracking are also discussed.

INTRODUCTION

The use of high-strength metals for critical components in
advanced high-performance ship structures such as surface-effect

ships and hydrofoils requires the designer to account for potential
crack growth and fracture. The fracture problem is more simple
than that of crack growth, in that several methods of designing to
prevent catastrophic failure exist and are being used. The most
direct means of fracture prevention is the use of a material which
will withstand excessively high loading at sharp crack locations
by yielding without rapid crack propagation; this method is often
criticized as overly conservative, but in many cases it is the only
practical approach for large complex structures. Crack growth is
more insidious than fracture, because subcritical crack growth in
a structure designed to be fracture-safe necessitates extensive
maintenance and repair, which can be as severe a problem as fracture.
Since these structures operate in seawater, prevention of crack
growth, as well as general corrosion, is a primary design factor.

     There are many micromechanisms by which cracks grow; however,
various aspects of macroscopic crack growth are most conveniently
classified according to loading and environmental aspects: stress-
corrosion cracking (SCC) - sustained load plus environmental
factors; sustained-load cracking (SLC) - no aggressive or unfavor-
able environment; fatigue - cycle load; and corrosion fatigue -
cyclic load plus environment. This chapter describes methods used
to analyze the significance of SCC, SLC, and corrosion-fatigue
properties of high-strength metals, as defined by fracture mechanics,
for structural applications which require prevention of subcritical
crack growth.

PROCEDURES FOR MEASURING CRACK-GROWTH RESISTANCE

     The utilization of linear-elastic fracture mechanics for de-
fining the resistance of structural metals to crack growth in
aqueous environments is well documented and is advanced to the point
where standardization of test procedures is imminent. As illustrated
in Figure 1, there are three phases of failure in structural components or in laboratory test specimens:

     1. Formation of a small pit by corrosion process
     2. Formation and rapid growth of a crack
     3. Final failure by fracture processes

The second phase of rapid crack growth is either SCC, when an environmental effect is present, SLC, when such an effect is absent, or
corrosion fatigue. In most structures the presence of small cracks
and other defects which initiate rapid crack growth eliminates the
necessity for considering the first phase. It is most important to
understand that all metals are not necessarily subject to either
phase 1, phase 2, or both.

     For several years, data have been reported for a variety of

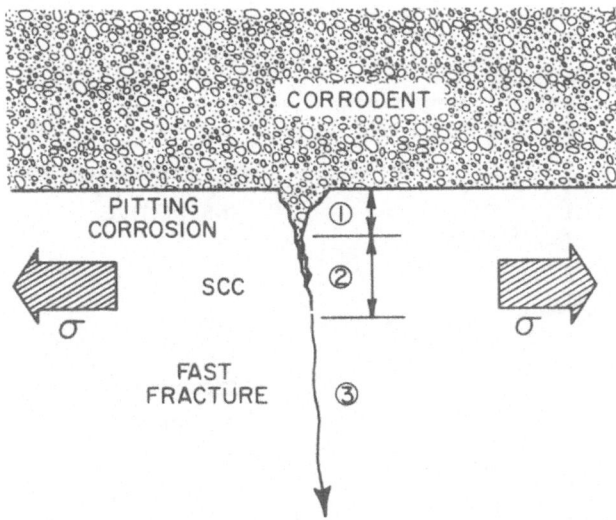

Figure 1.   Sequence of events in structural failure due to environmental effects. A given metal may not be subject either
to pitting or SCC or may be sensitive to both.

structural metals defining their resistance to the initiation of
SCC in terms of the threshold level of stress-liability, K_{Iscc}
[1-7]. The use of linear-elastic fracture mechanics methods that
were developed for describing brittle fracture to measure the
initiation of crack growth in metals has given structural designers
a potential for implementing "safe life" or other design concepts
based on crack-growth laws. Before the available data can be
incorporated into such design concepts, however, there are three
independent questions which must be considered:

1.   Is crack growth definitely involved in the reported result,
or does the data represent the result of a mechanical test conducted
in a wet environment?

2.   Are the reported data valid with respect to minimum
dimensions of the specimen so that the values are independent of
geometry effects?

3.   What do the reported data imply for the structure being
designed?

The objective of an SCC test is to measure the threshold
stress-intensity value K_{Iscc}, above which crack growth will definitely
occur [1]. This is done by testing fatigue-cracked specimens and
applying the calibrations of K_I developed for fracture toughness
testing; because of the physical similarities between test specimens,

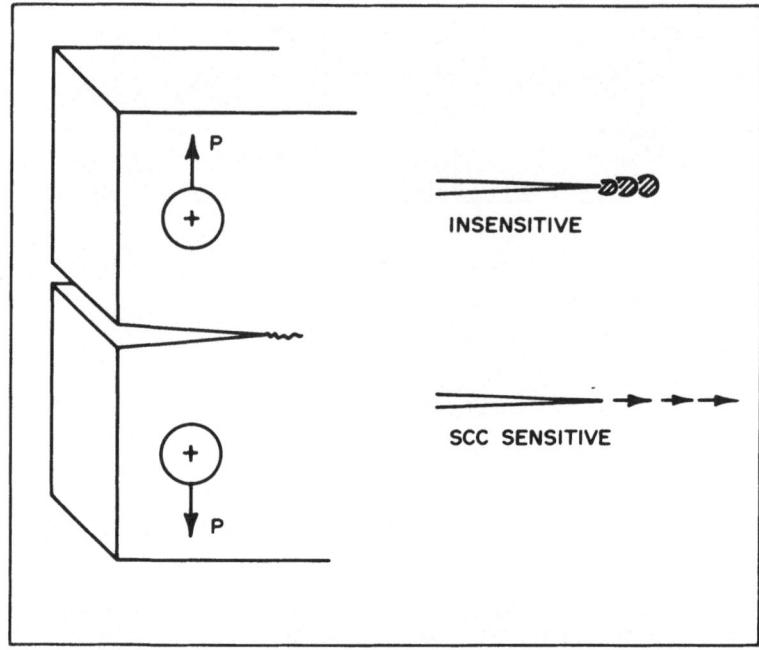

Figure 2.   Schematic illustration of laboratory tests for stress-
            corrosion cracking. The intent of such tests is to
            measure the sensitivity of materials to crack propagation.
            By SCC mechanisms.

the requirement for crack growth is often forgotten or ignored.
Figure 2 shows schematically the configuration of one type of test
specimen and illustrates the difference between the "insensitive"
case at the top and the "sensitive" case at the bottom. Tests
involving no crack growth constitute long-term fracture tests conducted in a wet environment; the results of such tests are not
significant for design to prevent subcritical crack growth.

    The validity of the threshold concept using stress-intensity
parameters for defining the resistance to SCC initiation has been
demonstrated [2]. There are two methods for applying fracture mechanics. The initiation method (Figure 3) employs a cantilevered, dead-
weight loading system so that crack growth results in increasing
applied K_I; for this reason a system of bracketing data derived from
specimens with and without crack growth is necessary to define the
threshold K_{Iscc} value [1]. The arrest technique utilizes a bolt-
loaded (constant deflection), modified wedge-opening-loaded specimen
that has a decreasing K field with increasing crack length [2-4].

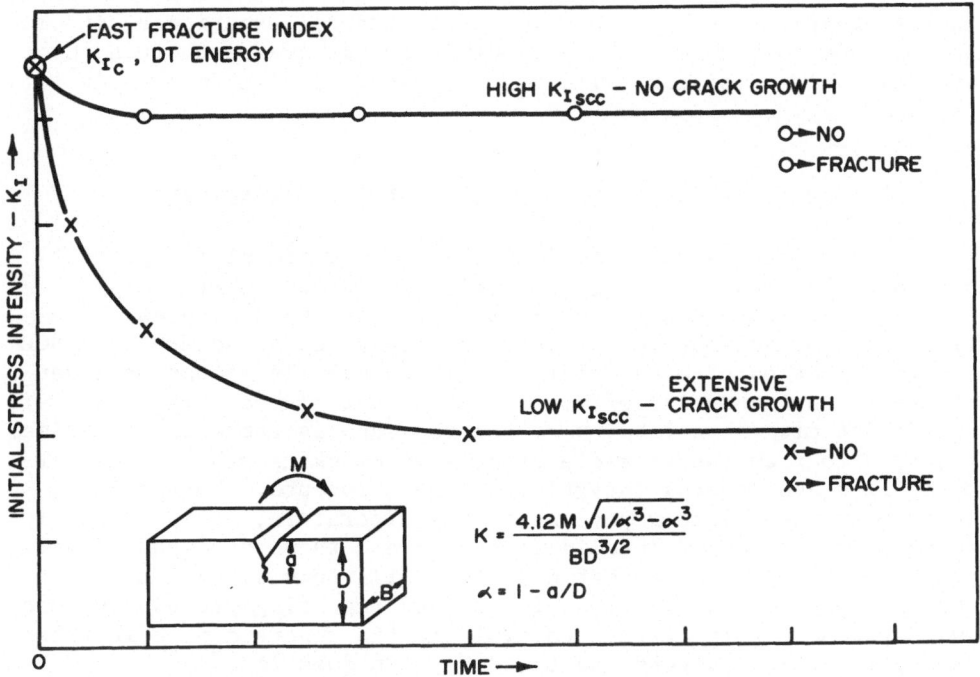

Figure 3.   Graph of K_{Iscc} determined by cantilever test method.
Purpose of test is to determine threshold level for crack
propagation.

In this method, the specimen is loaded to an applied K_I well above
the expected K_{Iscc}, and the crack is allowed to propagate until it
arrests at K_{Iscc}.

In both tests, there are criteria for assuring that the results
are independent of geometrical effects. These take the form of
minimum dimensions of the test specimen; i.e., crack length a and
thickness $B \geq 2.5$ $(K_I/\sigma_{ys})^2$. These dimensions are recommended until
more definitive results on geometrical effects are available.
Requirements for specimen dimensions and crack growth ensure that
the K_{Iscc} value is characteristic of the material for a specific
environment and that the analysis technology developed for fracture
mechanics can be applied.

It is important to note that validity criteria for specimen
size are not related to crack growth. It is certain that crack
growth due to environmental effects can be a structural problem
in cases where a completely valid K_{Iscc} could not be defined. The
fact that valid K_{Iscc} values could not be measured does not alleviate

the crack-growth problem; however, from general experience, it can
be observed that severe crack-growth problems are most often associated with dimensionally valid K_{Iscc} values, as will be seen in later
sections.

ANALYSIS PROCEDURES FOR SCC, SLC, AND FRACTURE

There are no direct implications for structural design that
are uniquely related to K_{Iscc}, as with other parameters, K_{Iscc}
must be evaluated in the context of other design variables. The
first line of defense for protection of a given structure from SCC
is the selection of a material that is completely immune or insensitive to environmental effects. Such materials do exist; however,
one is not always at liberty to use them and is therefore sometimes
required to face the possible effects of crack growth on the operating life and general integrity of the structure. Furthermore,
because of incomplete reporting in the literature, it is difficult
to separate valid and invalid test data and those data that involve
crack growth from those that do not. Until this situation is
rectified by standardization of test and reporting procedures, one
must systematically attempt to evaluate the reported data in terms
of the important factors and the design in question.

The procedure for making such analyses is to use existing
relationships of critical stress/flaw conditions calculated from
fracture mechanics equations and a general analysis capability
afforded by the Ratio Analysis Diagram (RAD) [8] with modifications
to include SCC [9]. Since the parameter K_{Iscc} describes the applied
K_I level for the beginning of crack growth, equations such as the
surface-flaw equation and the through-crack equation apply only for
the SCC initiation; nothing is implied regarding crack growth as a
function of time. The available analysis tools are the stress
intensity for SCC initiation and the final crack size for failure
as given by the fracture resistance parameter.

The surface crack equation (Figure 4) can be utilized to
demonstrate the type of analyses that can be made for either fracture
or SCC. Plotting the crack-depth parameter a/Q versus the ratio of
K_{Ic}/σ_{ys} or K_{Iscc}/σ_{ys} for various levels of applied stress reveals
some crucial factors. Figure 4 is divided into regions of high
ratio, low ratio, and intermediate ratio according to the following
rationale: A general inspectability limit for flaws and defects by
the usual methods is considered to be at an a/Q value of approximately 0.2; this translates to actual flaw sizes in the range from
0.18 inch deep x 1.80 long up to 0.28 inch deep x 1.12 long. The
range of K_{Iscc}/σ_{ys} ratios that is typical of the data reported can
be divided into three parts by the ratio values of 0.7 and 0.3.
Thus, the high-ratio region bounded by $a/Q \geq 0.2$ and $K_{Iscc}/\sigma_{ys} \geq$
0.7 is representative of conditions where high stresses and large

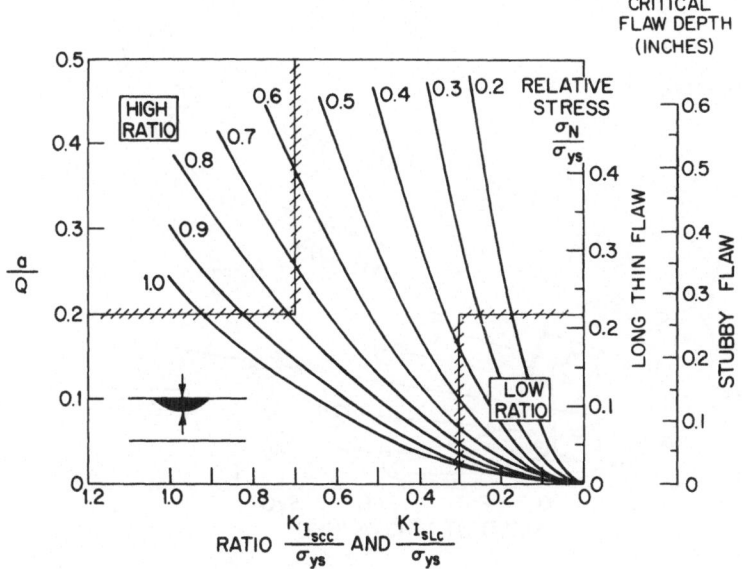

Figure 4.   Fracture-mechanics plot of surface-flaw equation.   The
            regions of "high ratio" and "low ratio" are related to
            limitations on detectability of flaws.

cracks are necessary to cause SCC initiation. Conversely, the low-
ratio region is representative of conditions where SCC crack growth
can initiate from very small defects at moderate to low levels of
stress. The low-ratio region is therefore one to be avoided because
of the present lack of capability to detect cracks of the critical
size. The intermediate, or transitional, region is one where the
combinations of high stresses and small flaws, low stresses and
large flaws, or intermediate stress levels and flaw sizes are
critical; these conditions are more tolerable than those in the
low ratio and constitute the range where highly refined application
of fracture mechanics technology is required for adequate design.

    The significance of the separations is more apparent from the
plot of Figure 5. In this figure, K_{Iscc} data zones for a variety
of steels are plotted against a grid of lines of constant K_{Iscc}/σ_{ys}.
The lines represent a fracture-mechanics-based plot of K_{Iscc} and
σ_{ys} and can be keyed to the flaw-size diagram of Figure 4, or to
other existing analyses. The location of the data points with
respect to the grid lines indicates whether each material is in
the low-, intermediate-, or high-ratio region.

    The dashed line in Figure 5 shows another factor: the transi-
tion in sensitivity to SCC with increases in yield strength. Data

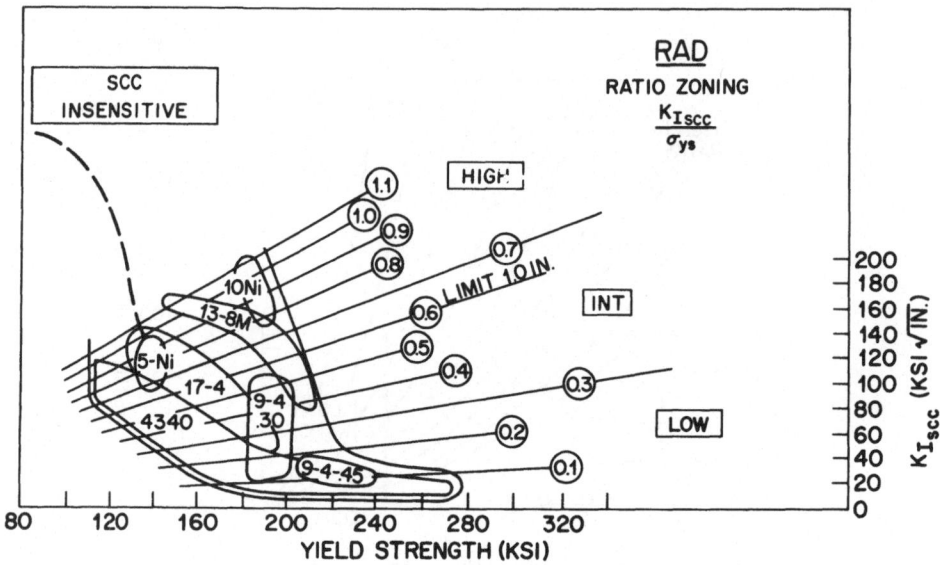

Figure 5. Characteristic data zones for various steels showing
their relation to K_{Iscc}/σ_{ys} ratio lines. These lines
reference combinations of flaw size and stress level for
the onset of SCC as shown in Figure 4. The dashed line
illustrates the SCC sensitivity transition with increasing yield strength. The levels of "high", "intermediate",
and low ratio are referenced to Figure 4.

for materials that are completely insensitive to subcritical crack
propagation are not represented on this SCC plot; however, the
notation of "insensitive" indicates that in the absence of environmental effects, the technology for guarantee of structural integrity
is concerned only with fracture and fatigue crack growth. It has
been observed for steels that a general trend does exist wherein
the yield-strength range beginning at approximately 120 ksi marks
the range where crack-growth problems being to become apparent.
This is somewhat lower than the yield strength for a similar transition in fracture resistance, but the effect is the same. It
appears that the metallurgical principles that are utilized to
increase strength contribute to crack-growth problems at the same
time.

     Plots of the type of Figure 5 can be overlaid on the RAD (as
shown in Figure 6) so that fracture and SCC properties can be
compared. Such comparison is necessary to interpret the severity
of subcritical crack growth on structural integrity. If the final

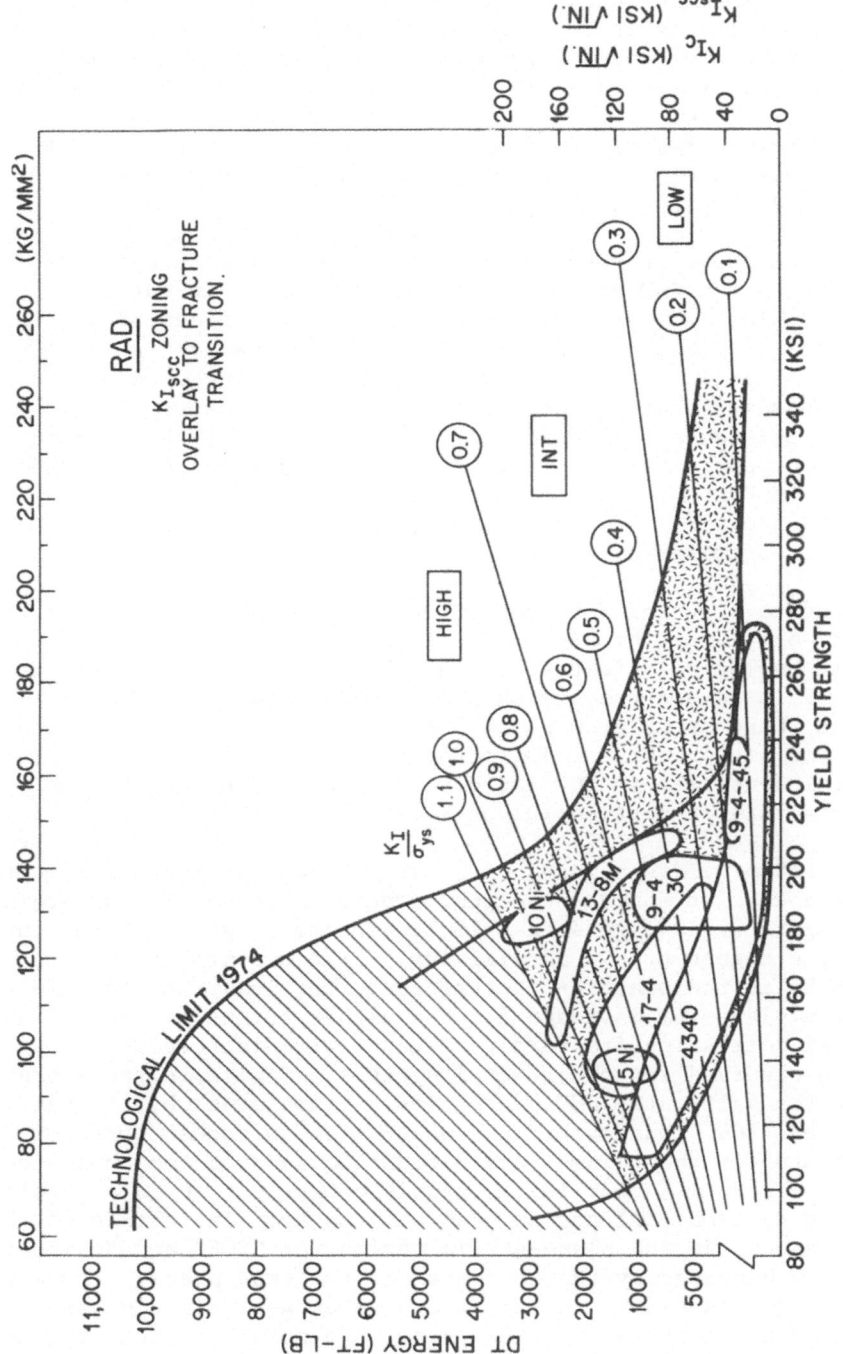

Figure 6. Ratio Analysis Diagram for SCC of various types of steel.

failure is by fracture, the life of the structure is determined by
the size of defect or crack that can be tolerated for a given
loading system; this in turn is defined by the fracture resistance
property. Structures designed with materials having high fracture
resistance have the capability of containing large flaws, so that
the problems of SCC are related to inspectability and maintenance;
however, those structures which are designed with materials of low
fracture resistance might suffer sudden crack extension after a
short period of crack growth.

For the above reasons, the comparisons of fracture and SCC
properties on the RAD are a most important factor in material
selection. The RAD system has two independent indices of fracture
resistance - K_{Ic} and dynamic tear (DT) energy - which apply to
brittle or elastic fracture, and to ductile fracture, respectively.
It is emphasized that K_{Ic} does not give a true measure of the
fracture resistance of ductile materials and cannot be utilized
for this purpose. The plane-strain limit is defined by the ASTM
Committee on Fracture [10] in terms of the thickness B, as
$B \geq 2.5 \ (K_{Ic}/\sigma_{ys})^2$. On the RAD, plane-strain limits for given
thicknesses of material are plotted as K_{Ic}/σ_{ys} ratio lines.

The data plots of SCC properties occupy the lower part of the
RAD (Fig. 6); i.e., the higher fracture resistance levels indicated
by the technological limit (TL) line are not attained for the case
of SCC. This is because the only SCC zones shown are those for
the cases where crack growth was present for linear-elastic loading;
thus by definition these fall in the range of ratio lines. Materials
that are insensitive to SCC would be represented on the RAD only
by the fracture properties (DT or K_{Ic}).

The RAD provides a means of examining several factors simultaneously; the case of a high-strength, precipitation-hardening stainless
steel (Figure 7) is a good example of effects of strength and environment on the degree of SCC sensitivity. Fracture properties are
shown to be in the plastic-fracture range for lower strength levels
and in the plane-strain range for the higher strength levels. Such
effects of decreasing fracture resistance with increasing strength
have been observed for other metal systems. Similar effects are
shown in the salt-water SCC properties of this steel by the data
zones coded "SCC" and "cathodic-couple". These zones represent
K_{Iscc} values for the open-circuit condition and for the condition
of electro-chemical coupling to metals commonly used in cathodic
protection systems - Al and Zn - as well as Mg. The shapes of the
data zones outside the plane-strain region for "SCC" and "cathodic
couple" are dictated by the thickness of the test piece. If very
thick sections should be tested, these zones would likely assume a
position parallel to the fracture zone. In point of fact, the data
zones are plotted with the understanding that crack growth was

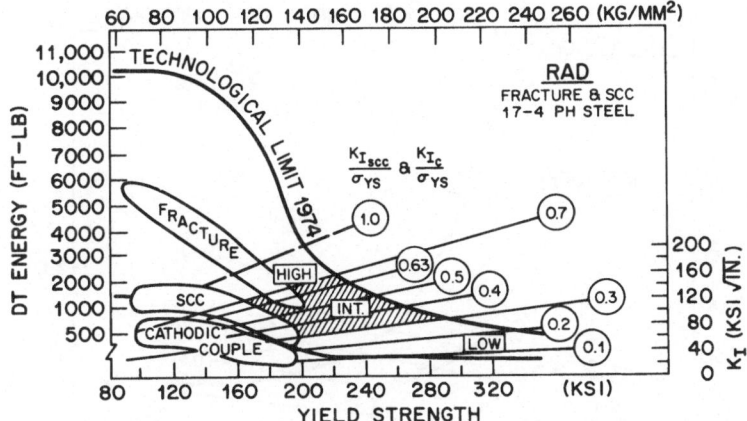

Figure 7.   Ratio Analysis Diagram showing the fracture and SCC
             properties for 17-4 PH stainless steel.

present on the specimens that are dimensionally invalid for the
values below approximately 140-ksi yield strength for SCC and 120-
ksi for "cathodic couple" condition.

     To use this high-strength stainless steel for applications in
a seawater environment, one must perform a very careful analysis of
potential crack growth. The most important single aspect is that
a maximum must be placed on the yield-strength value that will
absolutely preclude brittle fracture. The SCC properties dictate
that a maximum allowable yield strength less than that for fracture
would be necessary to absolutely prevent SCC under linear-eleastic
conditions. It was noted that crack growth in test specimens was
present for all strength levels regardless of the stress state –
linear-elastic or plastic loading. It is also apparent that
cathodic protection systems which depend on sacrificial zinc or
aluminum bars to prevent general corrosion aggravate the SCC problem by lowering the K_{Iscc} value.

     An SCC RAD for titanium alloys including some of the major
alloy families is shown in Figure 8. Data to form this RAD were
taken from the Damage Tolerant Design Handbook [5] and include only
the data which meet the thickness criteria. Whether or not crack
growth was present could not be determined from this listing. Some
data from previous NRL studies are also included. Two of the alloy
systems – 721 (Ti-7Al-2Cb-1Ta) and 811 (Ti-8Al-1Mo-1V) are well
known for sensitivity to crack growth in salt water. The other two
alloys – 6-4 (Ti-6Al-4V) and 6-6-2.5 (Ti-6Al-6V-2.5Sn) – are not
sensitive to salt water in all heat-treated conditions; accordingly,
the data zones may well be representative of either the fracture-
resistance properties or the SLC properties of these alloys.

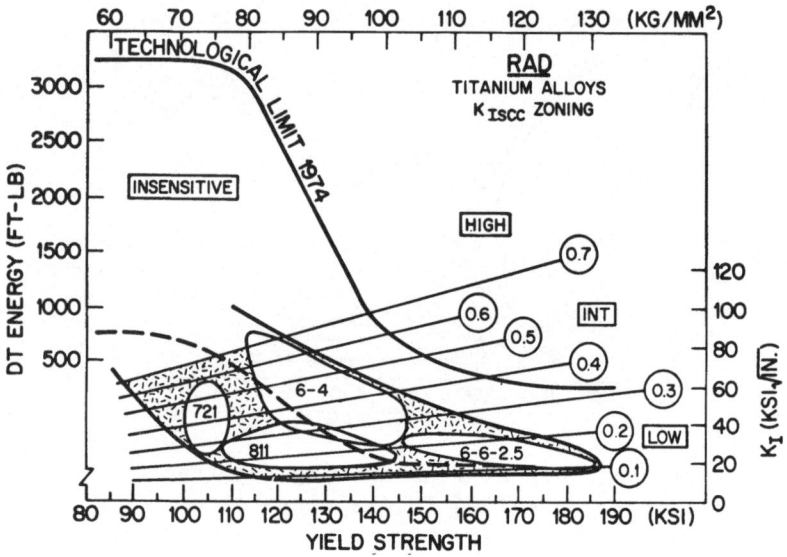

Figure 8. SCC RAD for several titanium alloy families.

        Sustained load cracking is a phenomenon that may affect
titanium alloys to a higher degree than other structural materials.
The failure mechanism is one of crack growth in an air environment
at moderate to high stress levels. Figure 9 summarizes results
for a sample of 6-4 [11] tested in a cantilever-bend configuration
(top) and as a part-through crack (PTC) tension panel (bottom).
The decrease in load-carrying capability for a PTC specimen with a
constant flaw size is indicated in the circles denoted "applied
stress/yield stress". It is noted that the absolute stress levels
would change for different crack sizes, but the decrease would
remain approximately the same.

        Results obtained to date indicate that the SLC phenomenon is
analogous to SCC and can be handled by the same RAD methods as
apply to SCC. The data points in the 6-4 zone of Figure 10 indicate
the SLC properties for selected samples of 6-4. The particular
samples of 6-4 included in this selection encompass a wide range of
processing variables and interstitial oxygen content (which is the
dominant factor in determining fracture resistance). Note that the
data points cover the range of SCC properties measured for the
Ti-6Al-4V system; this may be taken as evidence that the same
phenomenon might be responsible for the crack growth in both cases.

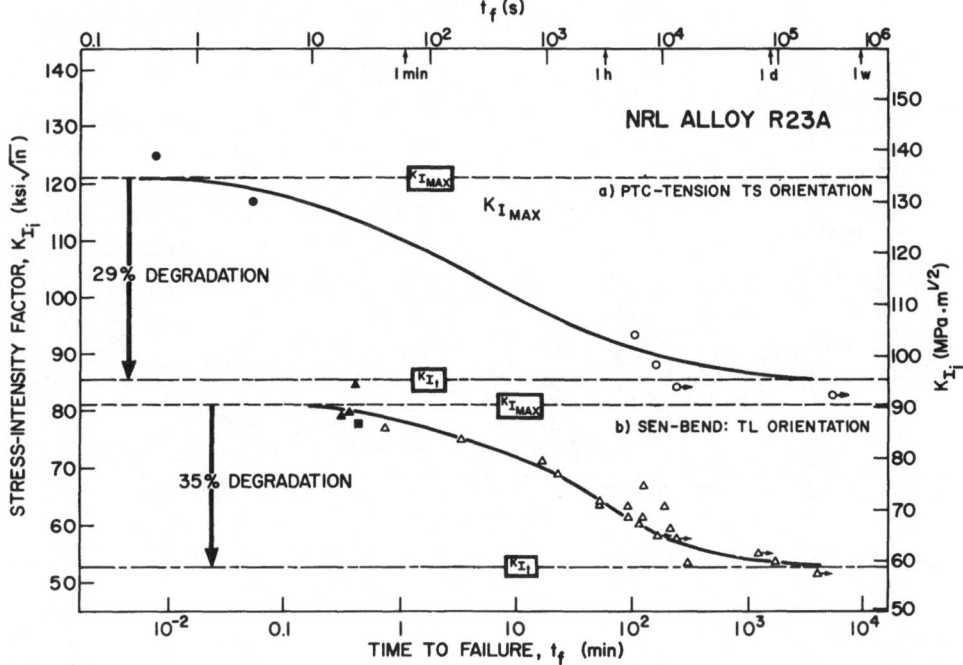

Figure 9. Sustained-load cracking of a commercial grade Ti-6Al-4V
           material in ambient air environment. Part-through
           crack specimens (top) and bend specimens (bottom) were
           used to measure the SLC properties for two crack-plane
           orientations.

Therefore crack growth in this alloy may not be caused by the salt-
water environment, although the K_{Iscc} values are completely valid
with respect to dimensions and crack-growth evidence.

CORROSION-FATIGUE ASPECTS

The application of fracture mechanics principles to fatigue
has permitted researchers to study crack propagation under strictly
defined crack tip mechanical conditions which duplicate structural
service. An important dimension to service simulation has been
added by conducting such studies in flowing seawater under freely-
corroding conditions and electro-chemically coupled (or potentiostated) to the potential of zinc or aluminum. This added degree
of service simulation is of particular importance in the characterization of new alloys being developed for application in critical,
high-performance structures.

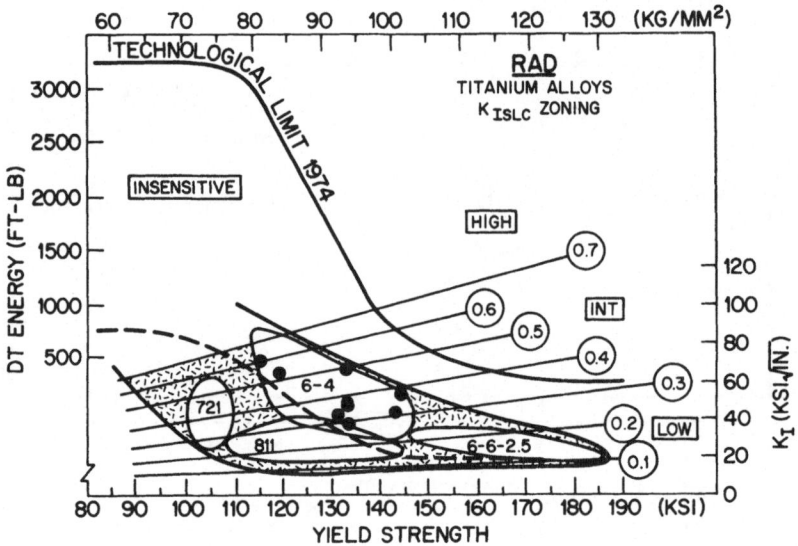

Figure 10.   RAD comparison of SLC properties of Ti-6Al-4V alloys
with the data zone for salt water SCC.

Most structural metals display a significant acceleration in
fatigue crack growth in the presence of a seawater environment.
The effects of corrosion fatigue tend to be greater in high-
strength materials, especially if coupled with stress-corrosion
cracking, thus can pose long-term problems to the reliability and
life-cycle costs of advanced high-strength ship structures.

The engineering technology for dealing with fatigue crack
growth in structural alloys rests on an empirical correlation
between the rate which a crack extends per cycle of repeated load
(da/dN) and the stress-intensity factor range ($K_{max}-K_{min} = \Delta K$) at
the crack tip.   This relationship takes the form of a power law
da/dN = $C(\Delta K)^m$ where C and m are material constants.   An example
of such a characterization is shown schematically in Figure 11.
The power-law relationship applies only to Region 2 of the plot
of da/dN vs. ΔK which in most cases represents the stress-intensity
factor range of engineering interest.

An example of the deleterious effect seawater can have on
fatigue crack growth is shown in Figure 12 for a 17-4 PH stainless
steel, an alloy of particular interest for fast surface craft
applications.   The curve for the material potentiostated to -650 mv
represents the electropotential condition for 17-4 PH stainless
steel electro-chemically coupled to aluminum in seawater, for
example, a condition that could arise in a hydrofoil if proper

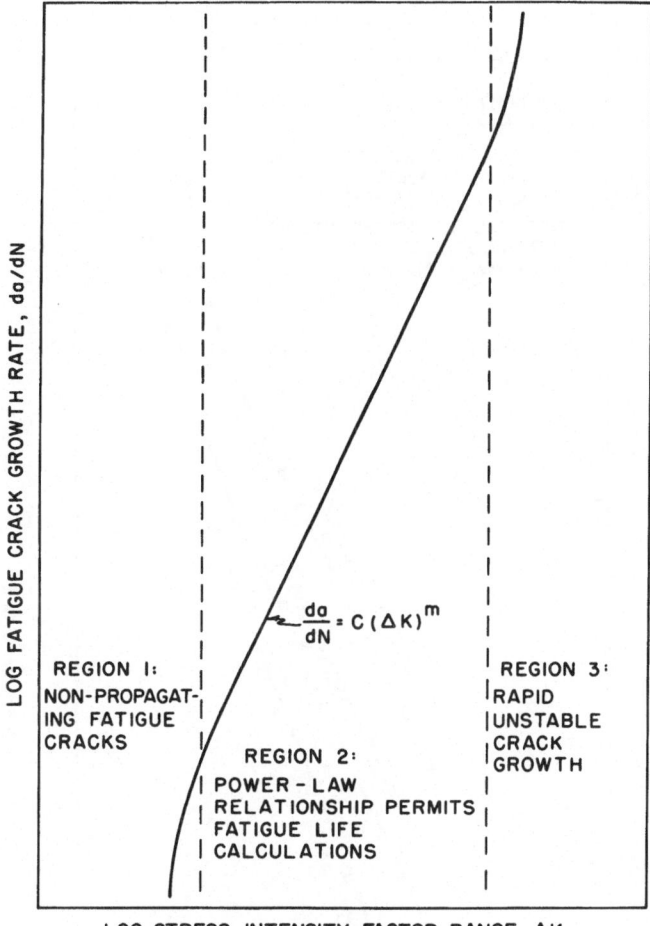

Figure 11.   Schematic illustration of the typical sigmoidal fatigue-crack-growth-rate curve for structural alloys.

electrical isolation is lost between the aluminum hull and 17-4 PH stainless steel struts and foils.

However, the fatigue-crack growth of some high-strength alloys appears to be essentially unaffected by the seawater environment as shown in Figure 13 for a high-strength titanium alloy. In this case, the conditions of freely corroding (-300 mv) and potentiostating to the electropotential equivalent to coupling the material to aluminum (-800 mv) and to zinc (-1050 mv) resulted in fatigue-crack growth rates obtained in air.

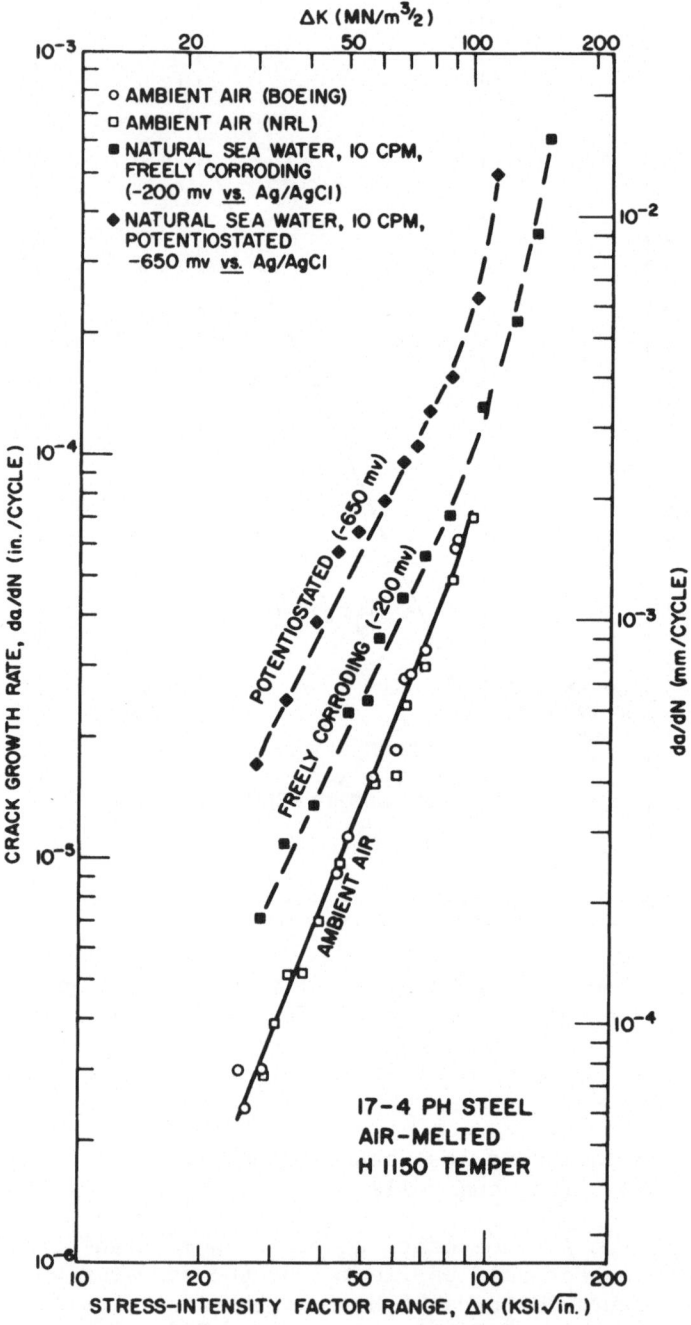

Figure 12.   Effect of seawater environment and electrochemical
potential on fatigue crack growth rate for a 17-4 PH
stainless steel.

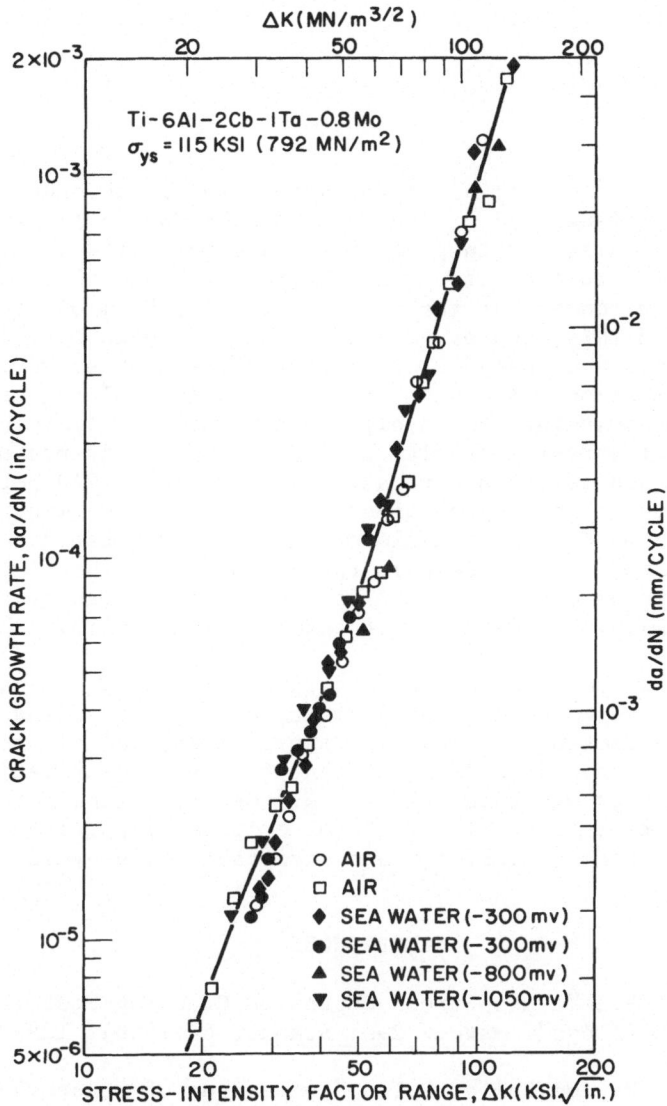

Figure 13.   Effect of seawater environment and electrochemical
             potential on fatigue crack growth rate for a high-
             strength titanium alloy.

        In actual practice, the analysis of fatigue crack growth in
structural situations is considerably more complex than what has
been presented here and involves consideration of such aspects as
residual stresses and loading spectra [12].   However, this discussion does show the importance of considering the influence of this

seawater environment, and particularly electro-chemical potential, on fatigue crack growth rate in design of reliable high-performance ships.

SUMMARY

The long-term reliability of structures is dependent on the tolerance for flaws and cracks that is inherent to the structural material. Problems of inspectability and maintainability can be as detrimental to the life of structures that must operate in a corrosive environment as the more apparent problem of catastrophic fracture. In this chapter, methods based on technology derived for prevention and analysis of fractures are adapted to facilitate material selection and design for situations involving crack growth under sustained and cyclic load. The significance of critical levels of stress intensity can be determined by use of a RAD format designed for this purpose. The three-part RAD permits determination of the degree of severity of crack growth problems in terms of the K_{Iscc}/σ_{ys} ratio or the K_{ISLC}/σ_{ys} ratio. The latter case (sustained-load cracking), rather than the dynamic fracture resistance properties, may well be a dominant design factor for applications involving highly stressed titanium alloys.

The growth of cracks due to fatigue is also a dominant factor in design of advanced high-performance ships. Depending on the alloy, the effect of the seawater environment, and electrochemical coupling can range from insignificant to severe degradation of fatigue crack growth rate properties. Detailed knowledge of these factors and their control will be of critical importance in designing for long-term reliability and acceptable life-cycle costs.

REFERENCES

1. Brown, B.F., "A New Stress-Corrosion Cracking Test for High-Strength Alloys", Mater. Res. Stand., 6 (1966), 129-33.

2. Novak, S.R. and Rolfe, S.T., "Comparison of Fracture-Mechanics and Nominal-Stress Analyses in Stress-Corrosion Testing", United States Steel Corporation, Monroeville, Pa., Naval Ship Engineering Center Contract Report No. ARL-B-63105, December 1968. (AD 846 125L)

3. Smith, H.R., Piper, D.E. and Downey, F.K., "A Study of Stress-Corrosion Cracking by Wedge-Force Loading", Eng. Fract. Mech., 1 (1968), 123-28.

4. Novak, S.R. and Rolfe, S.T., "Modified WOL Specimen for K_{Iscc} Environmental Testing", United States Steel Corporation, Monroeville, Pa., Naval Ship Engineering Center Contract Report, May 1968. (AD 836 310L)

5. "Damage Tolerant Design Handbook. A Compilation of Fracture and Crack Growth Data for High-Strength Alloys Including Data Sheets for the First Supplement", Metals and Ceramics Information Center, Battelle Columbus Labs., Ohio, Report No. MCIC-HB-01-Suppl-1, September 1973. (AD 772 810)

6. Judy, R.W., Jr., and Goode, R.J., "Stress-Corrosion Cracking of High-Strength Steels in Titanium Alloys", Weld. J., 51 (1972), 437s-48s.

7. Stress-Corrosion Cracking in High Strength Steels and in Titanium and Aluminum Alloys, ed. by B.F. Brown. Washington, D.C.: Naval Research Laboratory, 1972.

8. Pellini, W.S., "Criteria for Fracture Control Plans", Naval Research Laboratory, Washington, D.C., Report No. NRL-7406, May 1972. (AD 743 058)

9. Pellini, W.S., personal communication.

10. "Standard Method of Test for Plane-Strain Fracture Toughness of Metallic Materials", Designation: E399-72, in 1973 Annual Book of ASTM Standards, Part 31. Philadelphia: Am. Soc. for Testing and Materials (1973), 960-79.

11. Yoder, G.R., Griffis, C.A. and Crooker, T.W., "Sustained-Load Cracking of Titanium: A Survey of 6Al-4V alloys", Naval Research Laboratory, Washington, D.C., Report No. NRL-7596, August 1973. (AD 767 307)

12. Crooker, T.W., "Designing Against Structural Failure Caused by Fatigue Crack Propagation", Naval Eng. J., 84, no. 6 (1972). 46-56.

FRACTURE MECHANICS EVALUATION OF GENERATOR ROTORS

G. A. Clarke

General Electric Company, Schenectady, New York

ABSTRACT

Although a very few generator rotors have failed in comparison to the total number of turbine generator units presently in service, the amount of stored energy in a rotor when at rated speed is high enough to cause a catastrophic event if brittle failure does occur. With this in mind, the designer of generator rotors must use all the tools available, to design against the possibility of brittle failure.

By using the theory involved in fracture mechanics, the designer may input the design stresses to calculate their effects on fatigue crack growth and brittle fracture resistance due to the presence of internal discontinuities in rotor components. Due to the number of sonic indications found in many of the older generator rotors, a computerized approach is used to calculate the interaction effects of sonic indications within close proximity to neighboring indications and also within close proximity to the free surface of the prepanned center bore.

The computer program simulates the total number of cyclic start-stops that a generator rotor would normally see in its lifetime. The various cyclic stress states are applied to the sonic indications to estimate the crack growth of each indication. Once the crack growth is calculated, the fracture resistance capability of the generator rotor is found by comparing the maximum stress intensity of the internal discontinuities to the critical stress intensity found by measuring the fracture toughness of each rotor.

INTRODUCTION

Since 1953, there have been approximately four failure incidents in large steam turbine generator rotors which can be attributed to brittle fracture. A description of a number of the failures along with their causes has been outlined in two technical journals [1,2]. While this number of failures represents a very small percentage of the total units presently in service, the potential catastropic effects of such failures has created the necessity to design against any possibility of brittle failure.

In the mid 1950's a series of tests were developed by Winne and Wundt [3] to test for the fracture toughness of rotor materials. These tests, known as the Spin Disk tests, along with ultrasonic test results of the forgings, increased our confidence in the integrity of generator rotors. These results allowed the rotor designer using nominal design criteria to determine whether or not the forging could be applied. The theory of brittle failure required a plane strain fracture toughness value for comparison purposes. Since the plane strain constraint required very large specimen geometries, both the amount of material required, and the expense of running such a test, precludes plane strain fracture toughness tests (K_{IC}) on each rotor. A. Brothers et al. [4] developed a method for correlation of the 50% FATT value and K_{IC} for various classes of rotor steels. By taking charpy FATT tests on each forging, a quantitative assessment of all rotor forgings was possible.

Due to todays improved steel making techniques, improved non-destructive testing methods, and conservative design practices, a single cycle failure is highly unlikely to occur. However, the designer also must design to cyclic stresses which do not cause internal defects to grow to a critical size. To do this he must be aware of the crack growth characteristics of the material he is using. By using fracture mechanics, along with a knowledge of fatigue, the designer has the ability to fully understand the effects of the various stress states produced by his design geometry and of the environmental atmosphere on the internal discontinuities of the generator rotor.

THE FUNCTION DESIGN OF GENERATOR ROTORS

The primary function of a generator rotor is to produce a rotating electromagnetic field, which when passing a series of conducting stator windings produces an alternating voltage. The electromagnetic field is obtained by passing an excitation current through the copper conductor turns which are placed in a number of axial slots machined in the rotor. The magnetic flux (or total strength of the field) is dependent upon the magnitude of the

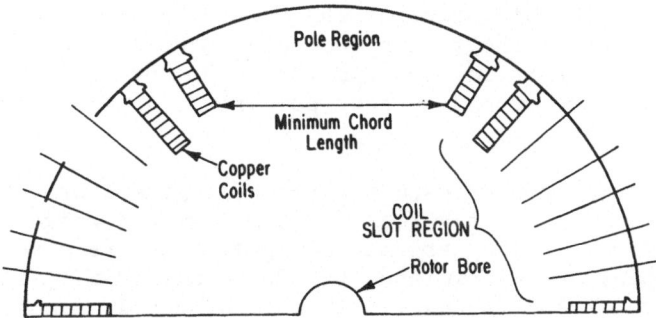

Figure 1. Minimum cord length between first coil slots in rotor pole region.

excitation current and the total number of copper turns placed in the rotor. The density of the flux in the rotor iron is directly proportional to the cross sectional area of the minimum cord length in the pole section. The maximum allowable flux density, for practical design purposes, that is acceptable in the rotor is dependent upon a material property called magnetic permeability.

While a more complex and thorough explanation is required to completely show why an increase in generator rotor diameter is necessary for an increase in power output, a simplified example is presented which will illustrate the point. Given a machine whose flux density is designed to the limit as defined by the magnetic permeability, then the more effective means of increasing the rating is to open the minimum cord length up in the pole areas, Figure 1, and increase the excitation current so that the magnitude of the flux increases. The easiest method of increasing the minimum cord length is to increase the diameter of the rotor. However, as the diameter of the rotor increases, so does the maximum stress state occurring in the rotor body. If we were to consider the rotor body as a cylindrical rotating beam, then the maximum stress (maximum tangential bore stress) occurs at the center of the body, for a solid cylinder, and at the bore surface for a center bored cylinder.

When two pole rotors have an extensive design history of about 70 years, they are designed to optimum diameters. These optimum diameters are pushing the design strength limits. There-fore, the quality and cleanliness of two pole rotors are of the utmost importance. The low pressure steam from nuclear power plants requires the larger annular areas of large turbine rotors, and, therefore, four pole generator rotors which run at half the speed of two pole machines. To obtain the required output for

half the speed, the diameters of the four pole units must be much
greater than that of the faster two pole rotors. The rated speed
of two pole generators is 3000 rpm for 50 Hz current and 3600 rpm
for 60 Hz while corresponding four pole rotor speeds are 1500–
1800 rpm.

The maximum diameter used for two pole generator rotors at
the present time is about 4 feet (1220 mm) with a weight of about
200,000 lbs (90MT). The largest four pole rotor received by the
author's company to date weighed 448,000 lbs (203 MT), with an
overall length of 662 inches (16815 mm) and a body diameter of
6 feet (1829 mm). The alloys used for generator rotor steels are
usually NiMoV and NiCrMoV.

STRESSES IN GENERATOR ROTORS

There are a number of components in generator rotors that
require careful analysis due to high stresses imposed by rotation.
The design of these components require both a static and dynamic
analysis of all critical sections to insure that the total number
and magnitude of cyclic stresses are below that necessary to grow
a possibly existing defect into a critical size crack. As an
example of the various methods employed in the design of generator
rotors, listed below is the function, and the type of analysis
used to calculate the stresses in some of the components.

Rotor Body

As discussed previously, the function of the rotor body is to
create a rotating electromagnetic field. The coil slots in the
periphery of the rotor body contain the copper coils which carry
the electric current, which in turn magnetizes the body. Due to
the difference in mass between the solid steel pole and the copper
carrying coil slot region shown in Figure 2, the non-uniform centrifugal force creates a non-uniform stress pattern which produces a
maximum tangential stress at an angle of 90° from the centerline
of the pole region on the bore surface of a center bored two pole
rotor. Important stresses that are calculated are: the maximum
and average tangential bore stress and the average bursting stress
across the section between the deepest coil slot depth and the bore
surface. Since once per revolution vibratory stresses will accumu-
late billions of cycles in the life time of a generator, all bending stresses are calculated to insure that the magnitude of these
stresses are well below the high cycle fatigue "endurance" limit
and that these stresses, if applied to the maximum size defect
which could possibly be missed by inspection limitations are low
enough to be well below the "threshold" stress intensity factor,
and hence small enough to prevent fatigue crack propagation.

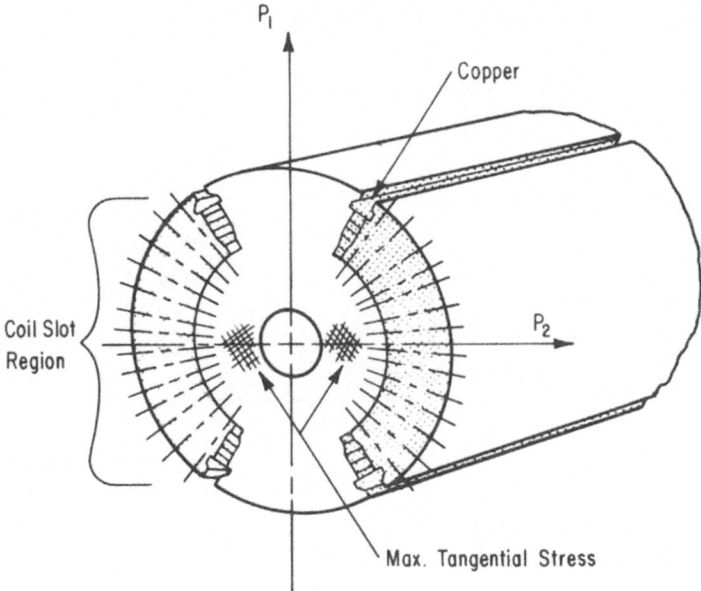

Figure 2.   Point of maximum tangential stress.

The nominal torsional stresses on the generator rotor trans-
mitted by the turbine are calculated at the various shaft diameters.
Transient stresses are also imparted to the rotor by faults relay-
ed back through transmission lines. These transient stresses are
calculated as cyclic stresses superimposed on the mean torsional
stress.

Rotor Body Teeth Between Coil Slots

Stress states at various heights on the rotor teeth are calculated to insure that they are able to carry the full copper load
in the coil slots. Due to the relatively small cross section of
these teeth, a careful analysis of the stress states must be performed along with a design against fatigue initiation in these
regions. An analysis of one of the failed rotors [2] showed that
the initiation site was close to the base of a coil slot tooth.
In that particular case, a hydrogen flake was found to be subjected
to a relatively high concentrated stress state at the base of the
coil slot.

Wedges Over the Coil Slot Copper

In general, the only cyclic stress that a coil slot wedge
receives is due to the centrifugal forces resulting from the stop-

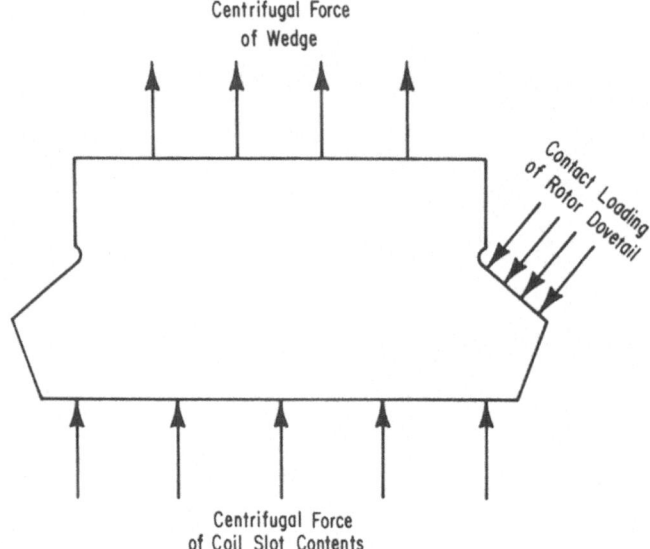

Figure 3.   Loading configuration of coil slot wedge.

start operation of the turbine generator set.   A simple stress
calculation is made such that a fatigue analysis may be performed
on the dovetail fillet radius of the wedge, as shown in Figure 3.
The axial bending stresses on the wedge due to the normal deflections
of the rotor fixed in its bearings is minimized by keeping the
wedge length to an optimum small size.   This effectively removes all
once per revolution cyclic stresses.

Figure 4.   Retaining ring and centering ring assembly.

Retaining Ring

The highest stressed component in the generator is the ring
that contains the copper coils where they cross the pole outside
of the rotor body slots at the end of the body length. One end
of the retaining ring is shrunk onto the end of the rotor body and
is stiffened by a centering ring at the other end to maintain
symmetry, as shown in Figure 4. The centering ring is shrunk
onto the internal diameter of the retaining ring. A complex stress
state exists in both the retaining ring, as seen in Figure 5, and
the rotor body portion on which the shrink fit is performed. A
careful stress analysis is performed to accurately predict the
stress at both the zero speed condition and at running speed. An
elliptical load due to the unsymmetry of the coils under the retaining ring creates bending stresses, whose effects are superimposed on
the tangential stress. The components of the above stresses are
well considered when conforming to normal yield design criterion
and also for fracture and fatigue analysis.

INSPECTION TECHNIQUES

The inspection of the materials used for generator rotor components is of the utmost importance, as failures of large rotors can
usually be traced to structural imperfections rather than design
inadequacies. The sensitivity of non-destructive testing equipment
used to inspect generator components has improved dramatically over
the last decade. With this increased sensitivity, the size of a

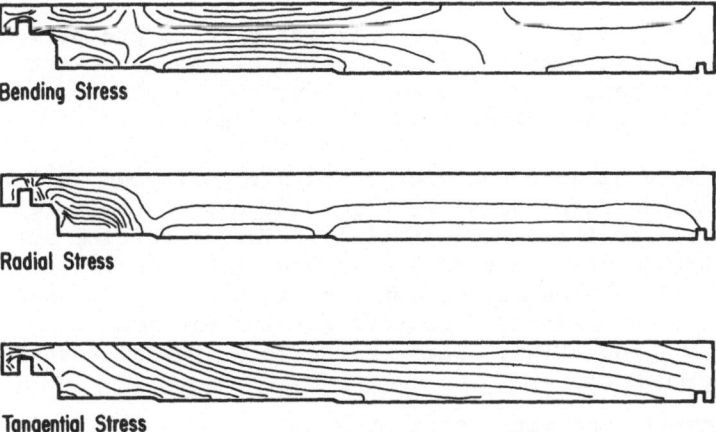

Bending Stress

Radial Stress

Tangential Stress

Figure 5. Typical stress contours found in retaining rings.

structural imperfection which could possibly be missed during an inspection is considerably reduced. However, the location and orientation of a defect is still a primary cause of inability to accurately determine all defect magnitudes.

To this end, the author's company requires that all rotor forgings have a trepanned bore hole through the center of the rotor. Not only does this enable the mapping of sonic indications from the bore and peripheral surfaces, it also removes the majority of the sonic indications in the rotor which are normally found near the centerline.

An ultrasonic inspection is made from the periphery of the rotor forging and compared to the results of a bore-sonic inspection, performed from the bore surface. All sonic indications from both tests are then matched by comparing their magnitude and location to reduce the probability of missing any internal defects. When all the sonic data is reduced to a suitable three dimensional map, they are evaluated from a fracture mechanics viewpoint by the designer to insure that in the presence of the maximum allowable design stresses they do not adversely affect the rotor integrity. The method used to perform this evaluation will be briefly described later in the chapter.

As the most critical area for sonic indications in the rotor forging is near the bore surface, a high sensitivity bore-sonic test is performed to accurately describe their magnitude. The sensitivity of this test is optimum for a range of 0.125 inches to 4 inches radially outward from the bore-surface. To aid in the description of the indications in the immediate proximity to the bore, a magnetic particle inspection is performed along the surface of the bore.

Both magnetic particle and ultrasonic tests are performed on all other critical rotating forged parts of the generator rotor. Also a red-dye penetrant is used on many components to check for surface imperfections after machining is complete.

While mechanical properties tests are performed in various locations in the rotor forging, Figure 6, one of the more important locations is near the bore centerline. The properties are general-ly lower at this point due to lesser consolidation of the forging at the centerline than at the outer periphery. The trepan speci-mens from the bore are used to test for the minimum mechanical and metallurgical properties. Some of the more typical tests performed in this region are:

1. Tensile and yield strengths
2. Percent elongation and reduction in area
3. Charpy impact energies and Charpy 50% fracture appearance transition temperatures.

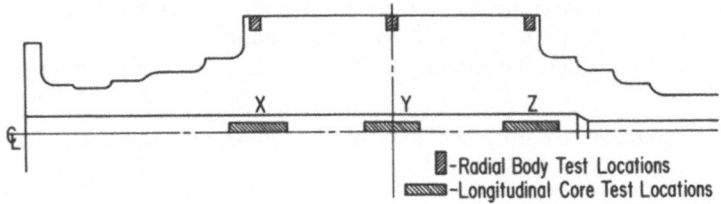

Figure 6.   Cross-section of NiMoV four pole generator rotor showing
            test locations

4.   Chemical analysis of the forging alloys

Experience has shown that safe design practices result from the use
of the material properties taken from the rotor trepanned core bar.

FRACTURE MECHANICS EVALUATION OF ROTOR COMPONENTS

Present day forgings have relatively few discontinuities when
compared to earlier manufactured generators. Due to the improved
efficiencies of the newer stations, the newest machines are primarily
used for base load with very few start-stop operations. In many
cases, the older units are used as "peak load generators" or daily
start-stop units, and consequently, receive more severe cyclic duty
than the newest generators. As these generators may have a large
number of sonic indications, a critical evaluation of these indications is important. It is a tedious job at best to evaluate the
interaction of these individual indications with its nearest neighboring indication, or the interaction of an internal indication with
the bore surface. It has become necessary, therefore, to computerize
our evaluation of these indications.

Once a detailed map of the sonic indications is made, they are
placed in a computer file with their various r, θ, and Z coordinates.
The computer program then simulates a prescribed number of stress
cycles on the forging to evaluate the growth rate and fracture resistance capability of the rotor. A list of the assumptions made
and the methods used in the computer program follows.

Assumptions

1.   Due to the size of the rotor forging, plane strain conditions prevail.

2.   Linear elastic fracture mechanics is applicable to the
material used for rotor components and hence small scale yielding
is assumed.

3.   The relationship between the excess temperature (operating temperature minus the 50% fracture appearance transition temperature) and the fracture toughness, K_{IC}, holds for rotor forging materials.

4.   The growth of cracks obeys the general form of da/dN = $C\Delta k^n$.

5.   All ultrasonic indications are treated as sharp cracks of a penny-shaped geometry whose orientation is in the radial direction normal to the tangential stress.

Computer Applications to Fracture Analysis

The primary function of our computer program is to simulate a cyclic stress applied to all of the sonic indications within the rotor body simultaneously. The interaction and linking of these indications with each other is considered throughout the cyclic process. If the linking of sonic indications does occur at a given number of cycles, then a new effective diameter of the linked indications is calculated and the remaining prescribed cyclic stresses are applied to the newly described map of sonic indications.

Following is a description of the techniques used to calculate the effects of sonic indications on the capability of the rotor forging to withstand brittle failure.

A. Mapping of Sonic Indications. A comparison is made between all indications found by a peripheral sonic inspection and boresonic inspection. After a comparison check is made, each indication is assigned a r,θ, z coordinate. The indications found by magnetic particle inspection are also used as input with their various coordinates. As the design stress distribution is known for each rotor, the stress state at each sonic location is calculated and assigned to each indication.

B. Calculation of Stress Intensity Factors. It is assumed that each sonic indication lies in a plane normal to the direction of the maximum tangential stress. For each single indication which is isolated from the effects of its nearest neighbor and any free surface, a stress intensity is calculated from the formula for a penny shaped crack:

$$K_I = \frac{2\sigma(\pi a)^{1/2}}{\pi} \tag{1}$$

and recorded in each indication record.

To calculate the stress intensity for indications close to a neighboring indication, we assume that each indication is in the same plane as its neighbor. Although this is a somewhat conservative approach, it must be remembered that a conservative approach is justified when dealing with the enormous amount of energy stored in a rotor at rated speed. The stress intensity of a co-planar tunnel crack can be found from [5]:

$$K_I = \sigma\sqrt{\pi a} \; \frac{2t}{\pi a} \; [\tan(\frac{\pi a}{2t})]^{1/2} \qquad (2)$$

where t is the distance between cracks. While the above equation is for a tunnel crack, appropriate adjustments can be made to calculate the stress intensity for penny shaped cracks [6].

To decide whether or not linking between indications takes place, a calculation can be made to determine the stress intensity of an effectively linked indication. If the value of this stress intensity is less than the value of the stress intensity between the two indications, then these two indications are considered to have linked. A second check must be continuously made to determine if any sonic indications break through to the bore surface to form a semi-elliptical surface indication. The method we use to calculate the stress intensity magnification factor between a penny shaped defect and a free surface was outlined by Gibbons et al. [7]. This method is essentially the result of an experimental program performed at the author's company based on a tunnel crack approaching a free surface. The magnification factor, M, applied to the stress intensity of a tunnel crack approaching a free surface is given by:

$$M = 1.08 \; (\frac{h}{L})^{0.13} \qquad [3]$$

where h is the length of the tunnel crack and L is the distance from the near edge of the crack and the free surface. A comparison of these results with previous work by Zwicky [8] and Shah et al. [9] show good agreement in the range of application as can be seen in Figure 7.

C. <u>Fatigue crack growth on sonic indications</u>. To simulate the growth characteristics of the internal discontinuities, the equation governing the fatigue crack growth curve of rotor materials is fed into the computer program. This equation is then integrated and rearranged in the following form [10]:

$$\frac{da}{dN} = C\Delta K^n \qquad (4)$$

where $K_I = \sigma\sqrt{Fa}$

and by integration:

$$N = \frac{2}{(n-2)C_oF^{n/2}\Delta\sigma^n} \left(\frac{1}{a_i^{\frac{n-2}{2}}} - \frac{1}{a_f^{\frac{n-2}{2}}}\right) \tag{5}$$

where F is a geometric function of the crack.

The values of the various crack growth material constants can be found from the slope and the intercept of a da/dN vs ΔK crack growth curve. Such a curve is shown in Figure 8 for NiMoV rotor steel alloy [11].

By increasing the total number of cycles incrementally, a calculated defect size, a_f, is found at the end of each block of cycles. At this point a check is made throughout the entire map of indications for linking, and subsequent effective diameters are calculated for the linked indications. This process continues until the total prescribed cycles are simulated by the program.

D. <u>Overspeed capability</u>. Upon completion of the total applied cycles, the stress intensity factors for each cyclically grown indication is compared to a critical size penny shaped crack at a stress state existing in the location of the indication. The critical size of this penny shaped crack is calculated by inserting the

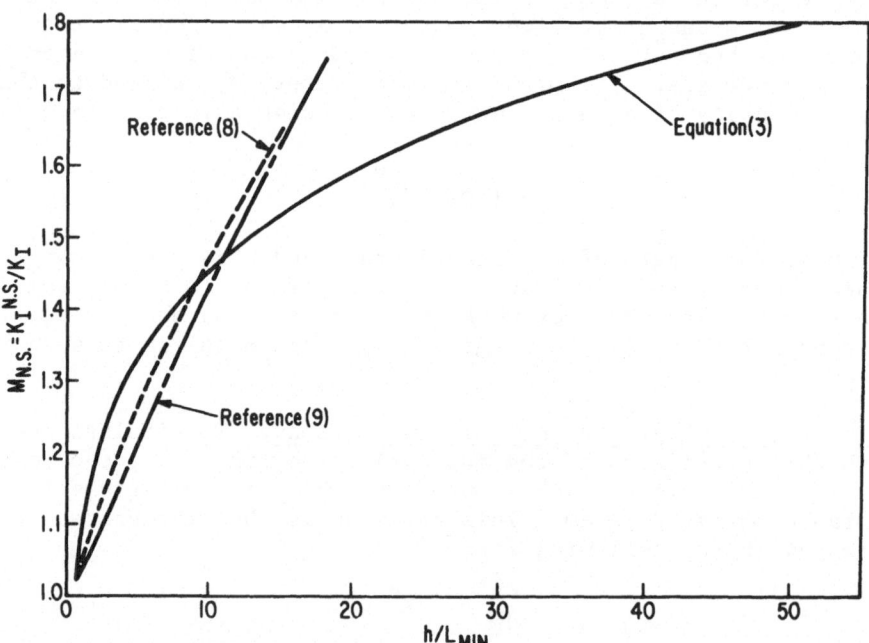

Figure 7.   Magnification factor of a tunnel crack approaching a free surface.

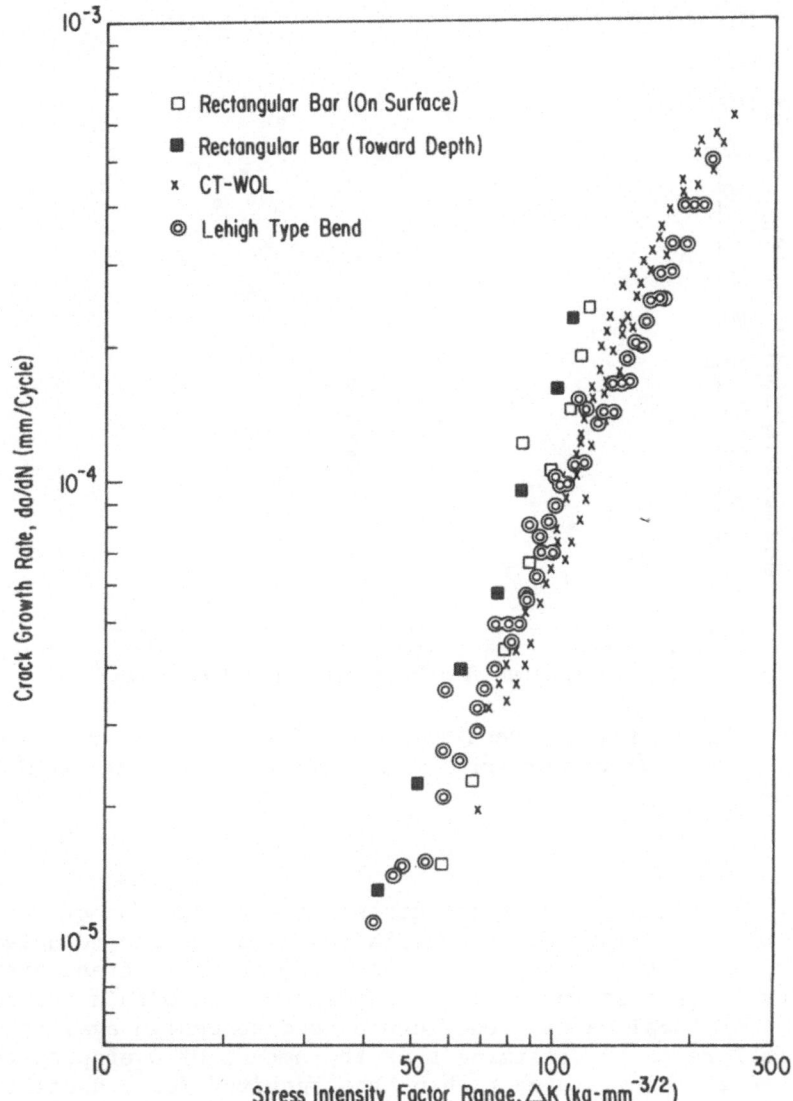

Figure 8. Fatigue crack growth rate as a function of ΔK for a
Ni-Cr-Mo-V steel forging.

fracture toughness in the equation:

$$a_{crit} = \frac{K_{IC}^{2}}{2\sigma}\,\pi \tag{6}$$

where K_{IC} is found from the curve in Figure 9 or Figure 10 for

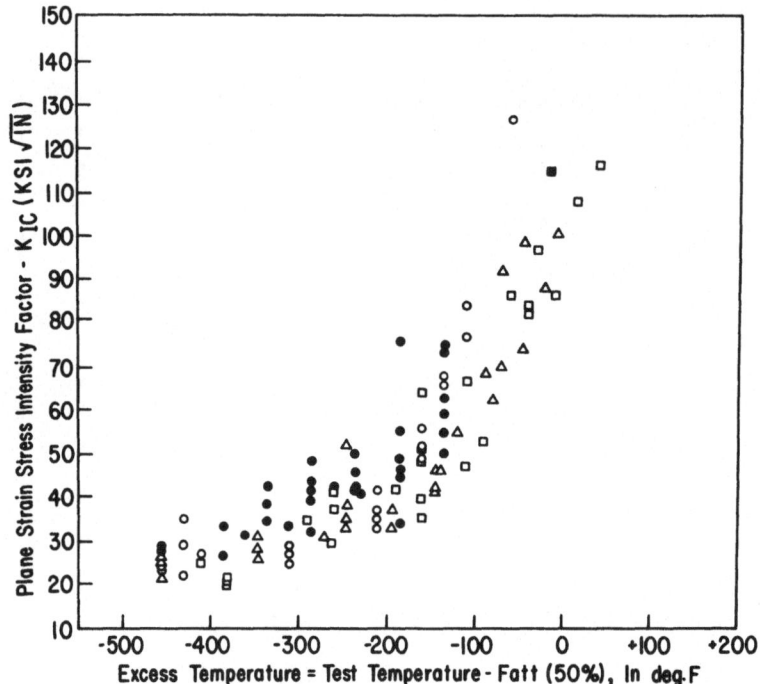

Figure 9. Plane strain stress intensity factor (K_{IC}) vs excess
temperature for four heats of nimo v forging steel.

various FATT values and σ is the stress at the location of the indication being considered. A calculation is also made to determine
the overspeed of the generator necessary to cause a stress high
enough for the stress intensity of any indication within the rotor
to reach a critical value. One method used to design against
brittle failure is to determine that the amount of overspeed required
for brittle failure is higher than that required for a ductile
yielding failure.

E. Updating of the computer program. A program of rotor component
and material testing continues at the author's company for an improved comprehension of the effects of environment and design. Tests
to define the fatigue crack growth curve in both air and hydrogen are
being performed and recorded for use as input for the computer
program.

The use of the J integral is also being investigated by the
author to augment the FATT method of calculating the fracture toughness of rotor materials. Various J-integral specimen configurations

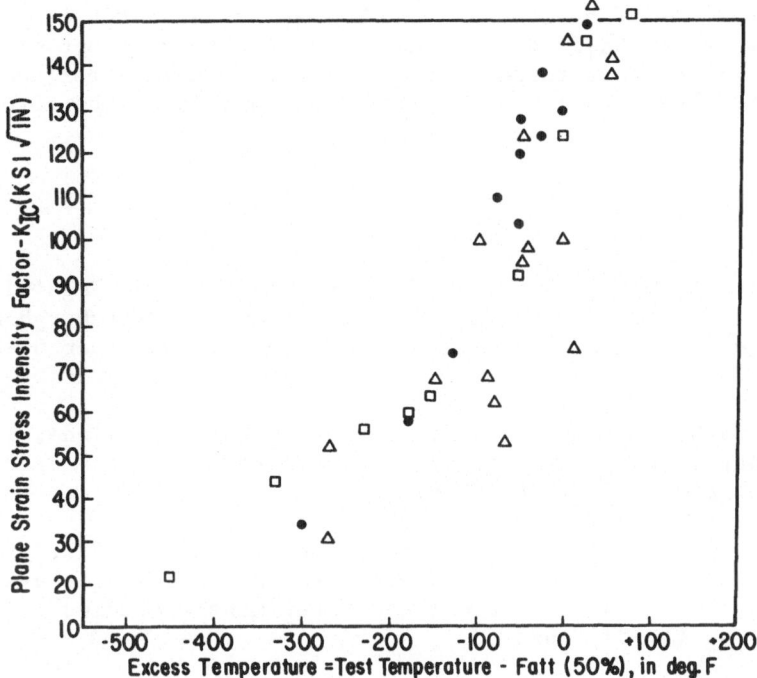

Figure 10.   Plane strain stress intensity factor (K_{IC}) vs excess
temperature for three heats of nimo v forging steel.

have been tested on rotor material with promising results. However,
it is believed that more basic research into non-linear fracture
mechanics is necessary before the J-integral technique receives
widespread acceptance.

CONCLUSION

A far more comprehensive overview on the effects of sonic
indications (non-metallics, porosity, or material segregation) on
the capability of generator rotors to withstand brittle fracture is
made possible by the use of a computer program to simulate the
crack growth characteristics and fracture resistance of large generator rotor forgings. Using the procedure outlined above, along
with the experience gained over many years of forging evaluation,
it is believed that present day rotor forgings have a negligible
probability of brittle failure.

Continuous effort is required, however, to evaluate various

adverse effects of environment and misuse of generator components. It must be re-emphasized that due to the enormous amount of stored energy of a rotor at rated speed, no apology is made for conservative approaches to a fracture mechanics analysis of the rotor design.

REFERENCES

1. De Forest, D.R., Schabtach, C., Grobel, L.P. and Seguin, B.R., "Investigation of the Generator Burst at the Pittsburgh Station of the Pacific Gas and Electric Company", ASME Paper 57-PWR-12, 1957.

2. Schabtach, C., Fogleman, E.L., Rankin, A.W. and Winne, D.H., "Report of the Investigation of Two Generator Rotor Fractures", Trans. ASME, 78 (1956), 1567-84.

3. Winne, D.H. and Wundt, B.M., "Application of the Griffith-Irwin Theory of Crack Propagation to the Bursting Behavior of Disks, Including Analytical and Experimental Studies", Trans. ASME, 80 (1958), 1643-58.

4. Brothers, A.J., Newhouse, D.L. and Wundt, B.M., "Results of Bursting Tests of Alloy Steel Disks and Their Application to Design Against Brittle Fracture", paper No. 93, presented at ASTM Annual Meeting, June 1965.

5. Paris, P.C. and Sih, G.C., "Stress Analysis of Cracks", in Fracture Toughness Testing and Its Applications, Special Technical Publication 381. Philadelphia: Am. Soc. for Testing and Materials (1965), 31-83.

6. Hale, L.R. and Kobayashi, A.S., "On the Approximation of Maximum Stress Intensity Factors for Two Imbedded Coplanar Cracks", Boeing Company Structural Development Research Memorandum No. 9, May 1964.

7. Gibbons, W.G., Andrews, W.R. and Clarke, G.A., "Fracture of Defects Approaching a Free Surface", Trans. ASME, Ser. J, J. Pressure Vessel Technol., 97 (1975), 270-77.

8. Zwicky, E.E., Jr., unpublished research.

9. Shah, R.C. and Kobayashi, A.S., "Stress Intensity Factor for an Elliptical Crack Approaching the Surface of a Plate in Bending", in Stress Analysis and Growth of Cracks, Special Technical Publication 513. Philadelphia: Am. Soc. for Testing and Materials (1972), 3-21.

10. Greenberg, H.D., Wessel, E.T., Clarke, W.G., Jr., and Pryle, W.H., "Fracture Toughness of Turbine Generator Rotor Forgings", paper presented at the International Forgemaster's Conference, Terni, Italy, 1970.

11. Kumeno, K, Nishimura, M., Mitsuda, K. and Iwasaki, T., "Defects and Fracture Strength of Large Rotor Forgings for Steam Turbines", ASME Paper 75-PWR-10, 1975.

PRACTICAL ASPECTS OF FRACTURE MECHANICS FOR STEAM TURBINE ROTORS

T. P. Sherlock

Westinghouse Electric Corporation, Lester, Pennsylvania

ABSTRACT

Various aspects of the use of a fracture mechanics approach
to steam turbine rotor design are covered in this chapter. The
importance of using proper operational procedure to prevent
brittle fracture is reviewed, and various methods of estimating
toughness properties from small amounts of material are discussed.
Some practical aspects of disc bursting are covered, and a probabilistic approach to design is introduced by way of an example
problem.

INTRODUCTION

One of the major applications of fracture mechanics technology
has been in the area of large rotating machinery for power generation. The massive forgings used in turbine-generator sets present
an increasingly complex problem for designers, for as unit size
increases, stresses and material strength requirements must also
increase. The forgings must be produced from extremely large
ingots (on the order of hundreds of tons), requiring close control
throughout the forging, heat treating and testing cycles to insure
a quality product. The trend toward larger, stronger, less ductile
forgings has made the use of fracture mechanics an important part
of the overall design of steam turbine rotors.

The number of failures of large rotors has been extremely small
in comparison to the number placed in service and the severe operating requirements imposed upon them. However, since reliability
and equipment availability are of paramount importance in the

251

power generation industry, a conservative design approach is
dictated. The use of a fracture mechanics approach to rotor design
has been covered in detail in two previous publications [1,2] and
these serve as an excellent introduction to the subject. The purpose of this chapter is to expand on a few details of a fracture
mechanics approach and to review some of the operational and design
features in light of this.

MANUFACTURE OF STEAM TURBINE ROTORS

The manufacture and materials of steam turbine rotors have
been the subject of two recent publications [3,4], and only a
brief outline will be presented here. Many of the rotors used are
of integral construction, that is, they are a one piece forging
made from an ingot several times larger than the finished product.
A typical manufacturing cycle is shown in Figure 1. Because of
segregation, forging and heat treating considerations, non-uniformity of properties may result, and a solid center core is removed
from the forging for subsequent testing purposes.

In the instances where size or quality restrictions prevent
the use of integral forgings, the turbine rotor is "built-up" using
disc forgings which are shrunk onto a central shaft by heating the
disc and allowing it to contract in place on the shaft (Figure 2).
The amount of interference fit is sufficient to maintain the disc-
shaft contact during operation.

Excess material is provided on each forging for testing purposes, which normally consist of tensile, impact and, in certain
cases, creep-rupture testing. Forgings are then inspected using
magnetic particle techniques for surface and near surface flaws
and ultrasonic techniques for possible flaws within the body of
the forging. If defects are found, the designer is able to make
a decision regarding the disposition based on estimated defect
size, material properties and intended service.

OPERATION OF STEAM TURBINE ROTORS

A typical fossil steam turbine arrangement is shown in
Figure 3. Rotation at 3600 RPM imposes local tangential (hoop)
stresses at the center bore which are on the order of half the
yield strength of the material. In addition, steady state and
transient thermal gradients can produce thermal stresses of considerable magnitude at the center bore and outer diameter of the
rotor. Thermal stresses are additive to the rotational component
when making a fracture mechanics calculation, provided elastic
conditions are maintained. The metal temperatures of the turbine

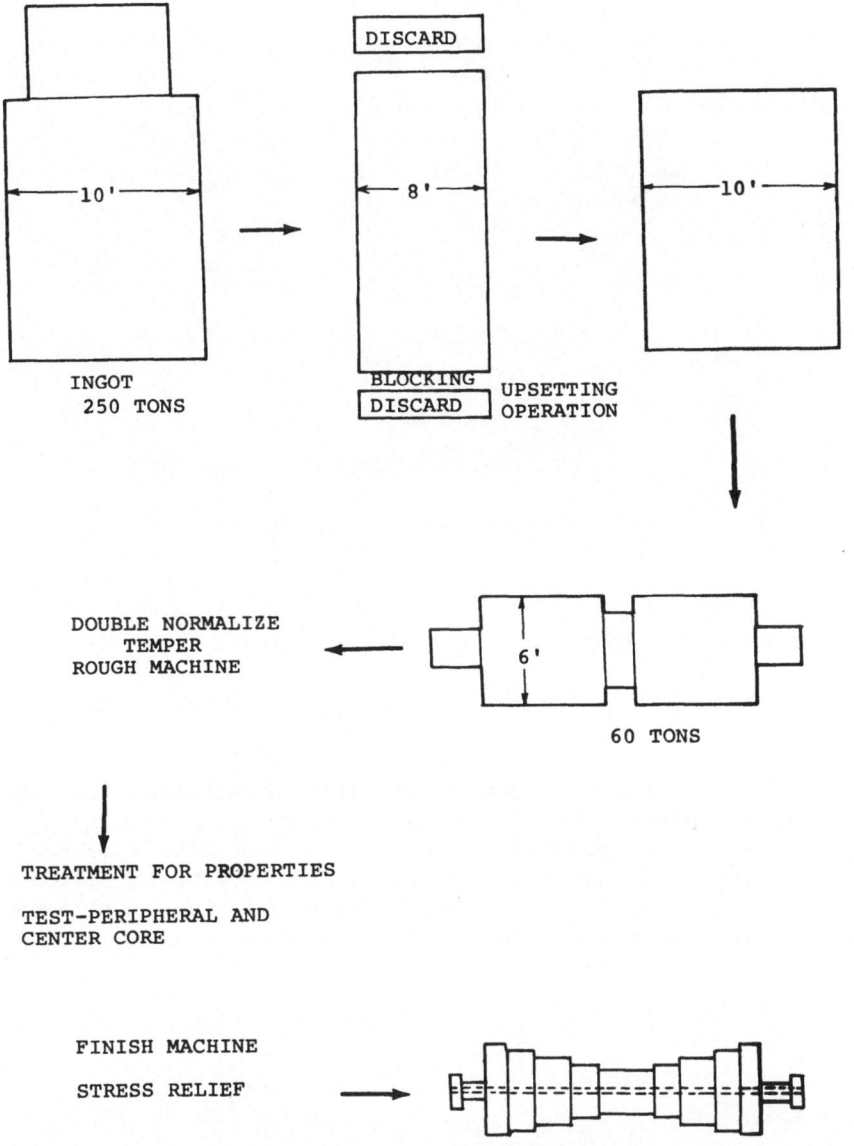

Figure 1. Typical production sequence for a rotor.

rotor vary from unit to unit, depending on the inlet steam temperature, unit size, operating conditions, etc. As the metal temperature
also dictates the value of fracture toughness to be used, accurate
determinations of steady state and transient metal temperatures are
required for a fracture mechanics analysis.

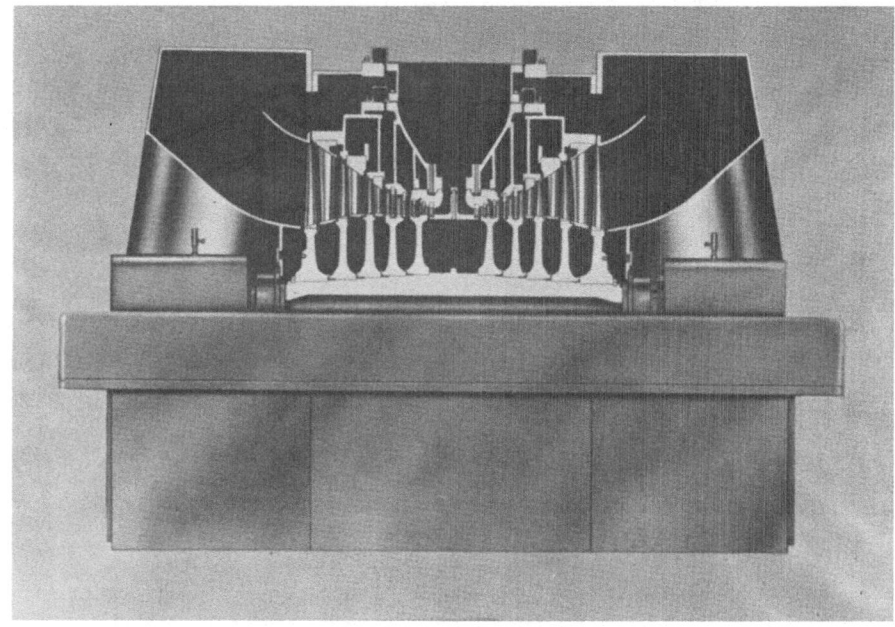

Figure 2.   Built-up construction for large rotors.

In general, the most severe operating conditions are imposed during a cold start, that is, when a turbine is nominally at room temperature and is gradually brought up to speed to synchronize with the electrical network.   During a cold start, the turbine is rolled at a low speed for several hours by passing small amounts of steam through the blade path.   The steam heats the rotor from the

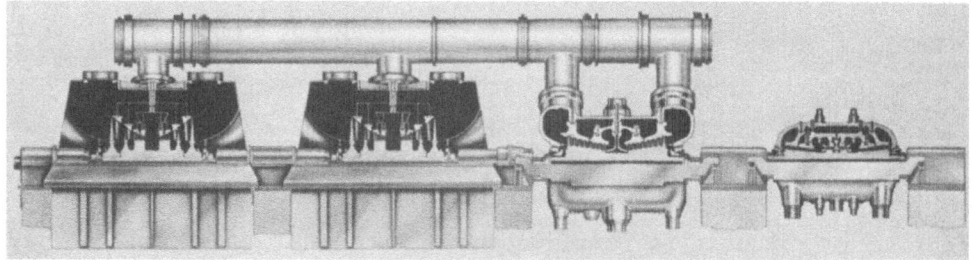

Figure 3.   3600 RPM fossil turbine arrangement.

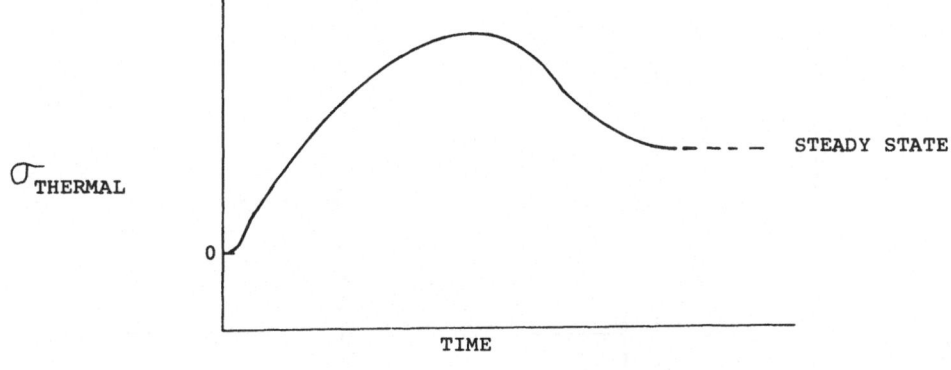

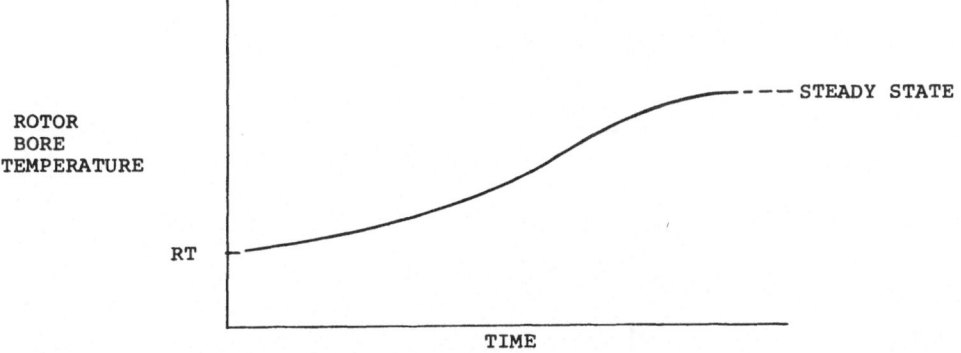

Figure 4.   Relationship between bore temperature and bore thermal
stress during start-up.

outer diameter, and tensile thermal stresses are developed at the
center bore. As the entire rotor heats up, the thermal gradients
become less severe, and the thermal stress peaks, Figure 4. As the
rotational stress varies with the square of the speed, judicious
operating procedures require that the turbine be rolled at lower
speeds until the thermal stress is past the peak value, and the
bore temperature (and toughness) reach a safe operating value.

Some of the steels used for steam turbine rotors are heat
treated to produce an upper bainitic microstructure which produces
good creep rupture properties, but results in material with a
fracture appearance transition temperature (FATT) well in excess of
ambient. In these rotors, the time-stress-toughness relationship
at the bore is complex, and a schematic example is shown in

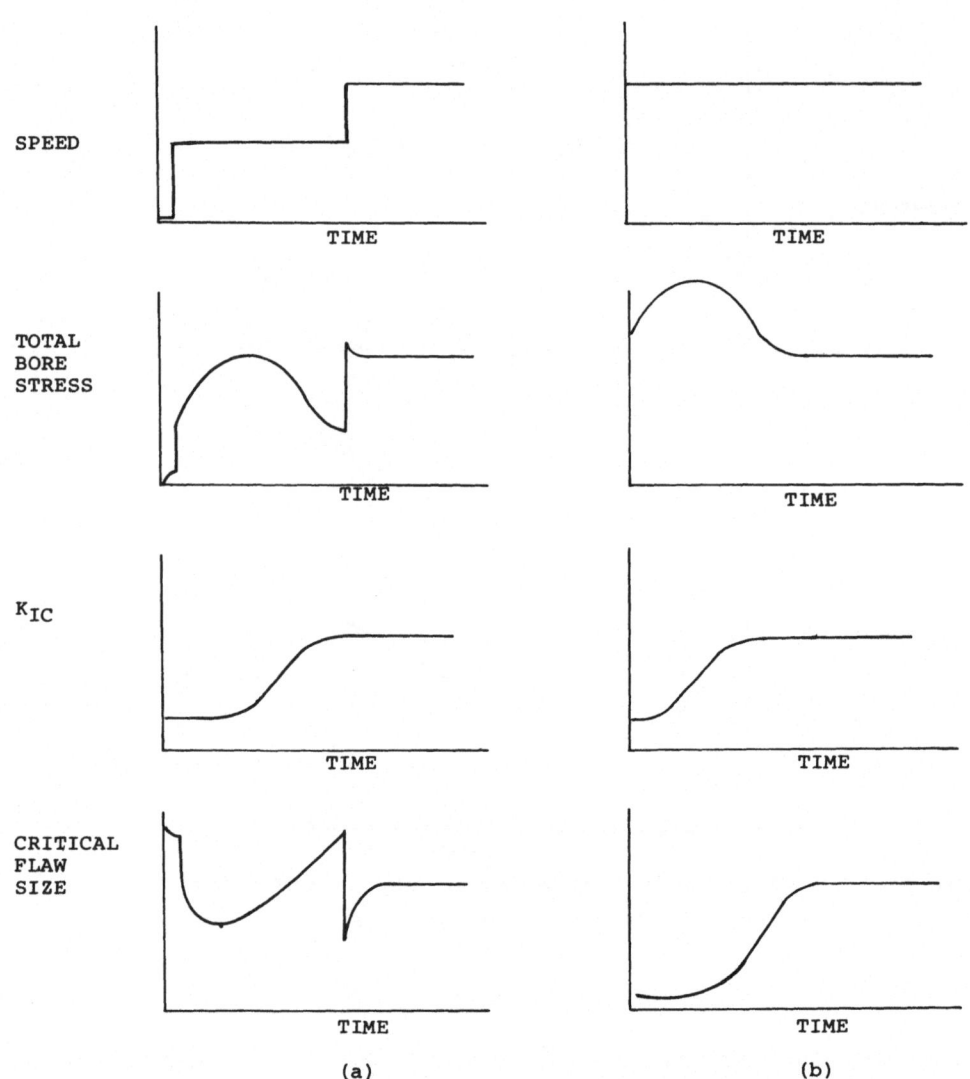

Figure 5.   Time-stress-toughness relationship for (a) slow roll
            condition, and (b) instantaneous full speed.

Figure 5. Two examples are given, the first using the slow roll
procedure prescribed earlier, and the second showing the effects
of attaining synchronous speed instantaneously. The variation of
the thermal stress component with time is assumed to be the same
for both cases for comparison purposes. In each case, the value
of the critical flaw size is calculated periodically during the

Figure 6.   Large flaw (dark area) contained in a fragment of a
             steam turbine rotor.

cycle, using the relationship

$$a_{cr} = C_1 \; (\frac{K_{IC}}{\sigma_{total}})^2 \qquad (1)$$

where a_{cr} is the critical flaw size, C_1 is a flaw geometry property,
K_{IC} is the fracture toughness and σ_{total} is a combination of rotational and thermal stress. In the first case, the minimum value of
the critical flaw size occurs when the turbine is brought to synchronous speed after the thermal stress peak. In the second case,
the total stress is higher and occurs at a lower value of K_{IC}, and
the minimum value of critical flaw size is much lower. When steady
state operating conditions are reached, the critical flaw size for
turbine rotors will be quite large. Figure 6 shows a fragment of a
rotor which ran for some time with a very large (>100 in.^{2}) flaw.
(Darker area on the surface.)

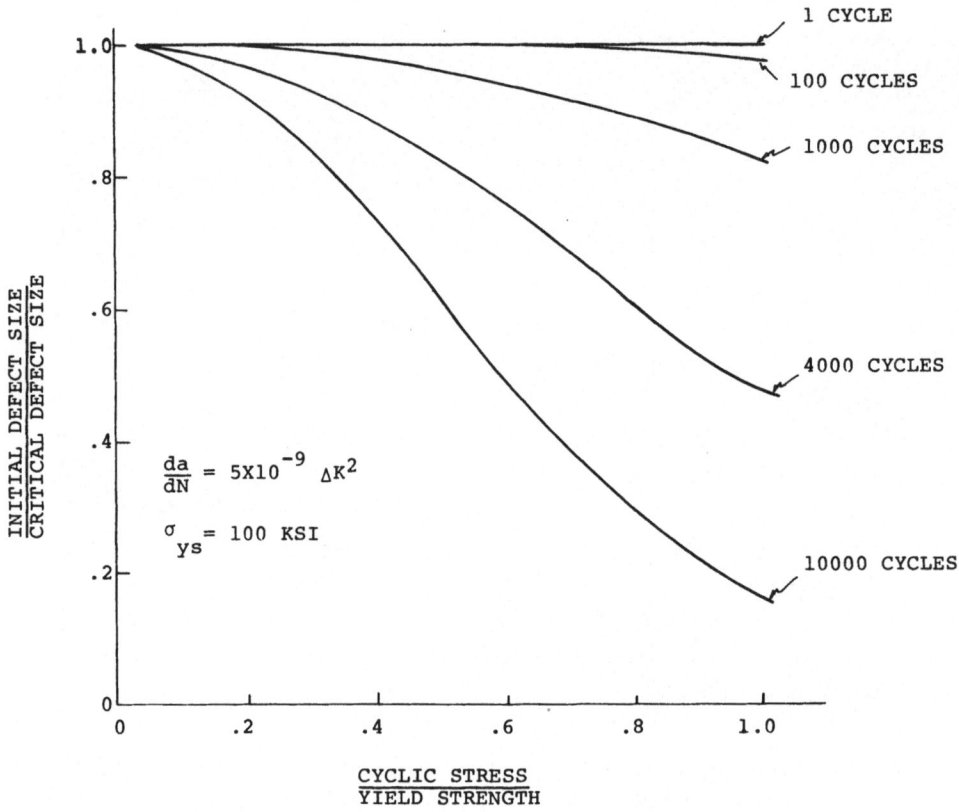

Figure 7.   Relationship between cyclic stress, cyclic duty and
            initial defect size.

        Cyclic stresses at the bore of a rotor are significant only
when a speed change (start-up and shut-down) is involved. Load
changes alone produce only minor variations in the bore stress when
compared to a full speed excursion (this is not true for the surface
stresses). Turbines used for base load duty experience relatively
few speed cycles in their lifetime. However, more and more turbines
are now being used for peaking operation, which involves being taken
off line each evening and restarted each day. In these units, the
cyclic crack propagation may become the critical factor in a brittle
failure analysis. Figure 7 shows the importance of cyclic stress
by comparing the ratio of the initial flaw size to the critical
flaw size with the number of cycles of stress and the relative
value of stress. In such cases where sub-critical flaw growth may
be substantial, periodic in-service inspection of turbine rotor
bores is recommended to insure that if any such flaws are present,
they are removed.

ESTIMATION OF TOUGHNESS PROPERTIES

Since the bore core bar removed from the forging is generally only a few inches in diameter, it is not possible to determine valid K_{IC} at temperatures of interest. Two methods of K_{IC} are available. The first method [5] is empirical and uses Charpy V-notch tests to determine the upper shelf energy which is converted to K_{IC}. An estimate of the lower shelf (100% brittle fracture in a Charpy specimen) K_{IC} is made from the yield strength, and K_{IC} at FATT is the average of the upper and lower K_{IC}'s. The second method [6] uses the J integral approach to elastic-plastic fracture and it has been shown that K_{IC} can be derived from J_{IC}, thus allowing the use of very small test specimens over the temperature range of interest.

At the present time, the empirical approach using Charpy specimens is used more extensively. The smaller size and lower cost per data point permits testing over a wider temperature range as well as for axial variations of properties along the bore bar. The other advantage is that the shift in FATT due to long time isothermal temperature embrittlement can be monitored using small Charpy blanks and the K_{IC}-T relationship can be predicted. Figure 8 shows an example of the use of the empirical correlation to predict the shift in K_{IC} after the FATT was shifted upward 165°F by step-cooling to embrittlement. A conservative estimate can be made by reducing the average line ~20-30%, depending on the material.

TURBINE DISC CONSIDERATION

In order to prevent the shrunk-on discs shown in Figure 2 from rotation relative to the shaft during transient operation, the discs are generally keyed to the shaft by some means. A common method is to use an axial shear pin(s) or key(s) at the disc-shaft interface. The keyway in the disc serves as a stress concentrator and also as a possible site for impurities in the steam to concentrate and cause stress corrosion cracking. Several years ago [7,8] a disc burst in Great Britain from a very small (~1/16" deep) flaw present in the keyway. More recently, the author's company experienced a similar incident on an older unit. The flaw which caused this incident was ~1/4" deep and located at the corner of the square keyway. The flaw is shown in Figure 9. In both cases the keyway was through the axial length of the disc, providing a steam leakage path. Although the discs were made to the best available steel-making practices at the time, the toughness values were only a fraction of those obtained with current methods.

Preliminary analysis of the latter cracking shows that the steam environment may have had a significant role in the initiation process. Modern designs employ a round, "blind" keyway between the

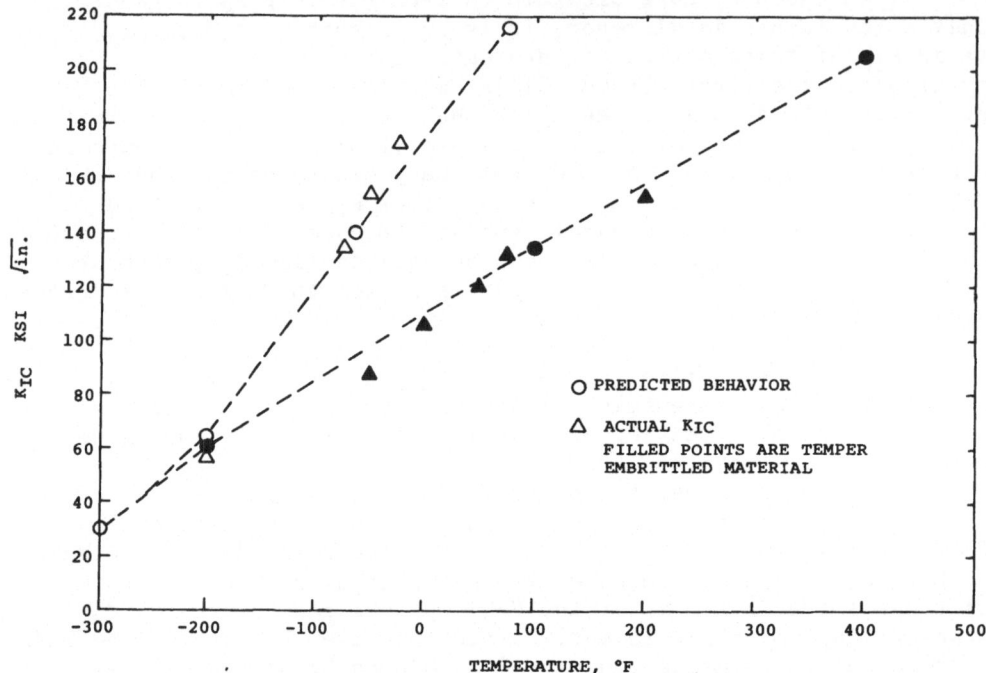

Figure 8.   Begley-Logsdon prediction of K_{IC} from Charpy tests for
embrittled and non-embrittled material.

anti-rotational pin and the disc, which limits the accessibility
of aggressive environments. The keyway introduces a local gradient
in the tangential stress which makes an exacting fracture mechanics
analysis a tedious task. Recent work [9] on this problem has
yielded experimental data to support analytically derived stress
intensity factors for rotating discs with central bores containing
notches. Figure 10 shows previously developed [10-13] analytical
solutions for stress intensity factors, while Figure 11 shows the
results of spin burst tests plotted on the same scale. The results
were obtained using the nominal stress calculated at the center of
a solid disc of the same dimensions. For most designs the average
tangential, or average bursting stress,

$$\sigma_{abs} = \rho\frac{\omega^2}{3}\ (R_o^{\ 2} + R_I R_o + R_I^{\ 2}) \tag{2}$$

Figure 9.   Critical defect size in a steam turbine disc.

is about equal to the nominal stress. R_O and R_I are the outer and
inner radii, ω is the angular velocity and ρ is the mass density
of the material.   Using the average tangential stress, the turbine
disc burst was analyzed and is plotted in Figure 11.   The good
correlation is an indicator that average tangential stress can be
used for both ductile and brittle bursting, even in the presence
of keyways.   Further work is under way in this area.

CONSERVATIVE VERSUS PROBABILISTIC APPROACHES

As previously stated, unit sizes and stresses are increasing
and the larger ingot sizes used result in generally poorer material
properties.   The more frequent cyclic loading associated with peaking application can result in higher rates of subcritical crack
growth.   Using a conservative approach to design based on fracture
mechanics technology sometimes leads to the result that the flaws

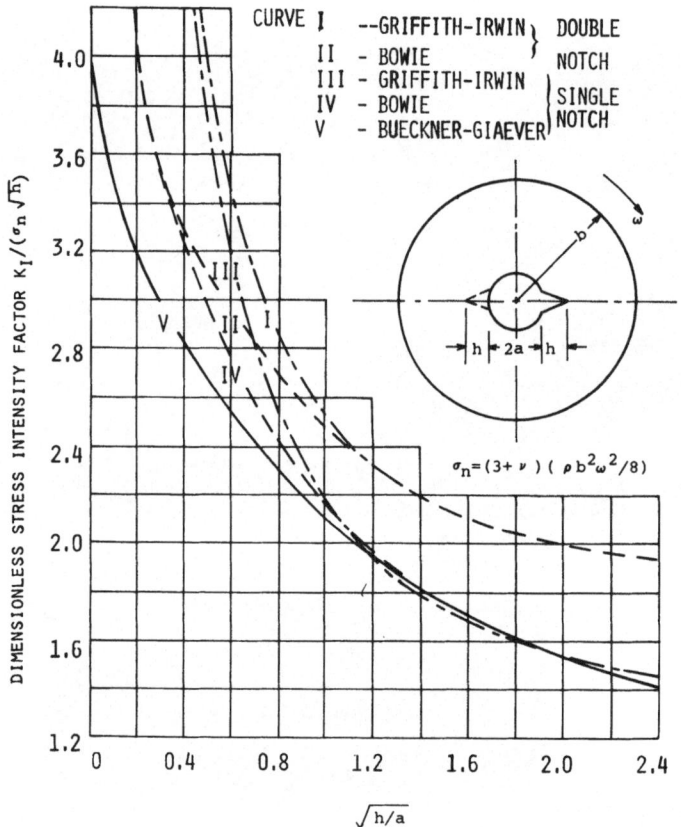

Figure 10.   Analytically derived stress intensity factors for a
rotating disc containing a center bore.

which could cause failure are approaching the limits of detectability.
The designer who uses the worst values of material properties,
assumes the highest values of stress to be operable on every cycle
and uses an extremely severe duty cycle may find that the flaw which
would cause failure may escape detection during routine inspection.
This approach is not consistent with the many thousands of machine-
years of successful operation. In order to provide a more rational
basis for design, it is assumed that input values for a fracture
mechanics analysis do exhibit a variability which can be described
mathematically using a mean and standard deviation. For example,
K_{IC} for a certain size forging might average 120 ksi$\sqrt{in}$ and a very
conservative estimate of a lower bound (<1% occurrence) of 80 ksi$\sqrt{in}$.
One reasonable distribution would be a normal distribution with an

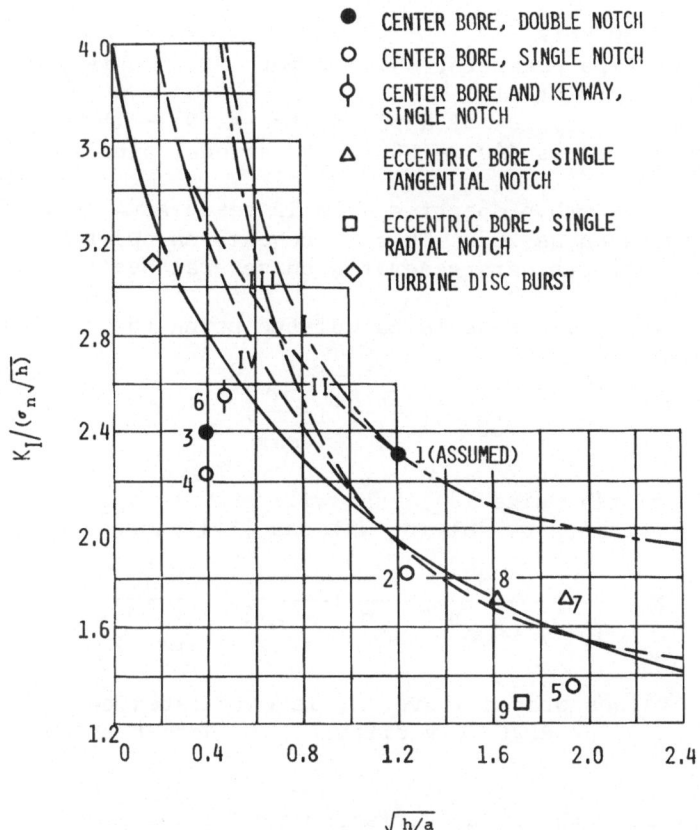

Figure 11. Comparison of test data with analytical solutions.

average of 120 ksi√in and a standard deviation of 13.3 ksi√in,
which would yield a lower bound (mean-3 x standard deviation) of
80 ksi√in. The same type of approach could be used for crack
growth rate, thermal stresses, flaw shape parameters, duty cycle,
etc. Random values of the variables are then selected from
these distributions and a value of the initial flaw size that would
grow to the critical size during the life of the turbine is calculated. Repeating this technique many times gives a distribution
on the initial allowable flaw size, which can be interpreted as
the probability that a given flaw size would cause failure. This
technique is best demonstrated with an example problem.

For the example problem, the following worst case values are
assumed:

1. Maximum stress (σ) - 66 ksi rotational plus 20 ksi thermal

stress.

2. K_{IC} - 80 ksi$\sqrt{in}$.

3. Flaw shape parameter (Q) - 1 for $\sigma \approx \sigma_{ys}$, this represents an a/2c ratio of .1.

4. Number of cycles of applied stress (N) - 16,000.

5. Cyclic stress ($\Delta\sigma$) - 66 ksi rotational plus 10 ksi thermal stress.

6. Crack growth properties - Co is taken as 4.56 x 10^{-9} and n is taken as 2.2. This represents an upper bound on crack growth data generated on several heats of rotor steel.

The critical flaw size is calculated using the equation for a surface flaw:

$$a_{cr} = \frac{Q}{1.21\ \pi}\ (\frac{K_{IC}}{\sigma_{max}})^2 \tag{3}$$

and the initial flaw size (a_i) that will grow to a_{cr} during N cycles of stress is obtained by solving Wilsons [14] expression:

$$N = \frac{2}{(n-2)\ Co\ M^{n/2}\ \Delta\sigma^n}\ [\frac{1}{a_i^{\frac{n-2}{2}}} - \frac{1}{a_{cr}^{\frac{n-2}{2}}}] \tag{4}$$

For the values stated above, a_i is calculated to be ~.008" deep, which would be extremely difficult to detect using standard techniques.

Assuming that the input variables can be described by the following distributions,

Variable	Mean	Standard Deviation
Maximum Stress	76 ksi	3.33 ksi
K_{IC}	120 ksi$\sqrt{in}$	13.3 ksi$\sqrt{in}$
Q	1.1	.05
N	10,000	2000
Cyclic Stress	70 ksi	2 ksi
Log Co (n = 2.2)	-8.67	-.11

a_i can be calculated by using the Monte Carlo technique of randomly selecting values from the distributions. This is done on a digital computer using a random number generator with a mean of zero and a standard deviation of 1. A plot of the frequency of a_i's for 100 trials is shown in Figure 12. To increase the confidence level in the output, the number of trials was increased to 100,000, and the results are plotted in terms of cumulative frequency in Figure 13.

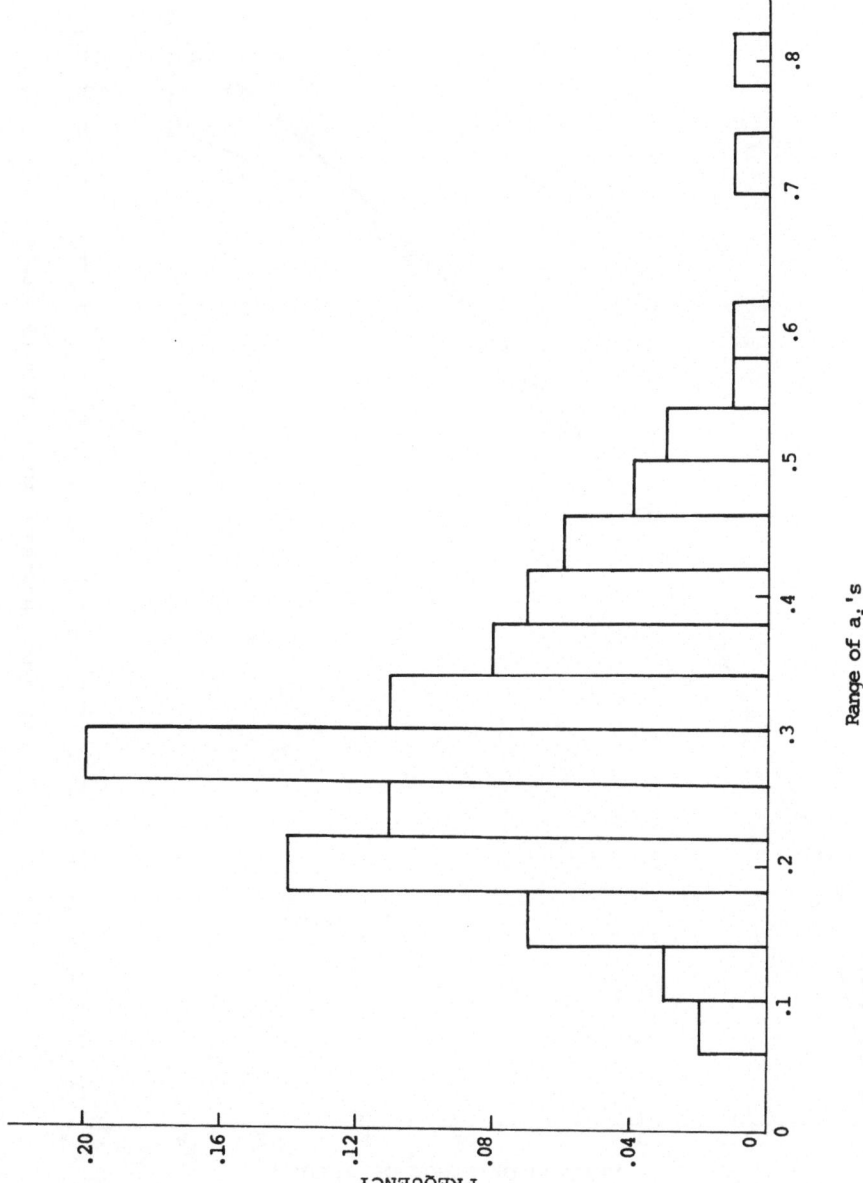

Figure 12. Distribution of 100 calculations for initial defect size.

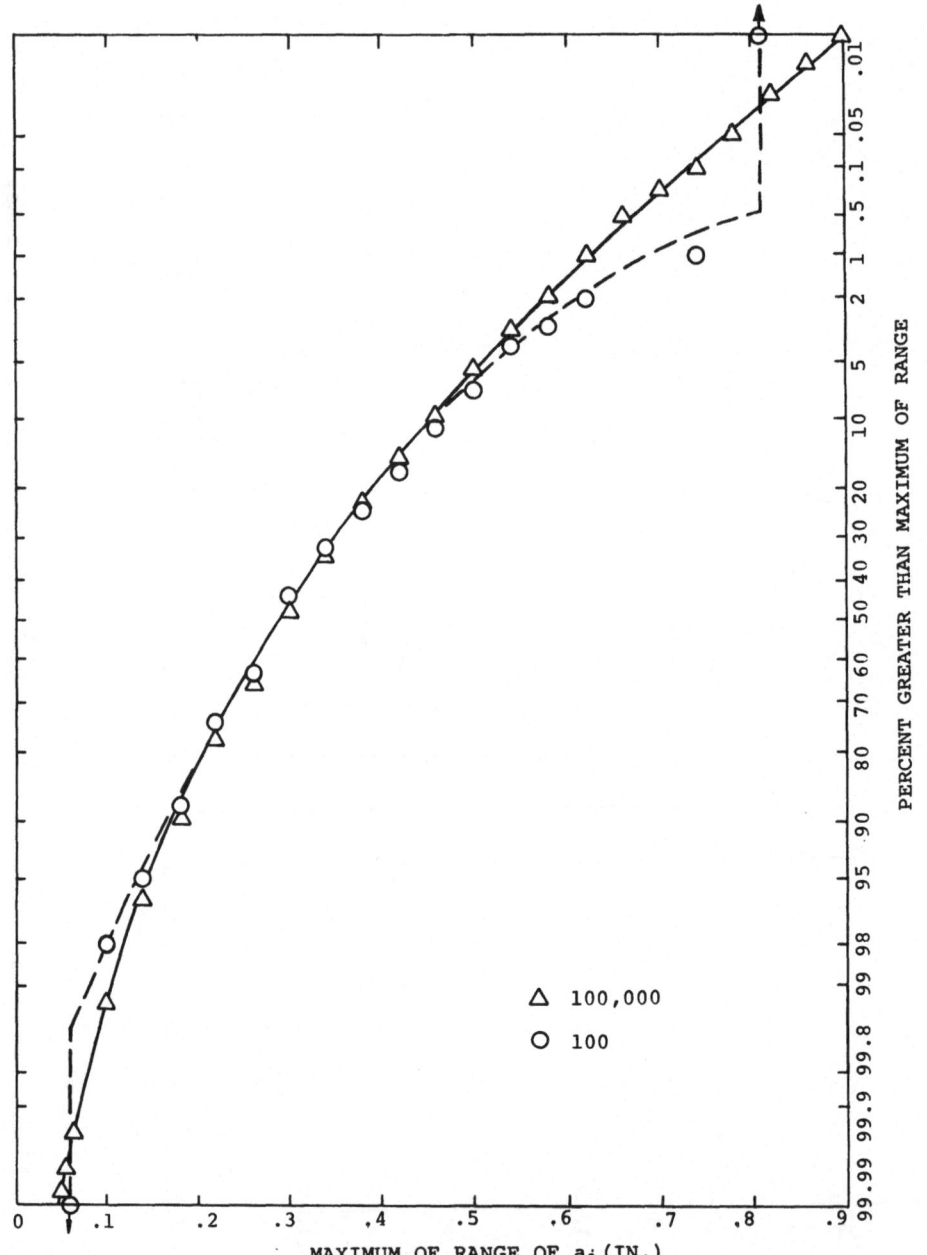

Figure 13.   Comparison of 100 calculations for initial defect size
with 100,000 calculations.

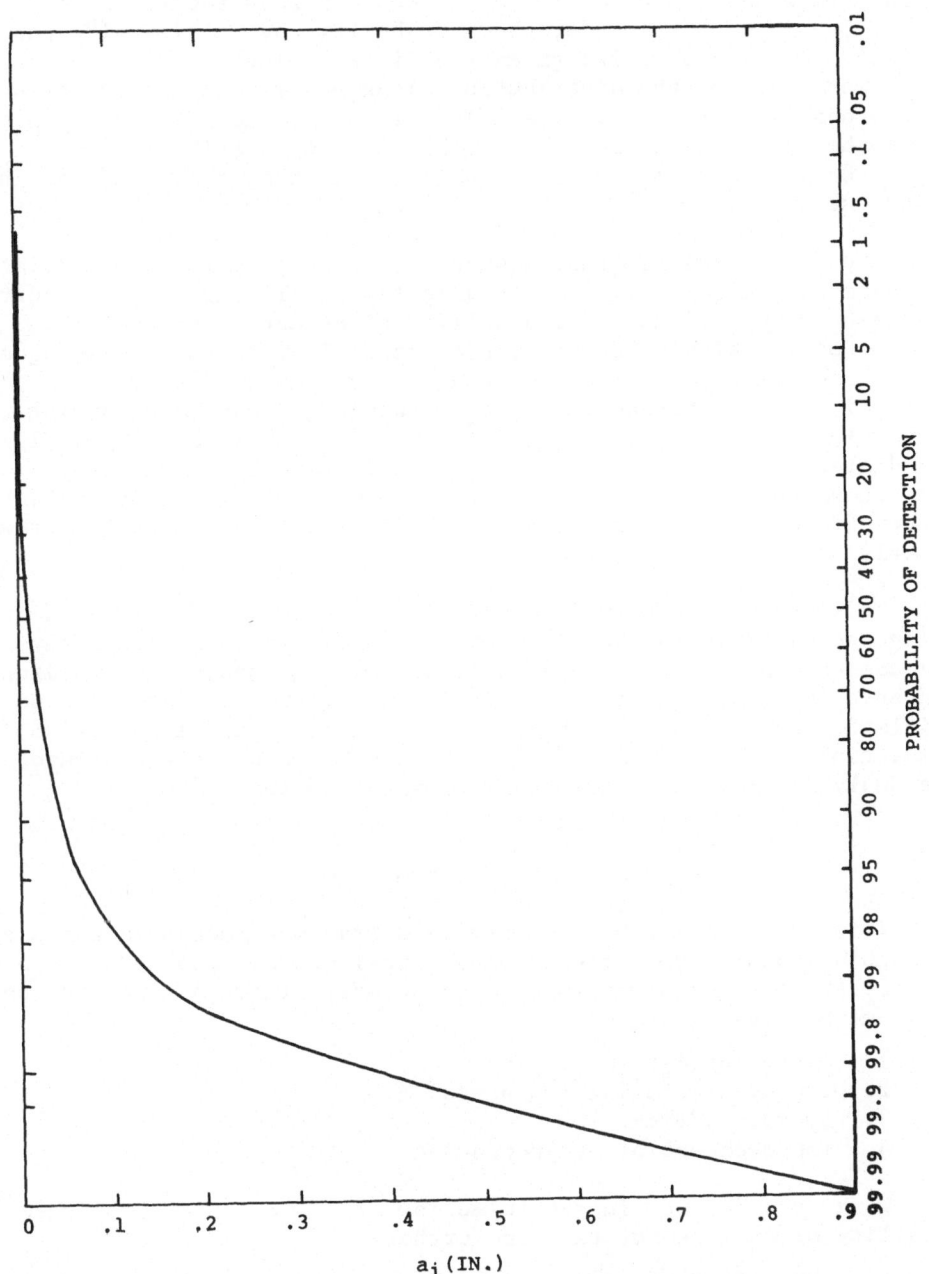

Figure 14. Hypothetical probability of detection versus flaw size.

From Figure 13, using the solid line, it is seen that for 99% of the trial cases run, a flaw >.1" deep was required to cause failure, or, in other words, the probability that a flaw .1" deep would cause failure is 1%, given the distributions of input variables. The lower limit of the distribution (mean - 3 x standard deviation or a probability ~.1%) gives a value of .075. The value calculated using the lower limit of each variable in the distribution was .088", or an order of magnitude smaller than the lower limit of the distribution.

One other variable that must be factored into the probabilistic analysis is the capability of finding flaws. In Figure 14, a curve is drawn which relates the probability of detection to the flaw size. For any given flaw size, the probability of failure is equal to the probability that the particular flaw would cause failure multiplied by the probability that it would go undetected during inspection. The probability of failure is calculated in Table 1 for different flaw sizes, and it is noted that the most likely flaw to cause failure is not at an extreme value of a_i, but at an intermediate trade-off between probability that the flaw will grow to critical and not be detected.

This example problem is meant to show some of the pitfalls of using too conservative an approach to a fracture mechanics analysis. Because so many variables are introduced in the analysis, building conservatism upon conservatism yields unrealistic output. A probabilistic approach has the benefit that the occasional low value of a material property or high value of stress can be factored into the analysis without unduly penalizing the design.

SUMMARY AND RECOMMENDATIONS

This presentation has dealt with a fracture mechanics approach to turbine rotors, both from an analytical and a hardware standpoint. Brittle fracture problems in turbine rotors, although rare, can be overcome by several means:

1. better analytical tools
2. improved material toughness
3. improved design
4. improved operating procedures

Work is under way in the following areas to improve the reliability of turbine rotors even further:

1. the effect of environment on material properties
2. improved inspection techniques for preservice and inservice use
3. Field determination of thermal transients to support

TABLE 1

PROBABILITY OF FAILURE FOR DIFFERENT FLAW SIZES

a_i (IN.)	Probability that a_i would cause bursting	Probability that a_i would go undetected	Probability of failure
.05	.0001	.1	1×10^{-5}
.075	.0015	.04	6×10^{-5}
.1	.007	.02	1.4×10^{-4}
.125	.025	.014	3.5×10^{-4}
.15	.06	.01	6×10^{-4}
.175	.12	.008	9.6×10^{-4}
.2	.19	.006	1.15×10^{-3}
.25	.36	.004	1.45×10^{-3}
.3	.54	.003	1.62×10^{-3}
.35	.7	.0025	1.75×10^{-3}
.4	.82	.0017	1.4×10^{-3}
.45	.89	.0013	1.15×10^{-3}
.5	.96	.001	9.6×10^{-4}
.55	.97	.0007	68×10^{-4}
.6	.985	.0005	4.9×10^{-4}
.65	.9925	.0003	3×10^{-4}

analytical work for start-up and load change procedure
4. determination of the distribution of the various input
    parameters for a fracture mechanics analysis
5. the use of finite element techniques for stress analysis
    due to rotational and thermal gradient

ACKNOWLEDGEMENTS

The permission to use the as yet unpublished work of
G. O. Sankey, M. M. Leven, J. Begley and W. Logsdon of the
Westinghouse Research Laboratories is gratefully acknowledged, as
is the permission of the Steam Turbine Division to publish this
chapter.

REFERENCES

1. Yukawa, S., Timo, D.P. and Rubio, A., "Fracture Design
    Practices for Rotating Equipment", in Fracture, Vol. 5, ed.
    by H. Liebowitz. New York: Academic Press (1969), 65-157.

2. Greenberg, H.D., Wessel, E.T., Clark, W.C. and Pryle, W.H.,
    "Critical Flaw Sizes for Brittle Fracture of Large Turbine
    Generator Rotor Forgings", paper presented at the Fifth
    International Forgemasters Conference, Terni, Italy, 1970.

3. Curran, R.M., "Progress in the Development of Large Rotor
    Forgings", paper presented at the Fifth International Forge-
    masters Conference, Terni, Italy, 1970.

4. Sully, A.H., "Progress in the Manufacture of Large Forgings",
    Proc. Inst. Mech. Eng., London, Pt. 1, 181 (1966-67), 877-99.

5. Begley, J.A. and Logsdon, W.A., "Correlation of Fracture
    Toughness and Charpy Properties for Rotor Steels", unpublished
    research, Westinghouse Research Laboratories.

6. Begley, J.A. and Landes, J.D., "The J Integral as a Fracture
    Criterion", in Fracture Toughness, Special Technical Publica-
    tion 514. Philadelphia: Am. Soc. for Testing and Materials
    (1972), 1-20.

7. Kalderon, D., "Steam Turbine Failure at Hinckley Point 'A'",
    Proc. Inst. Mech. Eng., London, 186 (1972), 341-77.

8. Gray, J.L., "Investigation into the Consequences of the Failure of a Turbine – Generator at Hinckley Point 'A' Power Station", Proc. Inst. Mech. Eng., London, 186 (1972), 379-90.

9. Leven, M.M., Sankey, G.O. and Bitzer, J.H., "Experimentally Determined Stress Intensity Factors for Notched Rotors", unpublished research, Westinghouse Research Laboratories.

10. Griffith, A.A., "Phenomena of Rupture and Flow in Solids", Phil. Trans. Roy. Soc. London, Ser. A, 221 (1920), 163-98.

11. Irwin, G.R., "Analysis of Stresses and Strains Near the End of a Crack Transversing a Plate", J. Appl. Mech., 24 (1957), 361-64.

12. Bowie, O.L., "Analysis of an Infinite Plate Containing Radial Cracks Originating at the Boundary of an Internal Circular Hole", J. Math. Phys., 25 (1956), 60-71.

13. Bueckner, H.F. and Giaever, I., "The Stress Concentration in a Notched Rotor Subjected to Centrifugal Forces", Z. Angew. Math. Mech., 46 (1966), 265-73.

14. Wessel, E.T., Clark, W.G. and Wilson, W.K., "Engineering Methods for the Design and Selection of Materials Against Fracture", Westinghouse Research Labs., Pittsburgh, Pa., Army Tank-Automotive Center Contract Report No. 66-9B4-315-R1, June 1966. (AD 801 005)

PRACTICAL APPLICATIONS OF FRACTURE MECHANICS TO TURBINE

ENGINE ROTORS

J. L. Price

Pratt & Whitney Aircraft, West Palm Beach, Florida

ABSTRACT

Representative applications where fracture mechanics theories have been effectively applied to gas turbine rotor components are discussed. The theoretical basis for each application is reviewed and the results are correlated with actual experimentally determined lives. The resultant capability realized by the utilization of fracture mechanics to extend the service life of parts which had exhausted their apparent useful life, is also highlighted. The current trend in fracture mechanics technology development, utilizing the Boundary-Integral-Equation (BIE) concept, to produce a simplified total life prediction system is outlined.

INTRODUCTION

The continual demand, by all turbine engine users, for higher thrust to weight ratio engines has resulted in the development of very light weight, highly efficient structures, which ultilize materials in the high strength superalloy class. These structures do not possess the "forgiveness" characteristically demonstrated by previous generations that utilized lower strength materials having significantly lower allowable operating stress levels. This trend results in hardware being cycled through a stress range much higher than previously experienced, to fully exploit the potential of the higher strength material. Furthermore, since the major stress cycles in turbine engines are caused by thrust level excursions, a significant number of cycles can be accumulated in a relatively short period of time, especially if numerous throttle

excursions are required per flight. The high number of cycles combined with the nominally high stress level causes inherent material
defects to become much more significant for crack initiation and
subsequent propagation to reach the critical crack length can be
extremely rapid. A rigorous treatment of these effects is of upper-
most importance to provide a reliable structural design. Therefore,
an improved cyclic life prediction system based on fracture mechanics
has been under development for the past several years.

The application of fracture mechanics to turbine engine hardware can be divided into two classes. One class treats the residual
cyclic life of hardware associated with the growth of a subsurface
crack propagating from an inherent material defect within the body
of the part, and the other treats the growth of a crack initiated
on the surface of a part by fatigue damage. The latter usually
occurs in a structural notch. In both classes, elastic fracture
mechanics correlation of the crack growth rate and the crack tip
stress intensity factor, which combines the crack length and the
local stress, is numerically integrated to determine crack size as
a function of the number of cycles applied.

The development of a crack propagation design system for the
first class (buried flaw) was straightforward, since the rigorous
solutions for subsurface elliptical crack in a uniform stress field
were known. The system only required a definition of the flaw size,
probability of defect occurrence, and the cycles required to initiate
the defect into a crack. This system has been incorporated at Pratt
& Whitney Aircraft and is now utilized to predict the life of all
cycle limited disk bores.

The second class of problems (surface crack) requires the crack
to propagate through a varying stress field and part geometry. The
limited number of exact solutions were inadequate to handle this
problem. A powerful tool based on influence functions developed
by J. R. Rice [1] and H. F. Bueckner [2] for cracked bodies has been
incorporated with slight modifications for treating both two dimensional (2-D) and three dimensional (3-D) problems. To date, applications of this tool are to planar cracks propagating through a
symmetrical stress field (Mode I loading). The solution procedure
utilizes boundary integral equations (BIE) or finite element solutions to develop influence functions for general geometries. Influence functions specifically adaptable to turbine engine hardware
have been computed and catalogued for automatic computer utilization.
These functions include part through surface crack, corner crack,
through the thickness crack and subsurface crack solutions. A new
criteria has also been developed to define stress intensity factors
(K). The K factors used are not point functions applied to the
crack tip locations alone, but instead it is an integrated (rms)
average K over the surface area, symbolized by $\bar{K}$. Most of the
background work required to treat the second class of problems has

been sufficiently evaluated to support its acceptance. Additional
work is under way to improve the reliability of the system by
improving the crack growth prediction during initial stages of
crack propagation where the crack growth may be affected by prior
plastic damage.

Except for the disk bore life prediction system the application
of fracture mechanics have been primarily limited to extending the
life of operational hardware which experience premature cracks in
service. These analyses were conducted by engineers schooled in
fracture mechanics techniques and usually involved "short-cut"
methods rather than an exact analytical fracture solution. The
internal defect system has been automated and is utilized by designers with little or no formal fracture mechanics training.
This system allows him to choose optimum materials, allowable
operational stress levels and nondestructive inspection standards
which are cost effective for each design.

Examples of typical applications in which the above fracture
mechanics life prediction techniques have been incorporated in the
turbine engine rotor are treated in this chapter. Specific examples
include residual life prediction of disk bores with internal flaws,
turbine blade root with high cycle fatigue initiated cracks, turbine airfoils with thermally induced low cycle fatigue cracks and
a general method currently being developed to treat surface cracks
in stress concentrated areas like disk bolt holes. Also a brief
review of the future planned improvements is presented.

APPLICATION OF FRACTURE MECHANICS TO DISK BORES

This section will treat the first class of problems requiring
the application of fracture mechanics theory; the growth of a subsurface crack. Subsequent sections will discuss applications of
fracture theory pertaining to surface initiated cracks.

The techniques developed to treat this problem are applicable
to many types of hardware with subsurface cracks originating from
many causes. In the turbine rotor field, disk rupture from inherent material defects is of paramount consideration and it has been
chosen to be addressed here.

Background

The leading cause of premature disk rupture, in the past, has
been crack propagation from a subsurface crack like material
defect (flaw). The flaw is often comprised of a single irregular
void but occasionally a chain porosity defect links together and
constitutes an equal failure source. These defects are usually

Figure 1.   An example of a fracture originating from a flaw.   FD 92287

surrounded by a brittle encasement caused by alloy segregation,
gaseous contamination, etc., which is a function of the alloy type
and fabrication methods used.   For a disk these flaws have a pre-
ference for the bore section.

    An example of such a failure is shown in Figure 1.   This disk
which was manufactured from air melted AMS 6304 steel, accumulated
about 10,000 cycles and was in the field for about ten years before
it ruptured due to the cyclic growth of the crack initiating from
a material defect.   The light circular area is the zone of LCF
crack propagation followed by a rapid tensile failure.

    As a result of this experience a fracture mechanics life pre-
diction system has been developed to preclude premature ruptures
within the design lifetime from the maximum allowable NDI permitted
flaw in the bore region.   The system developed for this material
will be briefly described to illustrate a fundamental application
of fracture mechanics theory.

                              Approach

Since the stress in the disk bore region is uniform, the

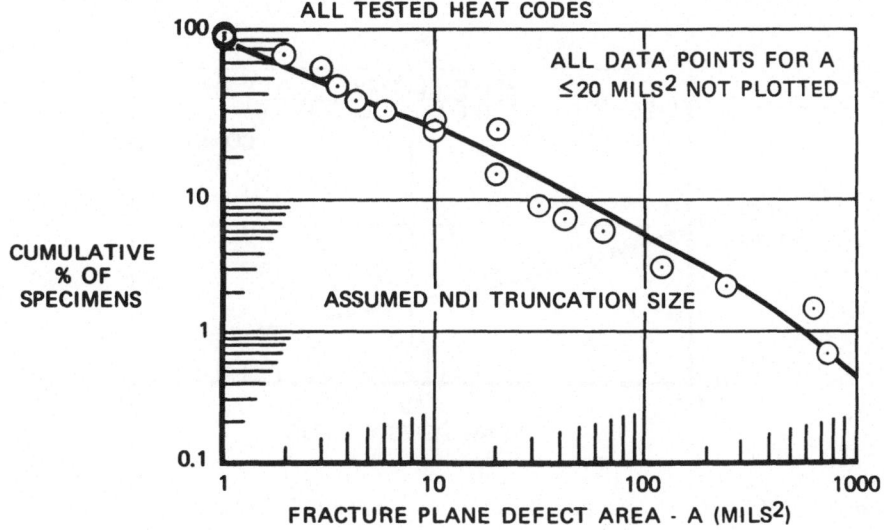

Figure 2.   Defect areas in ASM 6304 smooth specimens.   FD 92288

propagation life of an initiated subsurface crack can be accurately predicted by linear elastic fracture mechanics with the standard rigorous closed solution of a circular crack in an infinite body. The method developed for air melted AMS 6304 can be considered in four phases: the determination of the initial crack size and shape, material characterization, crack propagation model, and statistical life prediction. The stress analysis for the disk bore is a standard in-house symmetrical ring disk analysis.

Initial Crack Size and Shape

A combined NDI and fractographic analysis was utilized to characterize the flaws existing in AMS 6304 disks. NDI evaluation of representative disk forgings located defect areas for future isolation in smooth LCF specimens. Fractographic evaluations of the fracture faces of these specimens after failure defined the presence of flaws and measurements of their length and width were recorded. For simplicity, the flaw area was taken as the product of the length and width. If visible voids existed adjacent to the primary flaw they were considered as part of the flaw area. The resultant flaw size and frequence of occurrence is shown in Figure 2 as determined from NDI occurrence records and the fractographic size measurements. For this particular design application, NDI

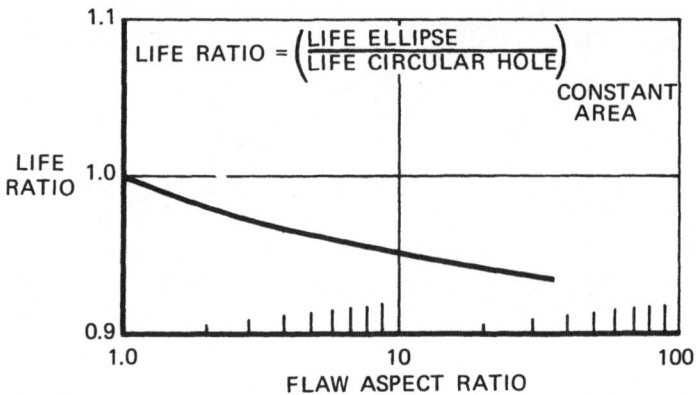

Figure 3.   Effect of aspect ratio on residual life.   FD 92289

permissible rejection rate was set to reject all disks with flaw areas greater than 100 mils^2. Therefore, the flaw distribution of Figure 2 was truncated at 100 in^2, to give the realistic design requirement. Furthermore, a fracture life analysis of actual disk ruptures indicated these flaws did in fact act like sharp cracks with little or no initiation time.

An analytical study of the effect of crack shape was conducted to help define the required shape parameter. This work has shown, for a uniform stress field, that the cyclic life of an elliptical defect is primarily a function of its initial fracture-plane area only and not its aspect ratio (see Figure 3). Therefore, the flaws of Figure 1 were assumed to be circular cracks, which remained circular in the fracture plane, from initial load application to critical length.

Material Characterization

Standard techniques for characterizing the crack growth pro-perties have been applied. The primary specimens were modified compact tension and standard center notch. A fatigue precrack is grown from an eloxed slot and then propagated with a low stress intensity until it clears the local stress field of the eloxed slot. The load is then raised and a crack length versus cycle is measured as the crack grows with a cathetometer.

The maximum scatter in crack growth rates anticipated for

actual engine hardware was determined from gang rig bolt hole specimen testing. This specimen was especially developed to allow attainment of a volume of cyclic load data economically. It is basically
a flat plate with a round hole at the center. A low cycle fatigue
crack is initiated at the hole and the crack growth is monitored to
critical. Sufficient data is obtained to evaluate the realistic
scatter inherent in the material for engine hardware. Results of a
statistical analysis of gang rig data for air melted AMS 6304 was
2.7 above the mean of 55 specimens.

Crack Propagation Model

The simplicity of this problem allows the rigorous closed form
solution for a circular internal void to be applied directly.

$$K = \frac{\sqrt{\pi a}}{\phi} \, \sigma \tag{1}$$

where

$$\phi = 1.571$$

Life is calculated by using Simpson's one third rule to numerically integrate the following equation from Reference [5]:

$$LIFE = \int_{ai}^{af} \frac{da}{c(\Delta K)^n} \tag{2} \quad \text{where } LIFE = N \text{ in } \frac{da}{dN} = c(\Delta K)^n \tag{2}$$

Statistical Life Prediction

A Monte-Carlo statistical treatment of the crack growth material
properties (see Figure 4) with data scatter included and initial
flaw size from Figure 2 were developed to determine the effect of
combining these two variables to set the mean life prediction. A
statistical computer routine was written to randomly select a crack
growth rate and flaw size combination for several arbitrarily selected disk bore stress levels. The cyclic life was then calculated
for each point selected as outlined above. Then, at each of these
stress levels, a histogram was conducted on the resultant life prediction as shown in Figure 5. A material life curve was then constructed by connecting the locus of points defining the maximum and
mean failure rate at each of the stress levels. This curve was then
used to predict the cyclic life of AMS 6304 disk bores.

The results of applying this system to known disk ruptures, from
internal flaws, are shown in Figure 6. Both steel and titanium
disks are included. The results shown are a comparison of predicted

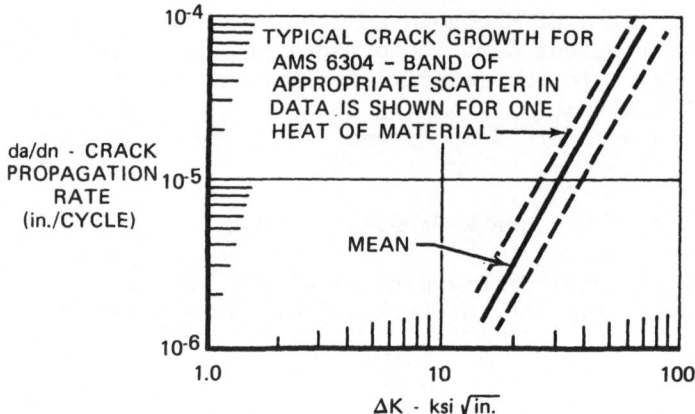

Figure 4.   AMS 6304 crack propagation data.   FD 92290

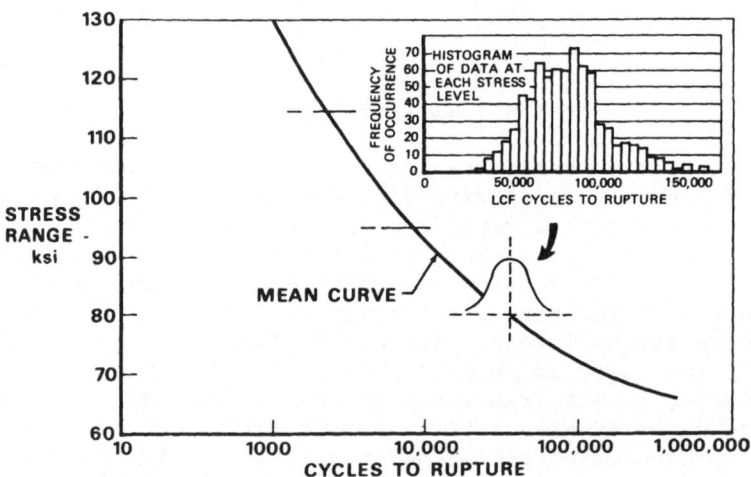

Figure 5.   Mean predicted rupture life for AMS 6304.   FD 92291

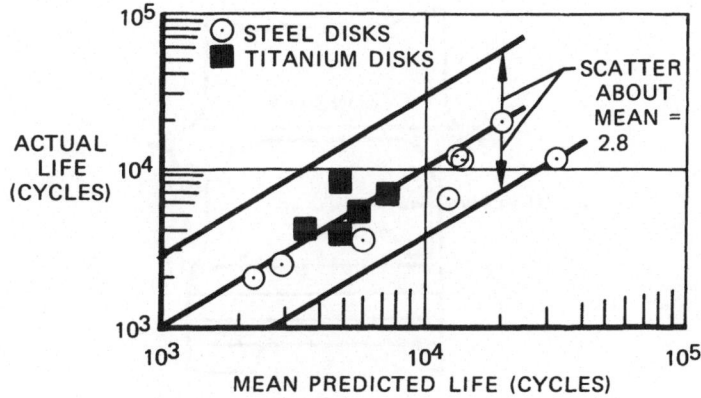

Figure 6.   Comparison of mean predicted residual life to actual
            life. FD 92292

life based on mean material data versus actual rupture life. If
the prediction was perfect all data points would fall on a 45 degree
line. Actually, all data points scattered along the 45 degree line
within a scatter factor of $\pm$ 2.8. This disk scatter indicates a
consistent agreement with inherent material property scatter of
$\pm$ 2.7 obtained in the gang rig crack growth data discussed above.

Conclusion

Based on fracture mechanics theory, a practical system has
been developed which will predict the residual life of disk bores,
with subsurface cracks, under low cycle fatigue loading.

APPLICATION OF FRACTURE MECHANICS TO TURBINE BLADE ROOTS

This and the remaining sections will treat the second class of
problems: growth of cracks on the surface of a structure.

Background

Premature cracking of a first stage turbine blade root attachment was observed on a production engine during service. The crack
was located as shown in Figure 7 and was confined to the top tooth

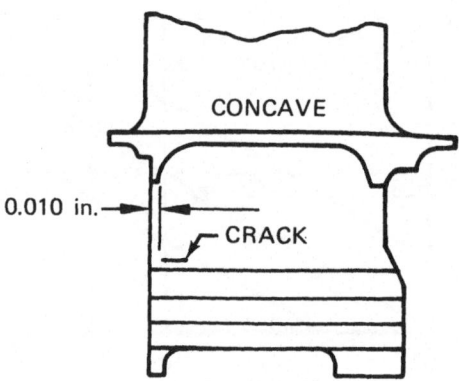

Figure 7.   Location of crack found in blade.   FD 92381

blend radius at the forward end of the attachment.   The material
was directionally solidified MAR M-200.   An experimental failure
analysis showed that crack initiation was caused by high cycle
fatigue (HCF) stresses induced at the normal steady-state operating
conditions.   Therefore the problem became one of crack propagation
under high cycle vibratory stresses.

A fracture mechanics analysis of the crack was instituted to
define quantitatively those conditions under which HCF cracks can
grow in this material and to establish the maximum allowable crack
size (inspection limit) which could be tolerated in the root when
subjected to both HCF and low cycle fatigue (LCF) loads expected
during normal operation.

The analytical program consisted of a detailed fracture mechanics stress analysis of the blade root, including the application of
integral equations and finite element techniques to evaluate the
blade root behavior in the presence of a crack.   The combined program established the maximum allowable crack length for a blade
root exposed to both HCF and LCF loading.   A design modification
which provided more efficient damping to reduce the vibratory
stress was implemented to facilitate continued engine operation.

Approach Taken

The program consisted of both an experimental and analytical
evaluation.   The experimental phase determined the threshold level
of stress intensity factor (ΔK_{TH}) below which cracks will not grow

in HCF, and to define the fatigue crack growth behavior in HCF. Both the LCF and HCF fatigue crack growth behavior as well as the HCF crack growth threshold were considered to fully investigate the extent of the blade root cracking problem under all operating conditions. Parameters investigated included temperature effects, mean stress level (R Ratio = $\sigma_{min}/\sigma_{max}$), and materials heat-to-heat variation.

The Experimental Program

Fatigue-precracked specimens were cycled at load levels which would duplicate crack tip stress intensity levels ($K_{mean}, \Delta K$) present in the blade roots under engine operating conditions. A high frequency fatigue rig was introduced to approximate the blade frequency encountered in the engine (1500 cps). A majority of testing was done at actual engine temperatures (1100°F).

Specimen Preparation

Center notched sheet specimens were used and precracking was normally done at the testing temperature to produce the minimum threshold stress intensity factor (ΔK_{TH}) for crack propagation. If the specimens were not precracked or if they were precracked at some other temperature, an effective ΔK_{TH} may be obtained which is higher than the true threshold value. During the precracking process the maximum stress in the specimen was maintained below the minimum stress for the test cycle. This was done to avoid introduction of significant residual compressive stresses at the crack tip.

K_{TH} Defined

To establish the threshold stress intensity factor, both the alternating and the mean stresses were incremented by five percent, every 10^7 cycles, keeping the R ratio (K_{min}/K_{max}) constant until crack propagation occurred. Once crack propagation began, the crack length was monitored as a function of cycles. A cathetometer was used to measure the crack length at the test temperature. Cellulose acetate surface replicas were also taken periodically and examined at magnifications up to 1000X to detect small changes in crack length. Because a crack extension as small as 0.001 inch could be measured, the threshold values ΔK_{TH} reported ensure that the crack growth rate is less than $0.001/10^7 = 10^{-10}$ in/cycle.

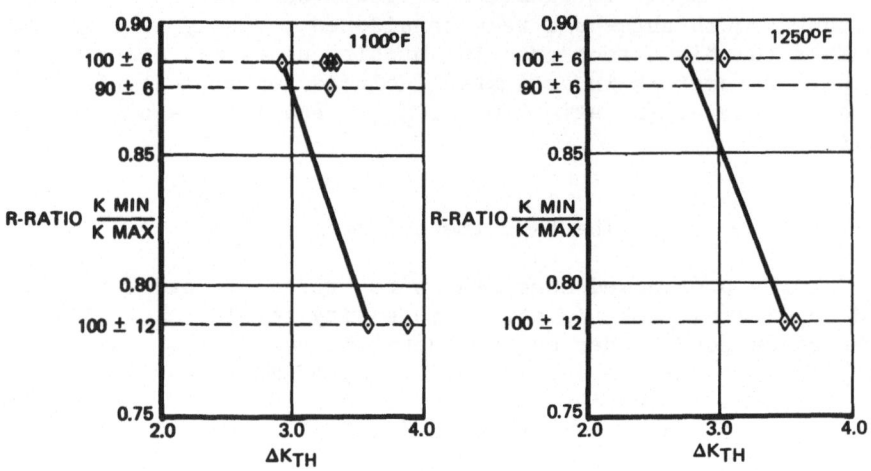

Figure 8.   Effect of R-ratio of ΔK_{TH}.   FD 92294

Test Results

Effect of R ratio and temperature on ΔK_{TH}. Figure 8 summarizes
the threshold stress intensity factors obtained for each temperature
and equivalent engine stress.   These data show that the threshold
stress intensity (ΔK_{TH}) increases with decreasing mean stress or R
ratio (R = K_{min}/K_{max} = $\sigma_{min}/\sigma_{max}$).   For the desired blade configuration, R = 0.875; (90 ± 6 ksi) the minimum threshold is ΔK_{TH} =
3.00 ksi√in at 1100°F and 2.85 ksi√in at 1250°F.

Effect of R ratio on crack growth. Figure 9 shows the crack
propagation rate as a function of ΔK at 1100°F and for various R
ratios.   Note that at higher R ratios the crack growth rate da/dN
is higher for equal ΔK values.   This effect is evident in both the
HCF and LCF crack growth behavior.

Effect of temperature on crack growth. Figure 10 compares
the crack growth rate vs ΔK relation at 1100°F with that at 1250°F
for R = 0.785; crack propagation is slightly more rapid at 1250°F.
The percentage difference in ΔK for a given propagation rate is
about the same as the percentage difference in ΔK_{TH}.

The effect of heat variation. Two additional heats of material
were investigated to determine whether significant variations might
occur from one heat of material to another within the stress and
temperature ranges of interest.   No significant differences in ΔK_{TH}

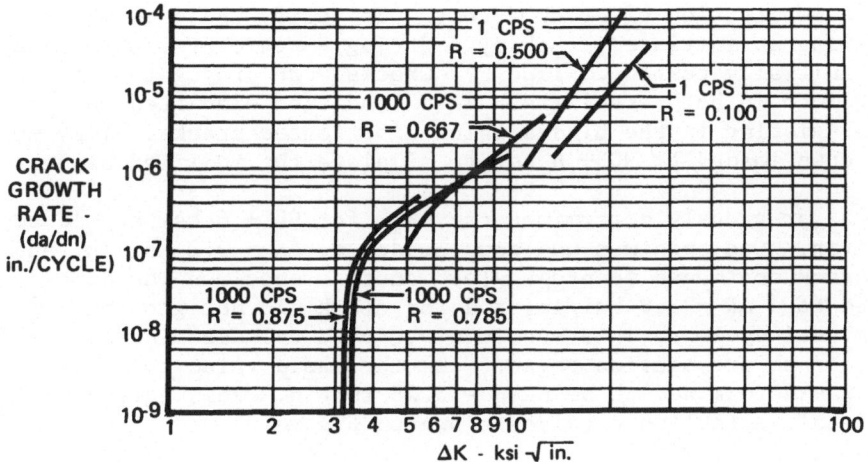

Figure 9.   Summary of fatigue crack propagation rate as a function
of stress intensity factor for PWA 1422L at 110°F and
various R-ratios.   FD 92301

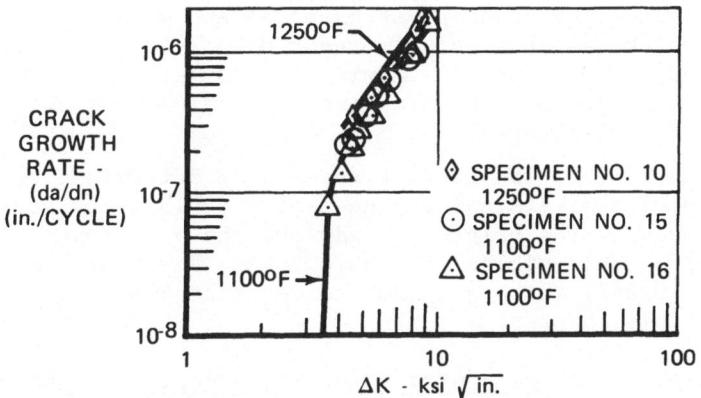

Figure 10.   Comparison of the fatigue crack growth rates at 1100°F
and 1250°F for R = 0.785.   FD 92303

or crack propagation rates were found in the tested specimens.

   Demonstration of crack arrest.   To verify the capability of
arresting an initiated crack by reducing R-ratio, two specimens
were initially tested at 1100°F.   Cracks were propagated about
0.100 inch in both specimens.   This condition simulated the conditions existing in the blades which experienced cracks.   They were
then down loaded to 90 ± 6 ksi to simulate the modified blade.

   As previously determined the ΔK_{TH} for 90 ± 6 ksi should be
3 ksi$\sqrt{in}$.   One specimen was down-loaded to $\Delta K = 3.2$ ksi$\sqrt{in}$, and
the crack continued to propagate.   Another specimen was down-
loaded to a $\Delta K = 2.9$ ksi$\sqrt{in}$, and the crack did not grow in 10^7
additional cycles.   This specimen was then uploaded incrementally
until crack propagation began.   The threshold value for this speci-
ment was determined to be $\Delta K_{TH} = 3.28$ ksi$\sqrt{in}$.   This agreed with
previously established values shown in Figure 8.

                        The Analytical Program

   Operating stress levels.   The stress analysis included finite
element solution techniques.   A three step procedure as shown in
Figure 11 was used to produce the desired results and is outlined
as follows:

   1.   A 3-D finite element analysis of the entire blade attachment under steady centrifugal loading was conducted to define the
gross stress boundaries in the attachment.

   2.   A detailed 2-D stress analysis in the vicinity of the
root radius with the 3-D displacements imposed was conducted.
Since the calculated boundary stresses agreed within 5 percent
the 3-D problem was successfully reduced to 2-D.   A detailed breakup
analysis was then conducted.   Maximum steady stress of 90 ksi was
predicted.   The proposed damping improvement did not change this
maximum predicted stress.

   3.   The fillet vibratory stresses were determined from a
combination of analysis and engine measured vibratory stress, 18
ksi and 6 ksi for the original blade and the new damper design,
respectively.   This yielded the 90 ± 6 ksi (R = 0.875) and 90 ±
18 ksi (R = 0.667) which are used to describe blade stresses
throughout this chapter.

   Calculation of blade stress intensity factors.   The existing
2-D stress analysis, with no additional breakup established K-
factors for through-the-thickness (K_{tt}) cracks.   Various crack
depths were introduced normal to the fillet surface at the peak
stress location.   Stress intensity factors (K) were then calculated
for crack depths from 0.005 inch to 0.030 inch.   However, the

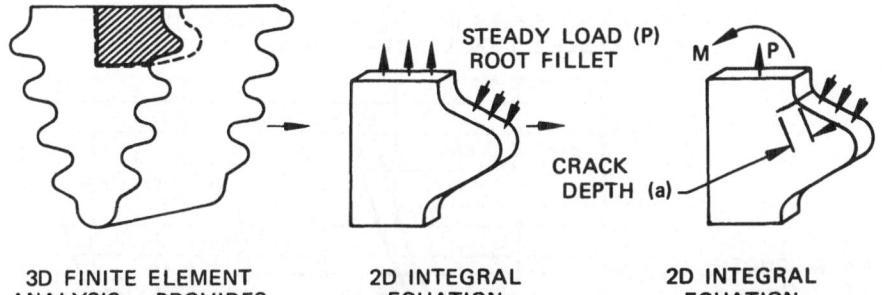

3D FINITE ELEMENT
ANALYSIS – PROVIDES
BOUNDARY
CONDITIONS FOR
DETAILED ANALYSIS

2D INTEGRAL
EQUATION
TECHNIQUES –
DETAILED STEADY
STRESS ANALYSIS
OF ROOT FILLET

2D INTEGRAL
EQUATION
TECHNIQUES –
OBTAIN STRESS
INTENSITY VS CRACK
DEPTH RELATIONSHIP

Figure 11. Flow chart of three part analytical procedure.
FD 92302

actual cracks in the blade grow with a length-to-depth ratio
(2b/a) of approximately six. The following equation was used to
calculate stress intensity factors ΔK (vibratory) and K_{mean}
(steady):

$$\Delta K = G(2b/a)\Delta K_{tt}^{(5)}$$ (5)

$$K_{mean} = \frac{\sigma_{mean}\Delta K}{\Delta\sigma}$$ (3)

$$K_{tt} = 1.1\sqrt{\pi a}\ f(a/w)^{(5)}$$ (4)

$\Delta K,\ K_{mean}$ = alternating and mean stress intensity factors
(ksi$\sqrt{in}$)

$\Delta\sigma,\ \sigma_{mean}$ = alternating and mean nominal fillet stresses (ksi)

a = crack depth

$2b$ = surface crack length

$\Delta K(a)$ = stress intensity factor (ksi$\sqrt{in}$) determined from
integral equation program for a through-the-
thickness crack of depth a (2b/a = ∞)

$G(2b/a)$ = part through crack correction function for uniform
stress fields. Throughout this chapter, it was
assumed that 2b/a = 6, and G 2(b)/a = G(6) = 0.896.

Figure 12 shows the calculated variations of ΔK with crack
length for the high and low stress blades. The respective curves

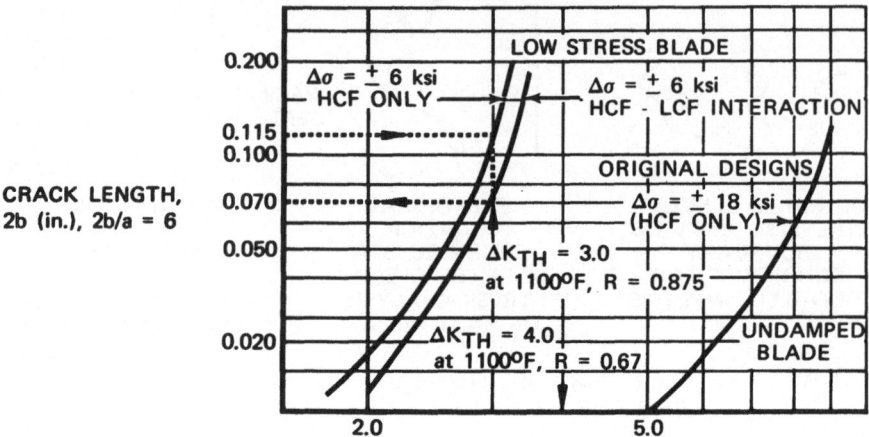

THRESHOLD CRACK LENGTH IN DAMPED BLADE IS 0.115 in.
ALLOWABLE CRACK LENGTH IS 0.070 in.

Figure 12.   Calculated variations of ΔK with crack length.   FD 92293

labeled "HCF only" shows that ΔK increases with increasing surface crack length (2b = 6a). For the low stress blade (90 $\pm$ 6 ksi) with a threshold value of 3.00 ksi$\sqrt{\text{in}}$, Figure 12 also shows that the crack length required for HCF propagation is 0.115 inch. For the original blade (90 $\pm$ 18 ksi) with a threshold value of 4 ksi$\sqrt{\text{in}}$, a crack as small as 0.010 inch may grow in HCF.

Calculation of blade allowable crack length. The maximum allowable crack length is a function of LCF stress excursions as well as vibratory cycling. Specifically, the crack was allowed to propagate in LCF until $\Delta K_{HCF} = \Delta K_{TH}$. At that size, the crack can propagate in HCF quickly ($< 10^7$ cycles) to failure.

LCF crack growth was calculated by integrating the appropriate material crack growth curves from Figure 9. For the defined LCF exposure, it was calculated that a 0.095 inch long crack in the 90 $\pm$ 6 ksi environment would grow to 0.115 inch in 300 hours at Military Qualification Test type endurance cycling.

Statistical results for the crack propagation behavior of other engine materials were then employed to extrapolate from typical to maximum (i.e., worst in 10,000 samples) crack growth rates. This analysis indicated that one in 10,000 cracks 0.070 inch long will grow to 0.115 inch long in 300 hours at MQT type endurance cycling.

Conclusion

Small cracks of a size which might not be revealed during inspection of the blade would not cause early turbine failure. Overhaul inspections at a time period equivalent to 300 hours of MQT type endurance cycling would reveal any cracks which may have occurred and no engine failure would be expected.

APPLICATION TO TURBINE AIRFOILS

Introduction

Traditionally turbine airfoil life analysis has been confined to using strain range and simple LCF strength of materials for predicting what consists mostly of crack initiation life.

The study reported by Linask and Dierberger [3] was required when cracking was unexpectedly observed in a first stage turbine blade after short time service operation. The cracks appeared to originate in the coating on the airfoil suction and pressure side walls, away from the critical leading or trailing edge locations usually associated with cracking, and propage into the base metal. (see Figure 13)

The novel application of fracture theory, developed to explain and correct this problem is now utilized as a standard design application. A brief review of this practical application will be made to highlight the techniques used.

Background

The blade was a hollow cast thin walled structure with short cylindrical pedestals structurally joining the suction and pressure sidewalls at the trailing edge as shown in Figure 14. Blade material was directionally solidified (DS) MAR M-200 superalloy, and the coating was CoCrAlY, an overlay type coating used for oxidation-corrosion protection at the blade operating temperatures. The coating thickness is about 0.005 in., typically five to ten percent of the airfoil wall thickness.

The initiation, in this case, was in the coating and was related to coating residual strain resulting from plastic deformation of the coating during normal engine operation. A brittle fracture in the coating occurred at shutdown conditions.

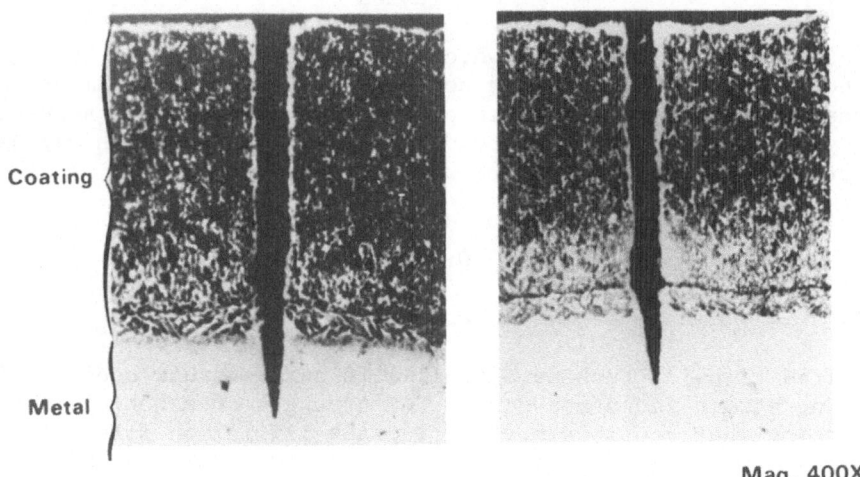

Mag. 400X

Figure 13.   First turbine blade surface cracks.   FD 92295

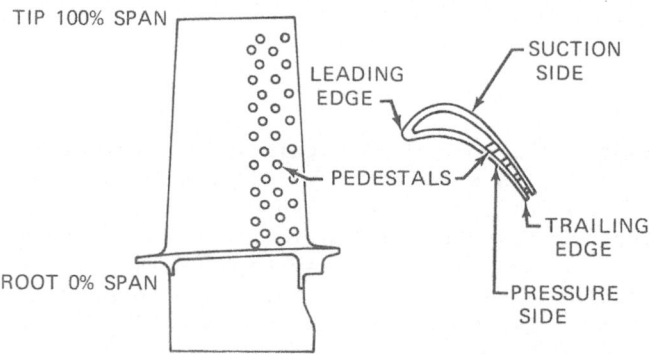

Figure 14.   Schematic of the blade analyzed.   FD 92296

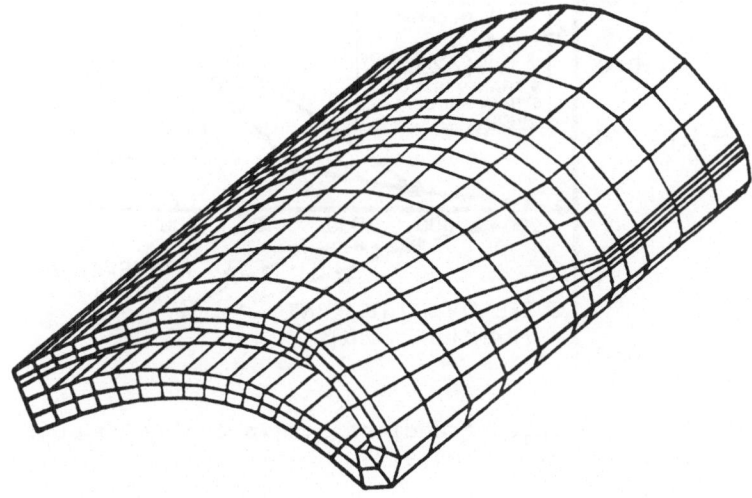

Figure 15.   Elemental breakup of the blade for ASKA analysis.
FD 92297

Approach Taken

Initiation occurred as a long crack in the coating along the
blade surface. Although some propagation occurred along the coating, the inward propagation was the principal concern.   The airfoil
life was controlled by base metal crack growth (da/dN) properties.
The surface crack propagation was neglected because its radial
length along the airfoil was very long and did not influence the
crack depth.   Therefore the coating effect was eliminated from the
crack propagation prediction.

A general three-dimensional finite element program (ASKA) was
used to model the directional properties of the airfoil for prediction of accurate strains through the thickness. A prismatic
isoparametric element was used as shown in Figure 15.

The thermal induced strains were the predominate loading.   The
centrifugal aerodynamic and vibratory loads were insignificant.
Therefore, the stress loading became one of thermally induced low
cycle fatigue (LCF) crack growth.   The ASKA results at the forty
percent span cross section, about which the cracks appeared to be
centered, are shown in Figure 16.   The surface cyclic strains
controlled the life to crack initiation in the coating and were
used to predict the coating crack occurrence.   The cyclic strains
through the blade wall were used subsequently in the propagation
analysis to predict the growth of the crack through the blade wall.

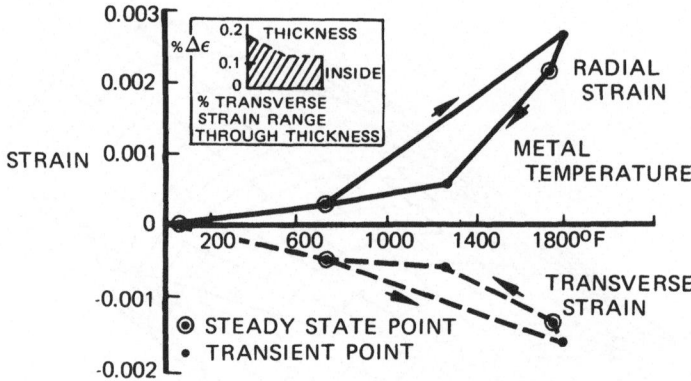

Figure 16. ASKA predicted transient strain cycles on outer surface
in base metal. FD 92298

Crack Growth Analysis

Propagation of the initiated coating cracks into the base
metal was modeled with linear elastic fracture mechanics. This
involves three steps:

1. Experimental definition of material crack growth properties
as a function of strain intensity factor.

2. Calculation of the crack tip strain intensity factor along
the projected crack path.

3. Propagation life prediction using the material property
and strain intensity information.

Two tests were made in order to assure that conditions necessary
for linear elastic fracture mechanics are satisfied:

1. The blade crack loading-strain range and crack size were
found within the criteria for linear elastic fracture mechanics [4].

2. The smallest size crack was found contained predominately
in elastic material, with the crack tip plastic zone site estimated
to be less than 0.002 in.

For the purposes of strain intensity calculation, the basic
fracture mechanics model was assumed to be that of a two-dimensional
edge crack, discussed by Paris [5], under pure in-plane bending and
tensile loading. Since the ASKA predicted strain range gradients
through the blade walls as shown in Figure 16 are nonlinear, a

computer method applicable to arbitrary load distributions was developed.

This method is based on a weighting function approach [6]. The edge crack is first assumed to be loaded by equal and opposite concentrated loads and influence coefficients are determined by solving the arbitrary unit load case.

$$K_I = \int_{ai}^{a} h_i(x,a) \ \sigma_{yy}(x) dx^{(10)} \qquad (5)$$

where the singular influence function is given by

$$h_o^2(x,a) = \frac{4a}{\pi(a^2 - x^2)} \ ; \ 0 \le x \le a, \ y = 0$$

and

$$\sigma_{yy} = \text{uncracked normal stress}$$

The total crack tip stress intensity is then found by numerical integration of the influence coefficients with the predicted stress gradients along the projected crack paths. This method could be simply extended to calculate the strain intensity because of the assumptions of linear elasticity noted previously.

Figure 17 shows a comparison of the resulting crack growth predictions and crack depths measured on the engine tested blades which were periodically removed and destructively inspected for crack size. The actual measured coating thickness ranges of 0.004 to 0.0065 in. and 0.004 to 0.0055 in. on suction and pressure side-walls, respectively, were used as initial crack sizes in the predictions. The final crack sizes shown include these coating thicknesses.

It is seen that the predicted crack propagation is in good agreement with engine measurements within the available data range. The suction wall cracks, which were predicted to proceed into the base metal, actually propagated nearly at the predicted rates up to a 0.012 in. total crack depth, while the pressure wall cracks essentially stop after initiation.

Conclusion

Further substantiation of the analytical model at still greater crack depth was not possible in the study because engine testing was terminated when the field service objectives were realized. However, a rational practical means to return parts to service which had

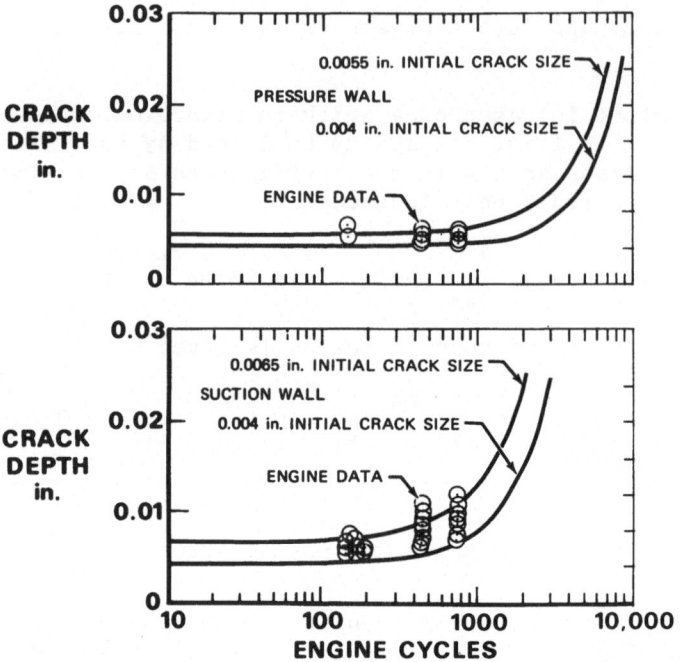

Figure 17. Comparison of measured crack depths with predicted
            crack behavior. FD 92299

already exhausted their apparent useful life had been developed
through the use of fracture mechanic theory.

APPLICATION OF FRACTURE MECHANICS TO
COMPLEX STRUCTURAL SHAPES

Introduction

The prior three examples were relatively straightforward problems because they could be closely approximated by 2-D applications
of stress and fracture theory. This section will deal with a 3-D
problem requiring a mixed mode crack propagation prediction.

Development of a fracture life prediction system to treat
surface cracks, subjected to steep stress gradients, in structures
such as disk rim slots, bolt holes, and cooling air holes is under
development at P & WA. The results of an initial disk corrobora-
tion test was reported by Cruse and Besuner [7]. The technique

developed is the basis for a general approach to a variety of crack
problems. The basic approach will be summarized here to illustrate
the future trend in practical applications of fracture theory.

The test structure was a low-pressure turbine disk removed
from engine operation with cracks emanating from the corners of all
twenty bolt holes. The material is Incoloy 901, a nickel-steel
alloy commonly used for disk forgings. As illustrated in Figure
18, each crack proceeded along the bolt hole surface through the
thickness (freedom a_y) as well as radially inward toward the disk
bore (freedom a_x). A high-temperature cyclic spin-pit experiment
was performed to ascertain the structure's actual residual lifetime
to a condition of brittle failure.

The primary difficulty in analyzing the growth of 3-D cracks
is that no one value of K may be assigned to characterize the
entire crack front; further, the stress state near the crack is
three-dimensional due to crack front curvature and complex local
geometry.

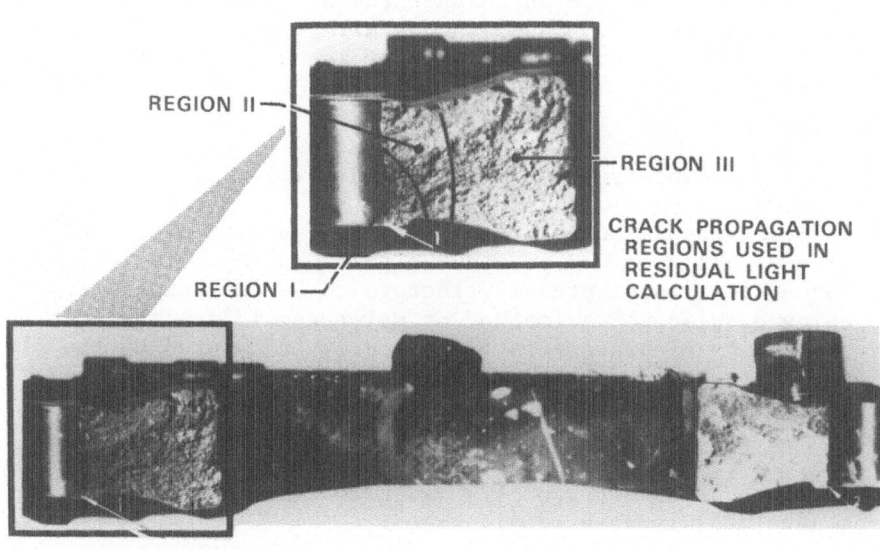

Figure 18.   Cross section showing primary failure origin and
             crack growth regions.   FD 92304

Approach

Two approaches may theoretically be employed to model this three-dimensional cracking problem. An engineering approach might consist of replacing the surface crack by an "equivalent" two-dimensional or line crack created mathematically by combining suitable analytical models with correction functions. Unfortunately, the correction functions have been found to be unique for each problem and often can be selected only when the answer (life) is already known. The other is to develop a special three-dimensional stress analysis model to reanalyze the crack geometry sequentially as it grows. This second approach requires a full three-dimensional solution for the crack at each increment in its growth history, and for each local stress distribution.

The residual life analysis procedure selected utilizes the latter approach. Recent developments in fracture mechanics analysis technology are employed for accurate 3-D stress prediction and crack propagation is accomplished through the use of an "equivalent" crack model. This work developed a technique based on the numerical implementation of a boundary integral equation (BIE) method of 3-D stress analysis.

The actual crack growth was simulated through the use of an influence function technique which uses the stresses in the uncracked structural detail. Thus, the details of both surface crack geometry and structural shape are directly accounted for in the residual life prediction.

Boundary Integral Equation (BIE)

The BIE method presented models only the surface of the body to be analyzed. This method provides significant solution accuracy to allow accurate stress intensity factors to be determined from crack opening displacements (method of Reference [8]). The necessary crack opening displacements are directly calculated by the BIE method without the need for interior solutions. The accuracy of this method has been evaluated relative to the closed form known solutions. The results of this comparison are shown in Table I. Five separate test cases with various crack geometries and loading are shown. The basic test geometry was a constant thickness plate. Excellent agreement is seen in all cases.

Residual Life Analysis

The features of the method proposed utilizing the developments for predicting crack propagation to critical length can be

Table I. Comparison of the Boundary Integral Equation (BIE) with the Literature Solution Stress Intensity Factor Prediction

Load Condition Crack Geometry	Nondimensional Stress Intensity Factor $K/\sigma_0 \sqrt{\pi a}$		
	BIE	% Dif. +	Literature Solution*
1. Internal crack constant def. = 0.01 a/w = 1/5	0.9935	0.4%	0.9896
2. External crack constant def. = 0.01 a/w = 1/5	1.1961	0.5%	1.2029
3. External crack point load a/w = 1/5	1.1817	2.2%	1.2104
4. External crack linear def. a/w = 1/5	1.1404	0.12%	1.1389
5. External crack linear def. a/w = 1/2	2.1496	2.5%	2.2059

*Error estimates for the literature solution (Reference 11) are estimated as being less than 2%.

+% difference between that result and the literature solution.

σ_0Average stress in uncracked solid at crack location $\sigma_0 = \int_0^a \sigma(x)\, dx / a$.

summarized in four steps. Step one is the procurement of appropriate material crack propagation data. Standard center notched sheet specimens machined in the proper orientation from representative disk forgings provided all the material data required. Standard data reduction techniques previously described were applied.

Step two is the stress analysis of the uncracked structure. Since the crack need not be included in the stress analysis, standard in-house programs are used for this step. Table II shows

Table II. Transverse Stress Concentrations Near the Bolt Hole

Distance from Bolt-Hole Surface (in.)	$\sigma_{zz}(x, y)/\sigma_{nominal}$	
	Disk Front Face	Disk Back Face
0.	2.20	1.96
0.0058	2.11	1.89
0.0115	2.05	1.83
0.0211	1.95	1.74
0.0307	1.84	1.67
0.0519	1.70	1.52
0.0709	1.56	1.40
0.105	1.45	1.28
0.137	1.34	1.20
0.212	1.24	1.11
0.287	1.18	1.05
0.437	1.15	1.03
0.587	1.18	1.05
1.037	1.34	1.20

σ_{actual} = Actual calculated local stress due to hole concentration.

$\sigma_{nominal}$ = Nominal calculated stress in disk web.

the complex nature and high gradients of the uncracked stress field with point estimates of the transverse stress concentration factor.

Step three is to model the propagating crack. Figure 18 defines three regions of crack propagation chosen for separate modeling. Region I defines crack propagation from a measured initial configuration ($a_x + a_y = 0.03$ in.) to a configuration three fourths ($a_y = 0.64$ in.) of the distance through the 0.85 in. thickness. Region II defines the transition crack growth to a one degree of freedom (DOF) through-thickness crack, while Region III defines the growth of the through-thickness crack to a critical dimension causing brittle failure.

The model chosen for Region I was found to be the most important part of the residual lifetime analysis. The acceleration of the crack growth as the crack length increased caused very fast propagation as the crack approached a through the thickness condition. The two-DOF corner crack model is chosen for estimation of the Region I. Advantage was made of the buried elliptical crack K analysis of Reference [9], the BIE was applied to compute crack opening displacements, influence factors, and $\overline{K}_x$ and $\overline{K}_y$ for a corner crack under arbitrary load.

The K_i was computed with influence function theory from

$$\overline{K}_x = \iint_A h_x (x,y,a_x,a_y) \, \sigma_{zz}(x,y) dA \qquad (7)$$

$$\overline{K}_y = \iint_A h_y (x,y,a_x,a_y) \, \sigma_{zz}(x,y) dA \qquad (8)$$

where σ_{zz} is the uncracked stress.

The influence functions h_x and h_y are computed from knowledge of the crack opening displacement $w = u_z$ as a function of position and geometry (a_x and a_y in the case of the corner crack model) for a single arbitrary uncracked stress, called a "reference stress". Reference [9] derived the key equations for the influence function theory as

$$h_x = [\frac{\partial(\iint_A \sigma_{zz}*w*dA)}{H \partial A}]^{-1/2}_{a_y} (\frac{\partial w*}{\partial A})_{a_y} \qquad (9)$$

$$h_y = [\frac{\partial(\iint_A \sigma_{zz}*w*dA)}{H \partial A}]^{-1/2}_{a_x} (\frac{\partial w*}{\partial A})_{a_x} \qquad (10)$$

where the superscript * denotes a single reference stress $\sigma*_{zz}(x,y)$ whose choice, in theory, does not affect h_x and h_y which depend on position and geometry alone. Therefore, accurate computation of h_x and h_y for any convenient $\sigma*_{zz}$ eliminates any future need to include the corner crack in stress analysis models. The BIE provided crack opening displacement ($w*$), the uniform reference stress $\sigma*_{zz}(x,y) = \sigma_o$ of the corner quarter-ellipse.

Thus the third step was completed by substitution of the stress data as shown in Table II, and crack dimensions (a_x, a_y) into a small computer subroutine that stores the BIE-generated influence functions and numerically evaluates the integrals in Equations (7) and (8). Initial values of a_x and a_y are not explicitly known, although their initial sum is 0.03 in. A procedure for initial value estimates, compatible with the applied model, was used to calculate the initial a_x and a_y values. The final Region I configuration is, as previously stated, given by $a_y = 0.64$ in.

Rigorous K computation for cracks in Region II was not required. Fortunately, the lifetime associated with Region II is small in this instance and was ignored in the analysis. The Region III crack growth lifetime is also small but was estimated with two-dimensional K-formulas and the accepted methods similar to those previously described in the turbine airfoil application.

The fourth step for Region I life estimates is the substitution of previously described estimates of $\overline{K}_x$, $\overline{K}_y$ and initial and final values of a_x and a_y into the crack growth rate determined from the material crack growth rate properties. For this material

$$da/dN = 1.3 \times 10^{-10} \, \Delta K^{3.1} \text{ in./cycle} \tag{11}$$

$12 < \Delta K > 0$ ksi$\sqrt{\text{in}}$ was used. Then

$$(da/dN)_x = 1.3 \times 10^{-10} \, \overline{K}_x^{3.1}$$
$$(da/dN)_y = 1.3 \times 10^{-10} \, \overline{K}_y^{3.1} \tag{12}$$

The final results of the life prediction and disk test results are shown in Figure 19. Both the mean of the twenty holes and the maximum hole crack growth are shown. A comparison of the mean measured crack-propagation life and the predicted life using median material properties shows good agreement. Slope correlation between the two median-based curves is very good with most discrepancy occurring at the smallest crack lengths. Causes for this discrepancy are currently investigated to further optimize the prediction.

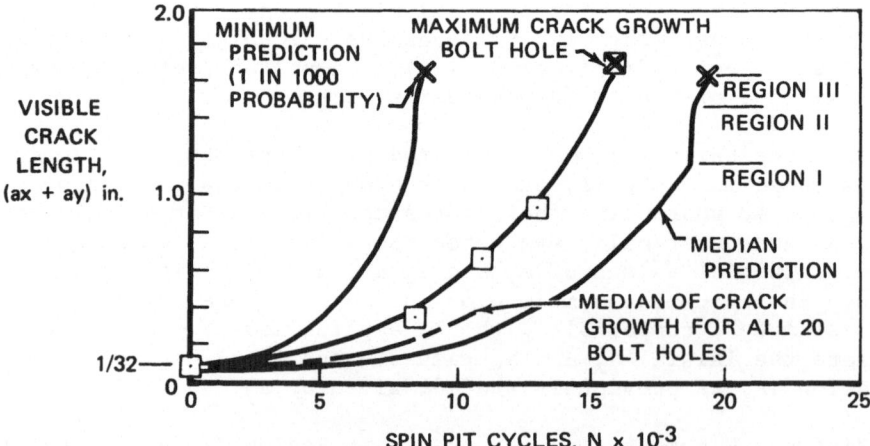

Figure 19. Disk spin test data to analytical prediction. FD 92300

However, from a practical design viewpoint, a comparison between
the crack growth observed in the fastest growing hole and the
minimum life projected and with the expected material da/dN
scatter included, would reflect a realistic design system prediction.
It can be seen from Figure 19 that the system proposed would give
a realistic minimum life prediction for the test disk.

Conclusion

The first step to a general purpose crack growth prediction
technique has been completed and with some additional substantiation
to verify its reliability, a viable design system is in the offing.

REFERENCES

1. Rice, J.R., "Some Remarks on Elastic Crack-Tip Stress Field",
    Int. J. Solids Struct., 8 (1972), 751-58.

2. Bueckner, H.F., "A Novel Principle for Computation of Stress
    Intensity Factors", Z. Angew. Math. Mech., 50 (1970), 526-46.

3. Linask, I. and Dierberger, J., "A Fracture Mechanics Approach
    to Turbine Airfoil Design", ASME Publication GT-79, 1975.

4. Rau, C.A., Jr., Gemma, A.E. and Leverant, G.R., "Thermal-
    Mechanical Fatigue Crack Propagation in Nickel- and Cobalt-
    Base Superalloys Under Various Strain-Temperature Cycles",
    in Fatigue at Elevated Temperatures, Special Technical
    Publication 520. Philadelphia: Am. Soc. for Testing and
    Materials (1973), 166-78.

5. Paris, P.C. and Sih, G.C.M., "Stress Analysis of Cracks", in
    Fracture Toughness Testing and Its Applications, Special
    Technical Publication 381. Philadelphia: Am. Soc. for Testing
    and Materials (1965), 30-81.

6. Bueckner, H.F., "Field Singularities and Related Integral
    Representations", General Electric Technical Report DF71LS162,
    Appendix II (October 1971), 92.

7. Cruse, T.A. and Besuner, P.M., "Residual Life Prediction for
    Surface Cracks in Complex Structural Details", J. Aircraft,
    12 (1975), 369-75.

8.   Cruse, T.A., "Numerical Evaluation of Elastic Stress Intensity
     Factors by the Boundary Integral Equation Method", in The
     Surface Crack: Physical Problems and Computational Solutions,
     ed. by J.L. Swedlow. New York: Am. Soc. of Mechanical
     Engineers (1972), 153–70.

9.   Besuner, P.M., "Residual Life Estimates for Structures with
     Partial Thickness Cracks", in Mechanics of Crack Growth,
     Special Technical Publication 590. Philadelphia: Am. Soc.
     for Testing and Materials (1976), 403–19.

10. Paris, P.C., Gomez, M.P. and Anderson, W.E., "A Rational
     Analytical Theory of Fatigue", Trend. Eng., Wash. Univ., 13,
     no. 1 (1961), 9–14.

11. Tada, H., The Stress Analysis of Cracks Handbook. Hellertown,
     Pa.: Del Research Corporation, 1973.

FRACTURE MECHANICS AND FAIL-SAFE DESIGN FOR HELICOPTER

ROTOR STRUCTURES

M. J. Rich

Sikorsky Aircraft Division of United Technologies
Corporation, Stratford, Connecticut

ABSTRACT

Fracture mechancis analysis is the key for design of fail-safe
structures. Controlled fracture offers a substantial improvement
for the design goal of increased safety and reliability of helicopter
rotor system components. However, to achieve an overall gain the
system must be inspectable and the fail-safe and safe-life features
must be integrated. Fracture mechanics analysis is currently being
used for many of the metallic rotor components with crack propagation time being the most important factor. The new advanced composites offer a substantial improvement in crack time but the
emphasis may well now be in static fracture strength. By proper
consideration of trade-offs in fail-safe and safe-life aspects
lighter weight and increased safety can be achieved.

INTRODUCTION

The safety and reliability of helicopter rotor system components depends to an important degree on the selection of materials.
The material properties that govern a lightweight reliable component design are at present fatigue resistance to the large number
of cyclic loadings and to a lesser extent the static strength.
Design practice has been to reduce mean fatigue data, either from
small specimens or from full scale tests, to assure that fracture
is extremely remote from the operating loads environment within a
specified replacement time. In addition, a factor of safety of at
least 1.5 is used for design limit loads to preclude static fracture
from the extreme peak conditions to be encountered in service.

An additional criteria, that has been growing in use by many of the helicopter manufacturers, is to consider damage tolerant design, wherein controlled fracture permits a significant safe flight time after the onset of initial cracking or other type of damage. The requirements by the military for ballistically damage tolerant design has accelerated the design concept of controlled fracture.

There appear to be two basic philosophies in design. The first, and most commonly used, is "Safe-Life", wherein safety is governed by providing sufficient design factors to preclude fracture within the replacement period of the component. The second, and now appearing to be gaining favor, is the "Fail-Safe" concept wherein damage modes are considered in the basic design of the component. Fail-Safety is specified in many ways, such as redundancy, controlled fracture, or some combination of the two. To achieve fail-safety requires that both the residual strength and life after damage be sufficient to preclude fracture within some specified inspection interval.

In essence the two philosophies are the same, i.e., a "life" is specified, either to crack initiation for "safe-life" or from crack initiation to fracture for "fail-safe". The major difference is that in theory no inspection is required for the "safe-life" approach, and inspection intervals are specified for fail-safe. In actual practice all rotor system components are inspected, even in the safe-life approach.

To achieve an overall safe, reliable design a combination of the crack initiation and crack propagation times is required. By proper combination the best gains in overall safety and reliability can be achieved. It is possible to achieve reduction of weight of the component and a reduction of cost through the economics involved through longer replacement periods and even, in some cases, replacement on condition only.

Fracture mechanics now provides the analytical tool to assess the residual strength and life of the rotor components, temper the inadvertent use of "higher" strength materials, and provide the assessment for required periodic inspections. Fracture mechanics also provides the means for the design and choice of materials for the most difficult problem of all for rotor components: the ability to withstand ballistic damage, particularly from higher threats.

DESIGN CRITERIA

The first and most important criteria is in the definition of fail-safety. Fail-safety in this chapter is defined as [1] "any structure whose characteristics are such that in the presence of

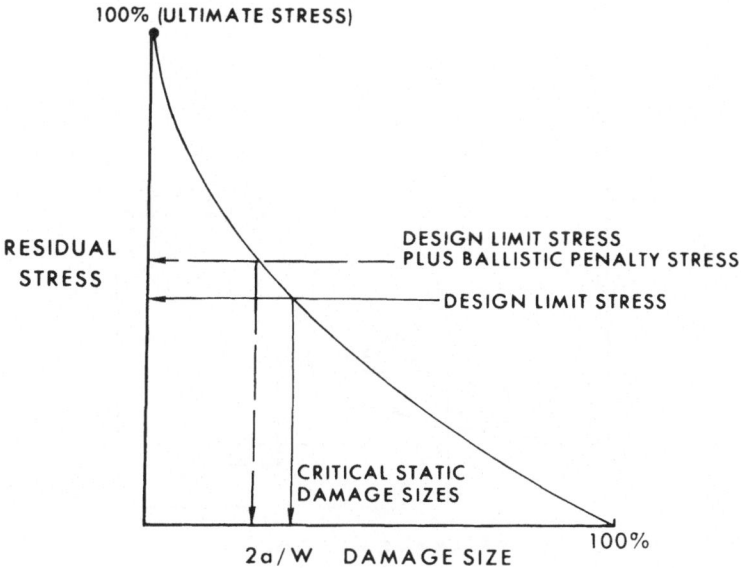

Figure 1.   Residual stress capability defines critical damage size.

abnormalities, such as fatigue cracking and/or physical damage or
deterioration, the probability of a catastropic failure prior to
detection of the abnormality is extremely remote". Similar definitions on fail-safety [2,3] have been presented in other papers for
fixed wing and helicopter structures.

Thus, to achieve a fail-safe design, for which we can be
assured that a catastrophic fracture is made extremely remote, we
need two factors: first, the structure can tolerate the initial
damage, and secondly, the progression of damage does not lead to
fracture within a specified inspection interval.

The size of the initial damage to the rotor system structure
encompasses a large range. An initial flaw or service damage may
be very small; however, the important aspect is "what is the size
that may be detected?" Therefore the method of inspection becomes
the controlling factor in assessing the residual strength and life.
Systems for main rotor blade inspection such as BIM[R] used by Sikorsky
or ISIS used by Boeing-Vertol have the advantages of early warning.
Thus initial damage and progression of such damage for specified
inspection intervals is well within the fracture strength of the
structure. Combat damage, particularly from higher threats, may
be large enough so that the residual strength problem may dominate
particularly when penalty stresses (at initial impact of projectile)
are high.

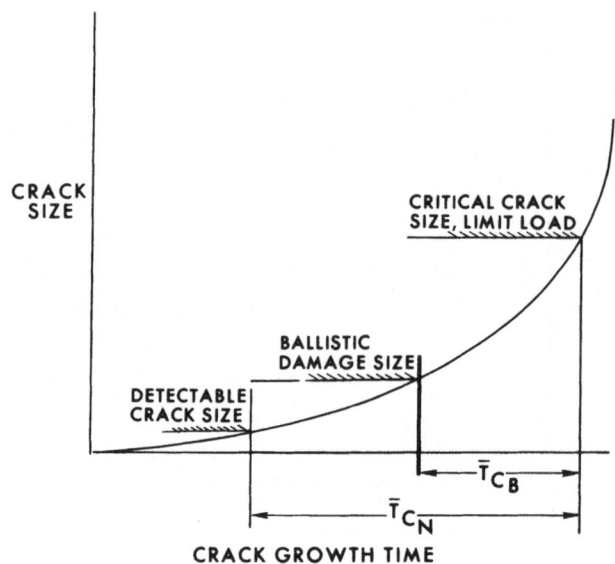

Figure 2.   Crack growth time depends on initial detection or
imposed damage.

    As illustrated in Figure 1, the residual stress capability
has two controlling damage sizes. One limiting damage size relates
to the ability to withstand fracture under the maximum (limit)
operating stress environment and becomes the upper bound for crack
propagation. The other size relates to an initial ballistic impact
wherein the penalty stresses of impact are present.

    Most current helicopter rotor system components are designed
by fatigue considerations from cyclic loadings. Therefore, the
residual stress capability is not usually a problem area due to
large static margins of safety for normal operations. However,
ballistic damage, combined with penalty stresses, presents a more
severe condition for survival at initial impact.

    The criteria for inspection intervals, or duration of safe
operating flight intervals with ballistic damage, depends on the
propagation rate and time of initial damage.

    As illustrated in Figure 2, the crack or damage propagation
time depends on either the initial size of crack detection or the
more extensive initial size of the ballistic damage.   The total

mean time to fracture is limited to the size of damage that pre-cludes the "static" fracture for peak load conditions.

Under dynamic loading the crack velocity will reduce the critical size for peak loadings as compared to the true static case. The importance of this aspect depends on the material and should be considered.

The mean crack growth time $\overline{T}_c$ for for fail-safe design is therefore dependent on the initial crack detection size or the imposed ballistic damage size and the extent to which the damage becomes fracture critical. The final critical fracture size is yet a question, but the conservative approach would be to limit crack growth to enable the structure to be safe for maximum (limit) loading.

A more important aspect is the question of what factor of safety to use in reducing the mean crack growth time to fracture for safe inspection or flight intervals. Various criteria for crack time have been used. For the H-46 main rotor blade Thompson and Weiss [4] proposed a minimum of 15 hours crack life for a rotor blade having a built-in inspection system ISIS that can be inspected prior to each flight. This suggests a life factor of about 4 for inspection intervals. Weiss and Zola [5] specified a criteria of thirty flight hours from detection to fracture for the UTTAS YUH-61A, and a specification requirement of 30 minutes flight time for 7.62mm damage with a limit load capability.

The question arises as to what benefits does controlled frac-ture buy in the overall component reliability. The answer is not a simple one, since there are three factors to be considered. These are:

(a) Reliability of structure for crack initiation

(b) Crack propagation time and its reliability between inspection intervals

(c) Reliability of the inspection

One method [6] of considering the interactions of the factors involved relates the overall probability of fracture as being:

$$P_F = P_{CI} (P_I + P_{CP} - P_I P_{CP}) \tag{1}$$

where P_F is overall probability of fracture

  P_{CI} is probability of crack initiation

  P_{CP} is probability of crack propagation to fracture within the given inspection period

  P_I is the probability of failure of the inspection method

For the safe-life concept the mean S-N data for the component
is reduced by some value to achieve a low probability of crack
initiation, one that service experience has proven to make fracture
extremely remote. In statistical terms the mean stress is reduced
by a given number of coefficients of variation in stress to arrive
at a working or allowable design fatigue stress.

$$S = \bar{S} \, (1 - n \, \sigma) \qquad\qquad (2)$$

where S is the allowable safe-life stress

$\bar{S}$ is the mean fatigue stress

n is the number of coefficients of variations

σ is the coefficient of variation in stress

The coefficient of variation in stress σ is known from test,
the value of n is specified from service experience. The result
is that one can relate a relative reliability in design. The
reliability is relative since given statistical values have been
proven to provide the necessary safety against fracture.

A comparative method can be used for crack propagation time,
i.e.,

$$T_C = \bar{T}_C \, (1 - n \, \sigma_C) \qquad\qquad (3)$$

where T_C is the permissible inspection period

$\bar{T}_C$ is the mean crack time

σ_C is the coefficient of variation in life for the crack
propagation data.

If the condition were that crack propagation has begun, thus
ignoring crack initiation aspects, then the life factor to be used
would be:

$$\text{L.F.} = \text{Life Factor} = \frac{\bar{T}_C}{T_C} = \frac{1}{1 - n \, \sigma_C} \qquad\qquad (4)$$

The life factor depends on the variability of the material,
σ_C, and the probability of fracture within the inspection interval.
As illustrated in Figure 3, the inspection interval would be
reduced where there is a large scatter in life for crack propagation.

The next factor, and the most important for the fail-safe
concept, is the reliability of the inspection method. For a given
inspection reliability the probability of failure within that
inspection period is:

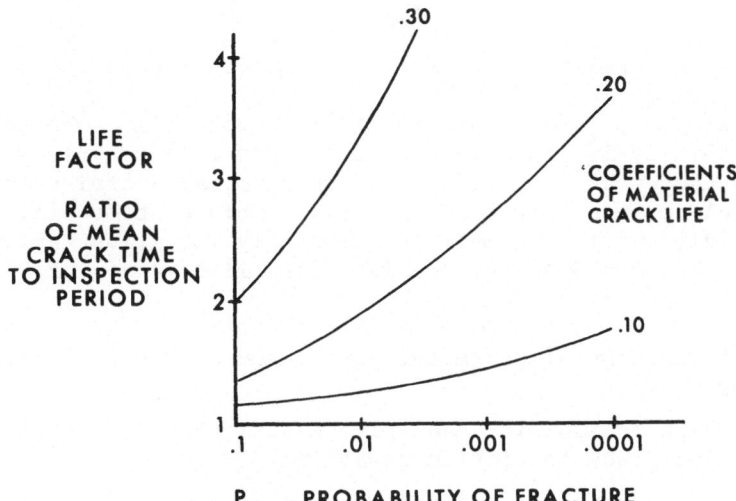

Figure 3.   The inspection interval is reduced for materials with large scatter in life.

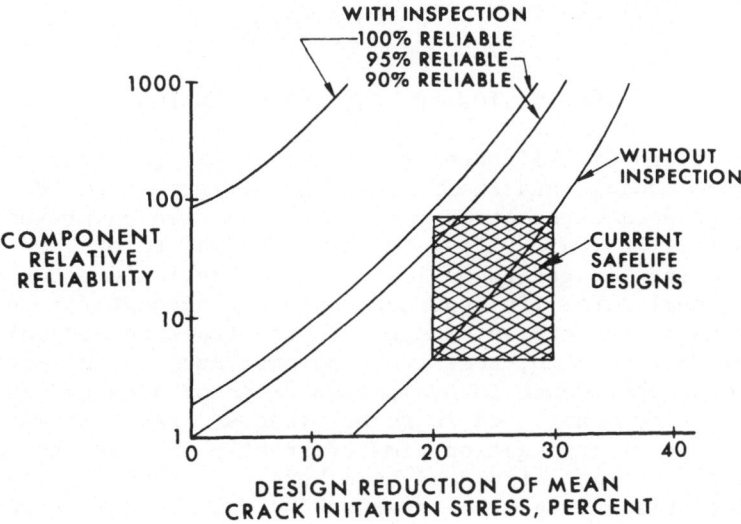

Figure 4.   Inspection provides reliability with residual life capability.

$$P_I = 1 - R_I \tag{5}$$

where R_I is the reliability of the inspection method.

An example of the combined effects on overall component relative reliability is shown in Figure 4. For this example a value of σ_C was chosen as being .20 based on a normal distribution of statistical crack propagation data [7]. This scatter will vary for materials and the example is shown only for illustrative purposes. For the 3 σ_C scatter in life a life factor of 2.5 is then used.

The comparison of parameters presented in Figure 4 illustrates the following:

(a) high reliability can only be achieved with consideration for crack initiation capability

(b) no increase in reliability can be achieved without inspection

(c) the reliability of the inspection method is the governing factor in achieving an overall high reliability

(d) by considering the crack propagation, with a reliable inspection method, a vast improvement can be made for the overall component reliability

(e) it is possible to achieve a lighter, more reliable component design by increasing the usual safe-life fatigue allowables and utilizing the crack propagation time with a reliable inspection method.

APPLICATION OF FRACTURE MECHANICS

The design of helicopter rotor systems is governed mainly by the cyclic loadings and their effect on fatigue strength. Prior to the last decade the primary emphasis has been on preventing crack initiation (Safe-Life). However, it was recognized [1-3] that while the fatigue technology produced reliable rotor structures that the usual safe-life approach would not account for such factors as produced by in-service damage. As the fracture mechanics technology became established, research programs (such as in Reference [8]) provided the confidence in evaluation of crack propagation for metal alloys commonly used in rotor systems. Early application of the use of crack propagation time to greatly increase the safety of rotor blades was introduced by Sikorsky Aircraft in the gaged pressurized blade inspection system BIM[R], as shown in Figure 5, and in a similar Boeing-Vertol blade inspection concept ISIS. Both systems rely on the concept of early crack detection, and sufficient crack propagation time to prevent an in-flight fracture.

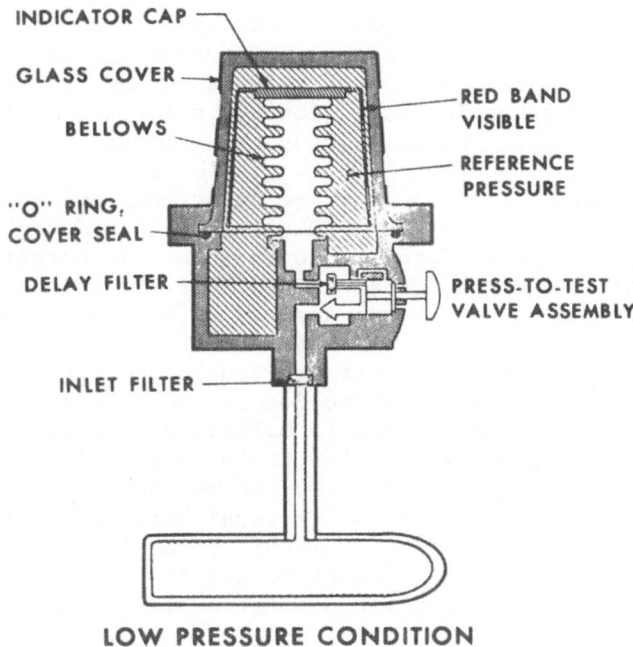

INDICATOR CAP

GLASS COVER

BELLOWS

RED BAND
VISIBLE

REFERENCE
PRESSURE

"O" RING,
COVER SEAL

DELAY FILTER

PRESS-TO-TEST
VALVE ASSEMBLY

INLET FILTER

LOW PRESSURE CONDITION

Figure 5.   Pressurized blade inspection method BIM[R].

Application of fracture mechanics for rotor blades [4,9] has
been to predict the survival time from the usual crack initiation
time and for the case of ballistic damage.

The major emphasis for fail-safety has been on main rotor
blades since their increase of structural reliability results in
the greatest performance gains of the helicopter. In addition,
ballistic damage considerations are most important for blade
structures since the latter represents a significant portion of
the vulnerable area for higher threats.

Increased safety for rotor blades has been proposed using
redundant design concepts [10] where the residual strength becomes
the important factor rather than crack propagation. New materials
such as composites may well offer increased crack propagation life
due to the inherent nature of fiber construction [11].

Basic concepts for using fracture mechanics to achieve fail-
safe designs for helicopter structures have been explored [12]
utilizing residual strength and crack propagation rates. Studies
such as these [12] have shown the need to select metal alloys for

other than the usual high static and crack initiation fatigue
strength. Fail-safe design has been incorporated into rotor controls [13] using a combination of redundancy and fracture strength
after damage. Further investigations of rotor and control components reveal that for typical helicopter structures a significant
residual strength and crack life is available after damage. The
studies also indicate if damage were detectable during flight that
long continued safe flight is permissible with reduced operating
conditions. The problem appears to be in early detection and
utilizing fracture mechanics to assure that the inspection intervals are adequate to assure the desired safety.

There appears to be a contradiction in the use of high
strength materials to achieve the desired goal of greater safety.
The application of higher strength materials, greater in static
and fatigue properties, gives rise to increased allowables for the
usual safe-life design concept. But these increased safe-life
stress allowables can reduce the residual strength and crack life
for fail-safe considerations. Fracture mechanics technology now
strongly enters into the choice of materials to temper the use of
materials and assure design methods for avoiding crack propagation
failures.

There appear to be two considerations to achieve the most
overall safe rotor design. The first is the consideration of
normal operation where cracks are initiated by cyclic loadings.
As presented in Equation (1), the overall probability of fracture
is the combination of crack initiation, detection, and crack
propagation. Thus to achieve a high overall safe design, one in
which the probability of fracture is extremely remote, it is
necessary to include low probability of crack initiation. The
second consideration is one in which crack initiation has started,
such as would be the case in ballistically damaged structures.
For this consideration residual strength and life properties are
dominant and inspection is only available from flight to flight.

The type of damage, as illustrated in Figures 6, 7 and 8,
shows the extensive damage incurred from .50 cal impact. Fracture
mechanics provides the analytical tool to predict the residual
strength of these types of metal structures used in the helicopter
rotor system.

An illustration of the use of fracture mechanics analysis is
the prediction [14] of the cracked S-N diagram for an aluminum
rotor blade as shown in Figure 9. The predicted values show a
good trend with the test results. However, it should be noted
that the difference in crack life is considerable and points out
that test verification is needed in order to specify inspection
intervals. Also the analysis is very conservative for the
ballistic damage, and fails to account for the local effects near

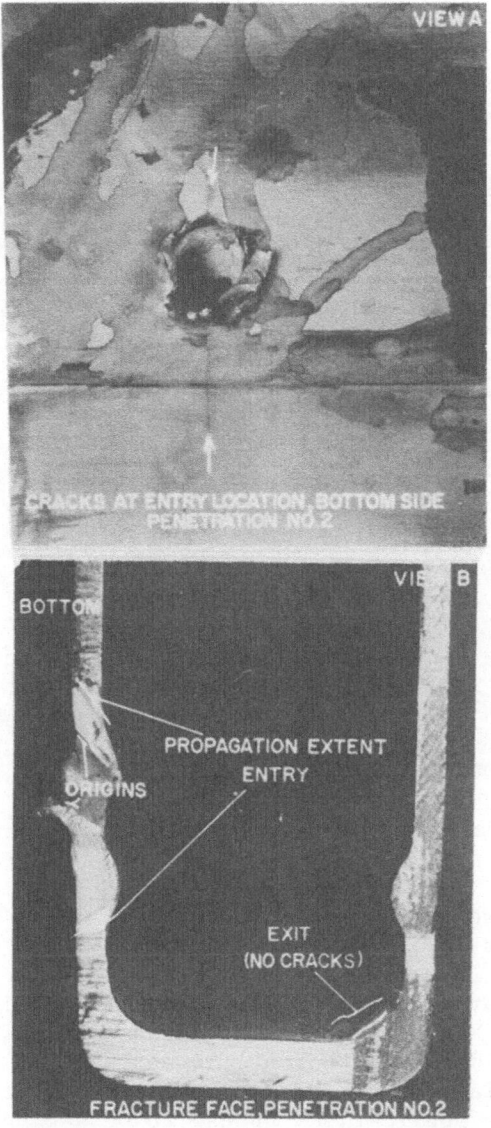

Figure 6. Ballistic damage to a main rotor spar.

the impact area. Nevertheless, the fracture mechanics approach
permits the designer to predict crack propagation to a reasonable
accuracy in terms of stress. In addition, the rotor structures
can be designed on a rational base and comparisons made as to the
most efficient damage tolerant design. For more complex structures,
such as rotor heads, the finite element analysis using a cracked

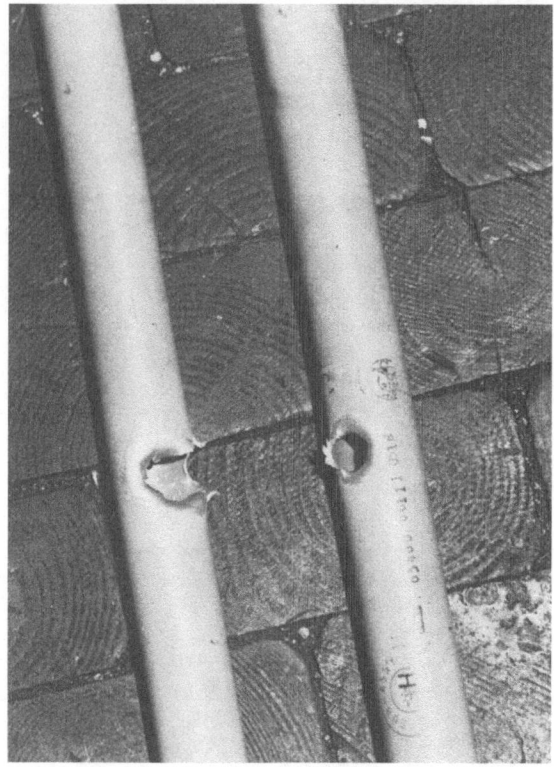

Figure 7.   Ballistic damage to main rotor control rod.

element [15] may well provide improved accuracy in predictions of
residual strengths.

    The use of advanced composites also promises improvement in
damage strength. Fracture mechanics appears to offer the means
of predicting the damage strength. Investigations [16-18] indicate
anisotropic materials obey a stress intensity relationship and
fracture mechanics applications appear to be a proper analytical
tool for composite materials.

    As illustrated in Figure 10, one composite material indicates
lower notch sensitivity to fatigue, particularly in the high cycle
regime most important to the rotor system structures. From the
experimental work [19] it appears from Figure 11 that the residual
strength to weight ratio is greater for composites with small
fractional damage, but is less than that for metals for large
damage. Also, as illustrated in Figure 11, note the importance
of ply orientation.

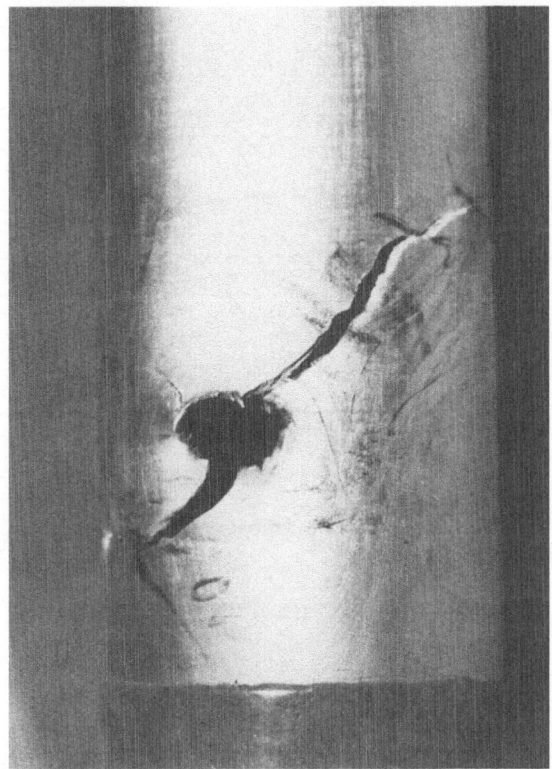

Figure 8. Ballistically damaged aluminum drive shaft survives three hours of test.

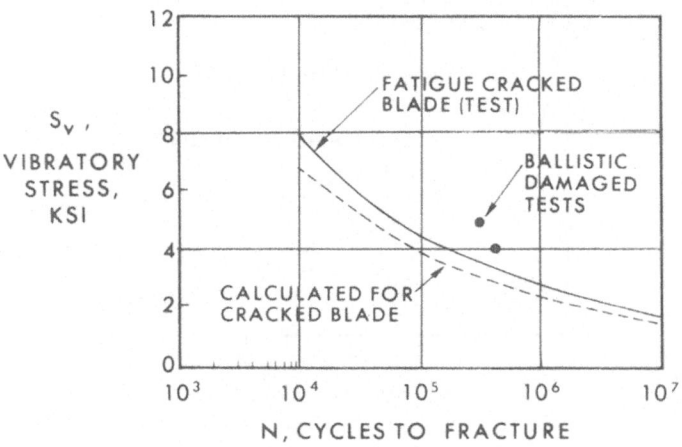

Figure 9. Fracture mechanics can predict damaged life of a rotor blade.

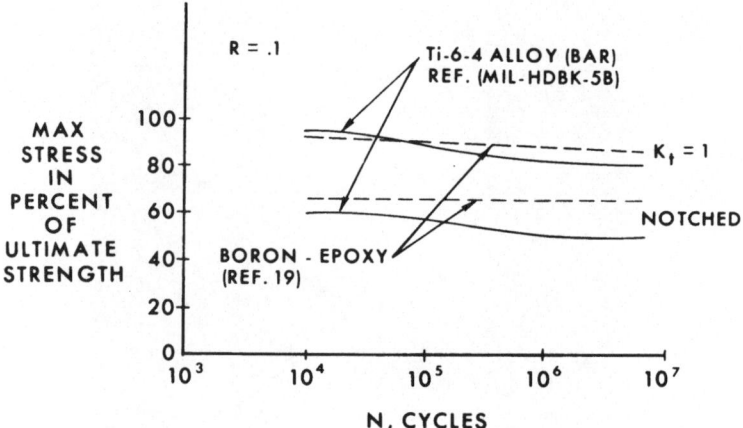

Figure 10.   Composite material shows less notch sensitivity in
              fatigue.

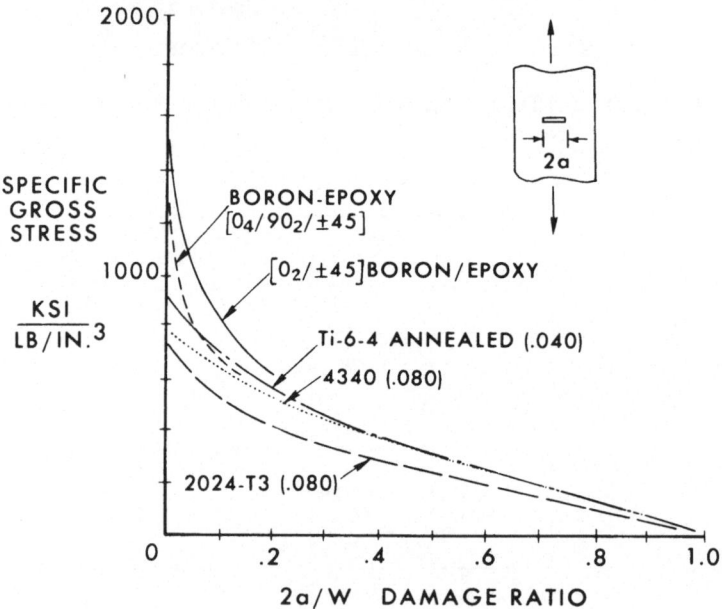

Figure 11.   Composite materials are more sensitive to damage.

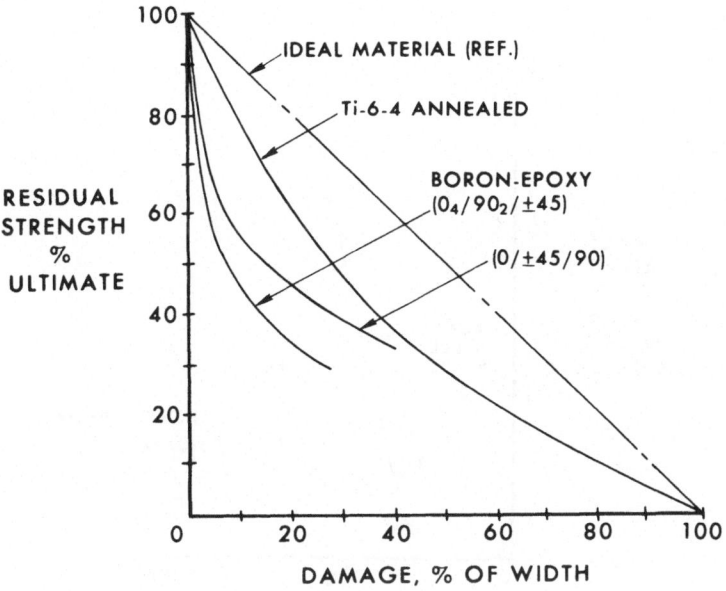

Figure 12. Composite materials may be limited by residual strength.

Another problem now appears when using composites which arises from the increased undamaged static strength and greater crack initiation time over that of metal alloys. For metal alloys the criteria of fatigue is dominant in the design of rotor system structures. For a good safe balanced design the effect of relatively low fracture toughness of the composites will have to be recognized. As shown in Figure 12, the lower fracture toughness of the composites (relative to their high static strength) can result in much lower residual strength than for metal alloys. The favorable aspect for composite materials, having taken into account the reduced residual strength properties, is the much lower rate of crack propagation as illustrated in Figure 13. Thus it is expected that fracture mechanics will be a most important aspect in selecting composite materials and the ply orientations for rotor system components.

SUMMARY

Fracture mechanics analysis is now being extensively employed in design for fail-safety of metal alloy helicopter rotor system components. The importance of fracture mechanics analysis has been to temper the use of high strength alloys by requiring properties

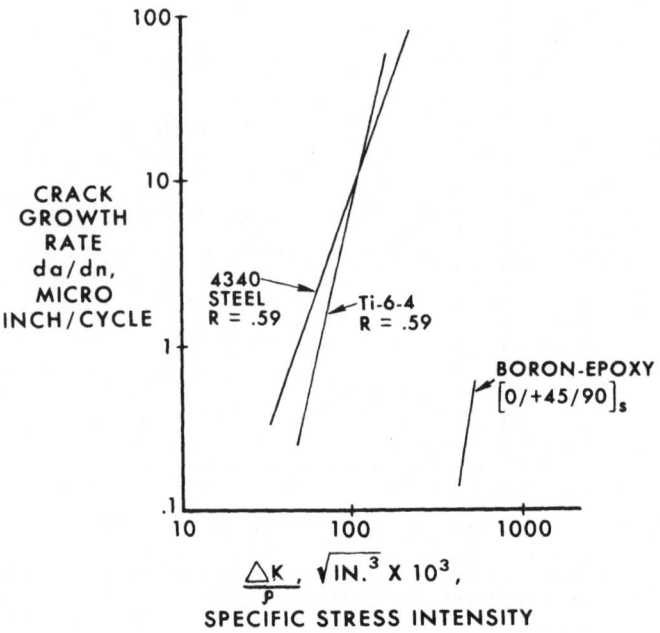

Figure 13.   Composites promise lower crack growth rate.

that provide a high residual strength and greater residual life. Controlled fracture, with reliable inspection methods, appears to be the key for fail-safe design. A combination of "safe-life" and "fail-safe" design philosophies offer a lighter weight solution for increased safety.

The new advanced composite materials appear to offer even greater reliability by virtue of their much lower rate of crack propagation. However, fracture mechanics analysis will now have to emphasize the residual strength aspect for composites due to their relatively lower fracture toughness.

REFERENCES

1. Jensen, H.T., "The Evolution of Structural Fail-Safe Concepts - Rotorcraft", in Fatigue Design Procedures, ed. by E. Gassner and W. Schütz. New York: Pergamon Press (1969), 433-86.

2. McBreaty, J., "Fatigue and Fail-Safe Airframe Design", SAE Trans., 64 (1956), 427-36.

3. Stratton, W.K. and White, R.F., "Fail Safety - What Is It?", Vertiflite, 14, no. 8 (1968), 4-10.

4. Thompson, G. and Weiss, W., "Fail-Safety for the H-46 Rotor Blade", Preprint No. 551, American Helicopter Society, May 1971.

5. Weiss, W. and Zola, J., "The Application of Fracture Mechanics to the Design of Damage Tolerant Components for the UTTAS Helicopter", Preprint No. 882, American Helicopter Society, May 1974.

6. Jensen, H., "The Application of Reliability Concepts to Fatigue Loaded Helicopter Structures", paper presented at the 18th Annual Forum of the American Helicopter Society, Washington, D.C., May 1962.

7. Jacoby, G. and Nowack, H., "Comparison of Scatter Under Program and Random Loading and Influencing Factors", in Probabilistic Aspects of Fatigue, Special Technical Publication 511. Philadelphia: Am. Soc. for Testing and Materials (1972), 61-74.

8. Degnan, W.G., Dripchak, P.D. and Matusovich, C.J., "Fatigue Crack Propagation in Aircraft Materials", Sikorsky Aircraft Div., United Aircraft Corp., Stratford, Conn., Army Aviation Material Laboratories Contract Report No. USAAVLABS-TR-66-9, March 1966. (AD 630 926)

9. Rich, M.J., "Crack Propagation in Helicopter Rotor Blades", in Damage Tolerance in Aircraft Structures, Special Technical Publication 486. Philadelphia: Am. Soc. for Testing and Materials (1971), 243-51.

10. Pociluyko, S., Griffen, C., Figge, I. and Blad, L., "Composite Material Geodesic Structures - A Structural Concept for Increased Helicopter Blade Survivability", Preprint No. 884, American Helicopter Society, May 1974.

11. Salkind, M., "The Twin Beam Composite Rotor Blade", Preprint No. 782, American Helicopter Society, May 1973.

12. Rich, M.J. and Linzell, L.E., "Damaged Static and Fatigue Stress Analysis of VTOL Structures", AIAA Paper 69-214, February 1969.

13. Barrett, L. and Mack, J., "Evaluation of Rotor Controls Designed for Increased Safety", Preprint No. 550, American Helicopter Society, May 1971.

14. Rich, M.J., "Vulnerability Considerations in the Design of Rotary Wing Structures", in Proceedings of the Air Force Conference on Fatigue and Fracture of Aircraft Structures and Materials, held at Miami Beach, Fla., 15-18 December 1969, Air Force Flight Dynamics Laboratory, Wright-Patterson AFB, Ohio, Report No. AFFDL-TR-70-144 (September 1970), 635-51. (AD 719 756)

15. Pian, T., Tong, R. and Luk, C.H., "Elastic Crack Analysis by A Finite Element Hybrid Method", Massachusetts Institute of Technology, Cambridge, Air Force Office of Scientific Research Contract Report No. AFOSR-TR-72-0752, December 1971. (AD 739 988)

16. Mandell, J., McGarry, F., Wang, S. and Im, J., "Stress Intensity Factors for Anisotropic Fracture Test Specimens of Several Geometrics", J. Composite Mater., 8 (1974), 106-15.

17. Holdsworth, A. and Owen, H., "Macroscopic Fracture Mechanics of Glass Reinforced Polyester Resin Laminates", J. Composite Mater., 8 (1974), 117-29.

18. Phillips, D., "The Fracture Mechanics of Carbon Fibre Laminates", J. Composite Mater., 8 (1974), 130-41.

19. Durchlaub, E. and Freeman, R., "Design Data for Composite Structure Safe-Life Prediction", Boeing Vertol Co., Philadelphia, Pa., Air Force Materials Laboratory Contract Report No. AFML-TR-73-225-Vol-1, March 1974. (AD 918 496L)

SOME RECENT IMPROVEMENTS IN ELECTRON FRACTOGRAPHIC ANALYSES

C. D. Beachem

Naval Research Laboratory, Washington, D. C.

ABSTRACT

Laboratory documentation of the manner in which interactions between environment, structure, and stress produce microscopic fracture surface markings is enabling more informative failure analyses. Cataloging of new features and calibration of the features with stress intensity promises to aid in failure analyses.

INTRODUCTION

Cleavage, microvoid coalescence, intergranular fracture and fatigue are some of the fracture mechanisms which were thought to be fairly well documented and specific but which are evolving into generalized families of mechanisms. This is due partly to the application of newer, more diacritical, analysis techniques and partly to the increasing number of alloys and cominbations of materials which enter service and break.

This brief chapter is intended to show how some recent fractographic mechanism studies may be used in failure analyses.

Intergranular Fracture

A preponderance of grain boundary facets on the fracture surfaces have for a long time been a useful indication of the cause of failures. It usually indicates a metallurgical embrittling process or stress corrosion cracking or hydrogen-assisted cracking process. Sometimes, even more can be learned by looking

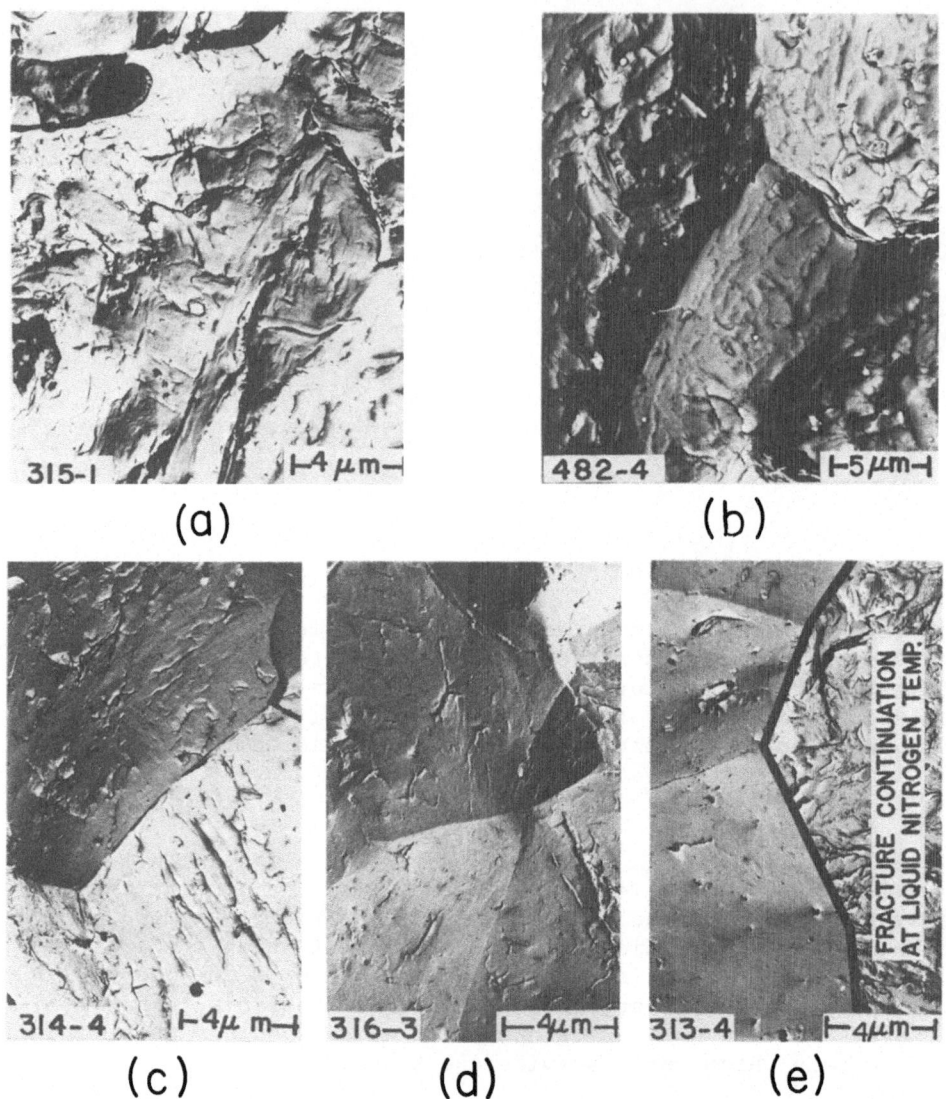

(a)   (b)

(c)   (d)   (e)

Figure 1.   Decreasing grain boundary plasticity with decreasing
            stress intensity in a specimen of AISI 4328 steel
            cracked as a cathode in salt water.   Examples of tear
            ridges, evidence of plasticity, are indicated by
            arrows.   [1]

Figure 2. Features which are present on many but not all fatigue
           fracture surfaces. The large arrow indicates the
           macroscopic cracking direction.

closely at the fine scale structures on the facets. Figure 1
shows how the sizes and numbers of plastically-produced tear
ridges decrease with decreasing stress intensity [1]. Little
attention has been paid to the details of these tear ridges.
With sufficient study their sizes might be correlated with K_I in
a useful quantitative way, such that K_I and perhaps the stress
operative in the cracked components in service could be deduced.

Cleavage

     The fractographic characteristics of cleavage which are most
well known are low-index facets, steps separating portions of
facets which are at different heights, river patterns composed of
steps converging in a "downstream" manner which indicates crack
propagation direction, and a renewal of step patterns as the fracture crosses grain boundaries. These features are frequently
found on simple overload fractures. However, cleavage has also
been found to occur as a result of fatigue, stress corrosion
cracking, hydrogen-assisted cracking, and sustained load cracking.

The fine scale features of these different forms of cleavage may
be sufficiently different to permit identification of specific
causes of failure.

Meyn [2] has recently shown the river patterns on Ti-6A1-4V
sustained-load cracked surfaces to consist of three different
types of tear ridge configurations depending upon the location of
local crack tip initiation sites relative to the a and B phases.
Davidson [3], and Newbury, Christ and Joy [4], using selected
area channeling patterns in the SEM have shown that deformation on
cleavage facets may be detected.

Fatigue

The standard in fractographic failure analysis of fatigued
components is the microscopic striation which often has general
features as shown in Figure 2. However, fatigue does not always
produce striations. Forsyth showed that striations can be ductile
or brittle, and that fatigue in aluminum alloys can lead to cleavage [5]. Hertzberg [6] recently showed that striations are present in only 1/3 of the crack growth range in a number of alloy
systems, with cleavage characterisitics being present in very low
crack growth rate conditions and overload fracture characteristics
present in very high crack growth conditions. Meyn showed that
striations are not formed in vacuum conditions in aluminum [7]
and titanium [8] alloys, with the fracture surface exhibiting slip
structures only.

Another indication of fatigue is the presence of features
called "tire tracks" [9] which may be present even if striations
are not. These are caused [10,11] by the fracture surfaces
shifting across one another as the crack opens and closes. It has
been found from service fractures, however, that these features
are occasionally formed on an origin caused by some other mechanism,
such as stress corrosion cracking, if the initial crack is opened
and closed by later fatigue stresses.

Microvoid Coalescence

Differences in microscopic dimple lengths on matching surfaces indicate macroscopic cracking directions. The shapes of
mating dimples indicate the local state of stress as well as the
crack propagation direction, as shown in Figure 3 [12]. The slope
of macroscopic ridges also indicates crack propagation directions
[13].

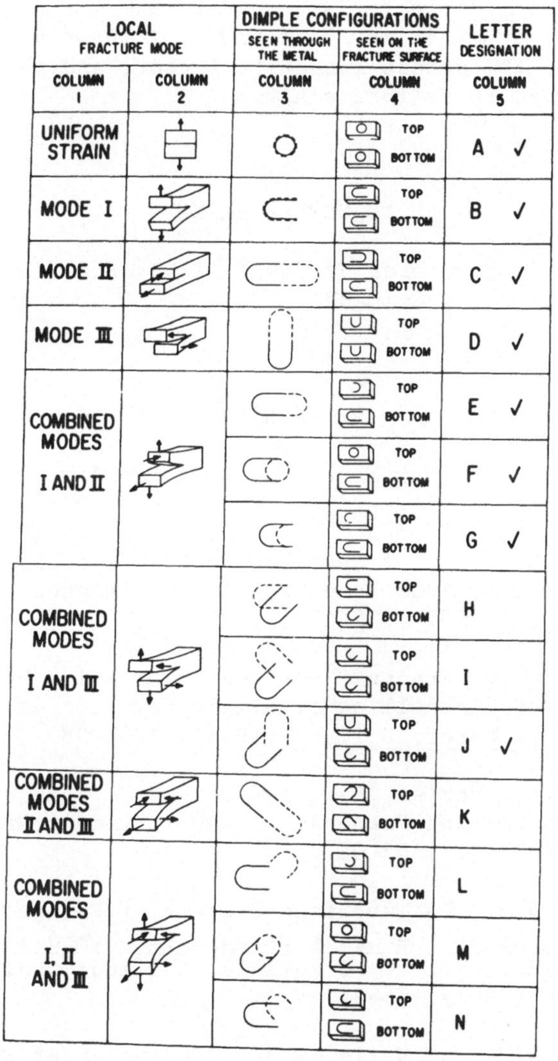

Figure 3. Eight observed and six probable dimple shapes depending upon local fracture mode, as determined by examining mating surfaces. [12]

Calibration of Mixtures of Fracture Modes

A large number of authors have shown that macroscopic fracture surfaces consist of mixtures of fracture modes, and that these modes change their characteristics with time and crack growth. With sufficient study it may be found that these mixtures may be correlated with the stress intensity factor, and that these correlations may be used in failure analyses.

ACKNOWLEDGEMENT

This work is sponsored by Naval Air Systems Command and Naval Facilities Engineering Command.

REFERENCES

1. Beachem, C.D., "A New Model for Hydrogen-Assisted Cracking Hydrogen ('Embrittlement')", _Met. Trans._, 3 (1972), 437-51.

2. Meyn, D.A., "Effect of Hydrogen on Fracture and Inert-Environment Sustained Load Cracking Resistance of $\alpha-\beta$ Alloys", _Met. Trans._, 5 (1974), 2405-14.

3. Davidson, D.L., "Fracture Surface Examination by Selected Area Electron Channelling of Single Crystals of Mo-15 At. % Re Alloy", _J. Mater. Sci._, 9 (1974), 1091-98.

4. Newbury, D.E., Christ, B.W. and Joy, D.C., "Relevance of Electron Channeling Patterns to Embrittlement Studies", _Met. Trans._, 5 (1974), 1505-08.

5. Forsyth, P.J.E., "A Two Stage Process of Fatigue Crack Growth", in _Proceedings of the Crack Propagation Symposium, Cranfield, September 1961, Vol 1._ Cranfield, England: College of Aeronautics (1961), 76-94.

6. Hertzberg, R.W. and Mills, W.J., "Character of Fatigue Fracture Surface Micromorphology in the Ultra-Low Growth Rate Regime", in _Fractography - Microscopic Cracking Processes._ Special Technical Publication 600. Philadelphia: Am. Soc. for Testing and Materials (1976), 220-34.

7. Meyn, D.A., "The Nature of Fatigue-Crack Propagation in Air and Vacuum for 2024 Aluminum", _ASM Trans. Quart._, 61 (1968), 52-61.

8. Meyn, D.A., "An Analysis of Frequency and Amplitude Effects on
    Corrosion Fatigue Crack Propagation in Ti-8Al-1Mo-1V", Met.
    Trans., 2 (1971), 853-65.

9. Whiteson, B.V., Phillips, A. and Kerlins, V., "Electron Fracto-
    graphy Handbook", Douglas Aircraft Company, Inc., Santa Monica,
    Calif., Air Force Materials Laboratory Contract Report No.
    ML-TR-64-416, January 1965. (AD 612 912)

10. Beachem, C.D., "Microscopic Fatigue Fracture Surface Features
    in 2024-T3 Aluminum and the Influcence of Crack Propagation
    Angle Upon Their Formation", ASM Trans. Quart., 60 (1967),
    324-43.

11. Koterazawa, R., Mori, M., Matsuiti, M. and Shimo, D., "Fracto-
    graphic Study of Fatigue Crack Propagation", Trans. ASME, Ser.
    H, J. Eng. Mater. Technol., 95 (1973) 202-12.

12. Beachem, C.D., "The Effects of Crack Tip Plastic Flow
    Directions Upon Microscopic Dimple Shapes", Met. Trans. A,
    6A (1975), 377-83.

13. Beachem, C.D. and Yoder, G.R., "Elastic-Plastic Fracture by
    Homogeneous Microvoid Coalescence Tearing Along Alternating
    Shear Planes", Met. Trans., 4 (1973), 1145-53.

INDEX